NanoScience and Technology

NanoScience and Technology

The series NanoScience and Technology is focused on the fascinating nano-world, mesoscopic physics, analysis with atomic resolution, nano and quantum-effect devices, nanomechanics and atomic-scale processes. All the basic aspects and technology-oriented developments in this emerging discipline are covered by comprehensive and timely books. The series constitutes a survey of the relevant special topics, which are presented by leading experts in the field. These books will appeal to researchers, engineers, and advanced students.

Sliding Friction
Physical Principles and Applications
By B.N.J. Persson
2nd Edition

Scanning Probe Microscopy
Analytical Methods
Editor: R. Wiesendanger

Mesoscopic Physics and Electronics
Editors: T. Ando, Y. Arakawa, K. Furuya, S. Komiyama, H. Nakashima

Biological Micro- and Nanotribology
Nature's Solutions
By M. Scherge and S.N. Gorb

Semiconductor Spintronics and Quantum Computation
Editors: D.D. Awschalom, N. Samarth, D. Loss

Semiconductor Quantum Dots
Physics, Spectroscopy and Applications
Editors: Y. Masumoto and T. Takagahara

Nano-Optoelectonics
Concepts, Physics and Devices
Editor: M. Grundmann

Noncontact Atomic Force Microscopy
Editors: S. Morita, R. Wiesendanger, E. Meyer

Nanoelectrodynamics
Electrons and Electromagnetic Fields in Nanometer-Scale Structures
Editor: H. Nejo

Single Organic Nanoparticles
Editors: H. Masuhara, H. Nakanishi, K. Sasaki

Epitaxy of Nanostructures
By V.A. Shchukin, N.N. Ledentsov, D. Bimberg

Nanoscale Characterisation of Ferroelectric Materials
Scanning Probe Microscopy Approach
Editors: M. Alexe and A. Gruverman

E. Kasper D.J. Paul

Silicon Quantum Integrated Circuits

Silicon–Germanium Heterostructure Devices: Basics and Realisations

With 263 Figures

Springer

Prof. Erich Kasper
Institute of Semiconductor Engineering
University of Stuttgart
Pfaffenwaldring 47
70569 Stuttgart, Germany
E-mail: kasper@iht.uni-stuttgart.de

Prof. D.J. Paul
Cavendish Laboratory
University of Cambridge
Madingley Road
Cambridge CB3 0HE, UK
E-mail: dp109@cam.ac.uk

Library of Congress Control Number: 2004116222

ISSN 1434-4904
ISBN 3-540-22050-X Springer Berlin Heidelberg New York

Springer is a part of Springer Science+Business Media.

springeronline.com

Printed in Germany

Typesetting: Data conversion by the authors using a Springer TEX macro package
Final processing by Frank Herweg, Leutershausen
Production: LE-TEX Jeloneck, Schmidt & Vöckler GbR, Leipzig
Cover design: *design& production*, Heidelberg

Printed on acid-free paper 57/3141/ - 5 4 3 2 1 0

Preface

For more than thirty years, progress in monolithic integration of transistors into integrated circuits (IC) has yielded an exponential growth of the performance and density of such circuits along with an exponential growth of sales. In microelectronics this rapid and longstanding exponential growth is referred to as Moore's law after one of the pioneers of the integrated circuit industry. The continuous shrinking of the lateral and vertical device dimensions is closely related to the increasing number of transistors on a chip, now approaching the tera scale (10^{12}). The scaling of complementary metal-oxide–semiconductor (CMOS) technology is defined by the lateral lithography rules which determine the future technology generations from the present 90 nm node to the 65 nm, 45 nm, 30 nm nodes and eventually down to some point which will be the ultimate scalability of the technology (or the final economically viable technology node). Vertical dimensions in devices are often a factor of ten finer than the lateral dimensions defined by optical lithography, due to sophisticated deposition and epitaxy methods.

Quantum electronics, therefore, more than traditional microelectronics frequently relies on the vertical structure of the device, as the vertical dimensions can be controlled at the nanometre scale more easily. In this context a definition of quantum electronics is required, because semiconductor physics and its application in electronic devices are completely based on the law of quantum mechanics. The transport and statistical behaviour of charged carriers in semiconductors must be derived from the first principles of quantum mechanics with statistical mechanics and are strongly influenced by the quantisation of charge and energy. Many of the basic semiconductor properties are uniquely defined by the quantum mechanics of the system. For instance the concept of holes, which – through the Pauli exclusion principle – states that a missing valence band electron can be considered as a positive charged hole, the existence of a band gap which forbids electron states in an energy range between the conduction band and valence band, and the effective mass approximation, which treats carriers near their energy minimum as quasi-free with masses differing from the free electron mass. The term *quantum electronics* is frequently only used when artificial man-made structures are small enough to allow the electronic or optical properties of devices to be strongly influenced by quantum effects. In particular the lowering of the dimensionality of a structure results in a change of the density of states along with quantisation of electron states into subbands (when larger than the thermal smearing ($\sim k_{\mathrm{B}}T$) of the system) which may frequently completely change the properties of devices. Typical structure dimensions for such quantisation in silicon with its relative large effective masses are below 10 to 20 nm. Frequently exploited quantum effects include the transmission of carriers through barriers (tunnelling), the bound states in quantum wells (quantisation) and the mini-band formation in multi-quantum wells and superlattices (artificial semiconductor). The general laws of quantum mechanics are the same for all semiconductor materials. Why do we think that silicon quantum electronics is a topic worthy of a whole book? Silicon-based quantum electronics is progressing in a significantly different way than for instance in group III/V materials where many basic studies were completed and early successes were obtained, e.g. with lasers and high electron mobility transistors (HEMTs). The reasons for the differences are related to the physics, the technology and the economics. Silicon is an indirect semiconductor with six degenerate conduction band valleys and the most important heterostructure SiGe/Si can be used to separate electrons and holes to opposite sides of the heterointerface (a so-called type II interface). A technologically stable heterostructure with equal lattice constants (such as Ga As and GaAlAs) is not available in a silicon-based system. As a consequence the SiGe/Si heterostructures are strained, giving additional freedom for material band structure designs but limiting the usable thicknesses. The widespread dominance of silicon substrates in all areas of microelectronics offers enormous market opportunities for quantum electronics, but also imposes strict manufacturing and operating conditions especially concerning high complexity integration and room temperature operation. This book is organised around three main topics. The first few chapters review the methods and techniques for creating quantum structures using combinations of heterostructures and conventional p/n-junctions; they also review the relevant semiconductor physics background (Chap. 3). Chapter 4 treats the influence of elastic strain on electronic structure and interface energies, in particular for the strained SiGe/Si system. Then follows a detailed discussion of devices, the examples having been selected by us with respect to their potential for integration and also operation under room temperature conditions. The standard silicon p/n junctions, bipolar transistors and MOSFET devices are reviewed in Chap. 5. The SiGe hetero-bipolar transistor (HBT) is now a mainstream production technology following its introduction onto the market in 1999 and now dominates many high-frequency applications (Chap. 6). The hetero field-effect transistor (HFET) is entering the competition for future CMOS technology generations, offering a variety of powerful solutions ranging from several tens of percent improvement to the ultimate symmetric n- and p-channel transistors (Chap. 7). Tunnelling and optoelectronic phenomena (Chaps. 8, 9) could be the key to novel and rapidly expanding system-on-chip (SOC) solutions. The important question of possible integration techniques is investigated in Chap. 10. The reader is referred to Chaps. 2, 3 and 5 when quick advice on materials science, technology, semiconductor physics and device principles is required. We have aimed this text at two groups of readers but hope that both these and many others will benefit from this focused treatment. We hope that the many engineers involved in fueling the exponential progress in micro- and nanoelectronics will be confronted with and increasingly utilise quantum effects. We also hope that researchers in device physics and modelling, nanoelectronics and semiconductor materials along with graduate students in electrical engineering and computer science, physics and material science will come to see silicon as the model system for strained heterostructures. During the writing of this book certain chapters were used as a manuscript for the lecture "Quantum Electronics" in a graduate course "Electrical Engineering and Information Technology" at the University of Stuttgart. We thank the students for their comments and are especially grateful for the help of the assistants Günter Reitemann and Jens Werner.

E.Kasper
D.J. Paul

Contents

1. Introduction

Quantum physics has recently had its 100th anniversary which can be traced back to the work of Planck with the development of the black body radiation formula. Planck showed that the spectral energy density $u(\nu, T)$ of the blackbody radiation is given by

$$u(\nu; T) = (8\pi h\nu^3/c^3)\left[\exp\left(\frac{h\nu}{k_B T}\right) - 1\right]^{-1} \tag{1.1}$$

where h is Planck's constant ($= 6.62617 \times 10^{-34}$ Js), ν is the frequency of radiation, c is the speed of light, T is the temperature and k_B is Boltzmann's constant ($=1.38 \times 10^{-23}$ J/K). This result was derived using thermodynamic considerations from Maxwell's equations on a dipole emitter/absorber and a finite amount of energy transfer. Planck postulated that the atoms in the walls of a cavity emit radiation in small packets or quanta. Einstein later concluded that these quanta of electromagnetic radiation of frequency, ν would have energy

$$E = h\nu \tag{1.2}$$

It was originally de Broglie who suggested that particles such as electrons could have wave behaviour. This wave-particle duality results in the energy of the photon being directly related to the wavelength, λ (where λ is the de Broglie wave length) of the particle wave. A similar universal relationship belongs between the momentum, p and the wave number, $k = 2\pi/\lambda$

$$p = \hbar k = h/\lambda \qquad (\hbar = h/2\pi) \tag{1.3}$$

One of the early successes of quantum physics was to explain correctly a number of phenomena including the binding energy of electrons on an atom and the emission of light from gases. It is in the field of semiconductors where quantum theory has predicted many new effects, all later to be demonstrated experimentally, which has transformed modern life. Examples include the internal photoelectric effect which explains the fundamental absorption,α of a semiconductor which is related to the bandgap E_g of the material through

$$\alpha = (h\nu - E_g)^n \cdot Const. \tag{1.4}$$

where $n = 1/2$ for a direct semiconductor and $n = 2$ for an indirect semiconductor.

Indeed the whole of modern semiconductor physics is based on band structures (Sect. 3.2) calculated from quantum physics and the development of new concepts from quantum mechanics such as holes and the effective mass approximation (Sect. 3.2.1) where the ability of an electron to be transported through a lattice is described as quasi-free electron waves. Chapter 3 will describe many of the quantum mechanical developments and results which can be applied to semiconductors and are key for an understanding of the present day microelectronic devices to be reviewed in later chapters.

The de Broglie wavelength λ (k_BT) of an electron with a thermal energy k_BT is given by

$$\lambda(k_BT) = h\sqrt{2mk_BT} \tag{1.5}$$

which results in de Broglie wave lengths of room temperature electrons typically between 5 nm and 25 nm depending on the effective masses. These wave lengths are far below the lithographic structures used up to now in microelectronics (Table 1.1). As the gate lengths become smaller in the future, quantum mechanical effects will play an ever increasing rôle in the modelling and understanding of transistors.

Table 1.1. Historical evolution of the metal oxide silicon (MOS) transistor data. Shown are the gate length L_g, the gate oxide thickness t_{ox}, the supply voltage V_{dd}, the number n of electrons in the channel (on, off) and the number N of atoms in the channel region (Si and dopants)

Year	$L_g (\mu m)$	t_{ox} (nm)	V_{dd} (V)	n		N	
				on	off	Si	Dopant
1970	20	120	20 (PMOS)	10^7	300	10^{13}	10^5
1980	2	15	5	10^6	30	10^{11}	3×10^3
1990	0.6	10	3.3	10^5	5	3×10^9	10^3
2000	0.18	5	1.5	5×10^3	2	10^8	250
2004	0.053	1.5	1.2	300	1	10^7	70
(2009)	0.025	0.7	1.0	150	1	5×10^6	35
projected							
(2018)	0.010	0.5	0.7	30	1	10^5	5

The complementary metal oxide semiconductor (CMOS) circuit architecture has dominated the market for logical circuits since 1985. This dominance is due to CMOS being the lowest power architecture and will be reviewed in Sect. 10.1. Before the CMOS dominance, bipolar circuits along with PMOS or NMOS architectures were dominating. Now, the market share for bipolar circuits is between 10% and 20% which is mainly for analogue circuits, high current drivers, high frequency circuits and input/output integrated circuits

(ICs). At present, the fastest demonstrated silicon circuits use more advanced SiGe heterostructure bipolar transistors which will be reviewed in Sect. 6.4. The high speed and analogue benefits of bipolar can be mixed with the low power and high density of CMOS to produce the mixed technology of BiCMOS (Sect. 10.6). In the past the functionality of ICs improved rapidly, e.g. the transistor density, the chip area and the clock frequency by factors per decade of 20, 2 and 5 respectively.

1.1 Microelectronics and Optoelectronics

The microelectronics industry has only been around since the 1950s. The first bipolar transistor was demonstrated at Bell Laboratories in 1948 and the first field effect transistor (FET) appeared in 1960. Since that date, the microelectronics industry has been growing at an exponential rate. This was first analysed and commented on by Gordon Moore, one of the founders of Intel, in 1965. Moore showed that the size of a transistor is halved every 18 months, a trend which is now referred to as Moore's law. This halving in size has been driven by economics. The smaller the transistor gate, the faster the transistor can switch, the less power it can consume and the larger the number of transistors which can be integrated onto the one silicon chip. The increase in numbers of transistors and the associated higher manufacturing yields reduces the cost per transistor to smaller and smaller sizes. Even 3 decades after Moore's prediction, the continued scaling of the MOS transistor has not just halved every 18 months, in the last few years the decrease in size is actually faster than Moore's original prediction.

With Moore's law, both the number of transistors on a silicon chip and the number of silicon chips being manufactured has also been increasing at an exponential rate. The enormous market for microelectronic circuits has been increasing strongly over the last five decades as the costs are reduced and this increase is predicted to continue. No saturation is expected at present within the next ten years (Table 1.2). This increase has been fuelled by the reduction in cost per transistor over the last five decades with the resulting enormous increase in the number of available transistors.

Table 1.2. Estimated market data for microchips

Year	market (10^9 US\$)	number of chips (10^9)	Si area (10^6 m^2)	Operations / person GOPS
2000	180	60	2	0.04
2010	800	250	10	1
2020	3000	1000	40	50

With over 300 million transistors per silicon chip on present day microprocessors, the main drive in the future is the reduction in cost per function on a chip. Single chip or few chip solutions are significantly cheaper to manufacture than multiple chips for complete systems and therefore the drive is towards system-on-a-chip. Systems-on-a-chip will be realised where heterogeneous functions as analogue, digital, high frequency, power, sensors, actuators and optical or microwave transmission are all combined. The personal communicator is a popular example where microphones, image sensors, and displays are combined with logic, transceivers and energy conversion/energy storage (Fig. 1.1).

Quantum effects can already be observed in existing micro- and optoelectronic devices. While lithographic dimensions are still much larger than the de Broglie wavelength of electrons, the vertical device dimensions, however, are up to a factor 10 smaller than lateral dimensions. Carrier abruptness at surfaces or interfaces is only fundamentally limited by the Debye length, L_D which for electrons is given by

$$L_D^2 = \tau_{Rel} D_n = \left(\frac{\varepsilon_r \varepsilon_0 V_t}{q N_D} \right) \tag{1.6}$$

with the thermal voltage $V_t = k_B T/q$, the donor doping level N_D, the relative dielectric constant ε_r, the permittivity of a vacuum, ε_0, the electron charge, q and the temperature, T. The relaxation time τ_{Rel} and the diffusion coefficient, D_n are given by

$$\tau_{Rel} = \frac{\varepsilon_r \varepsilon_0}{\sigma_n} \tag{1.7a}$$

$$\sigma_n = q \mu_n N_D \tag{1.7b}$$

$$D_n = \mu_n V_t \tag{1.7c}$$

with σ_n the specific electrical conductivity, and μ_n the electron mobility. The Debye length in Si at $T = 300\,\mathrm{K}$, is smaller than 10 nm for carrier levels above $2 \times 10^{17}\,\mathrm{cm}^{-3}$.

One common example of quantum effects can be given by considering the hundreds of millions of metal contacts per integrated circuit. The current from the metal contact to the semiconductor has to overcome the depletion layer beneath the metal contact. The technological routine solution is based on a highly doped contact formed by implantation which reduces the depletion width to such a small dimension that quantum mechanical tunnelling of electrons through the barrier dominates the current. Indeed, every transistor is composed of the inner transistor and three terminal tunnelling diodes connecting the device to the interconnect metallisation. With proper technology the voltage drop V_c across the tunnelling diodes is small and may be expressed by $V_c = R_c \times J$ with current density, J and the specific contact resistance, R_c proportional to the tunnelling probability

Fig. 1.1. A cartoon illustrating a vision of the future with a person wearing an independent, multifunctional, ubiquitous computing system

$$R_c \sim \exp\left[\frac{2\Phi_{Bn}}{q}\sqrt{\frac{\varepsilon_r\varepsilon_0 m^*}{\hbar^2 N_D}}\right] \tag{1.8}$$

where Φ_{Bn} is the Schottky barrier height and m^* the effective mass of the electrons. If the contacts are doped in the $10^{20}\mathrm{cm}^{-3}$ range, specific contact resistances may be realised with values as low as $10^{-7}\,\Omega\mathrm{cm}^2$ to $10^{-8}\,\Omega\mathrm{cm}^2$. Even with extremely high current densities of $10^6\,\mathrm{A\,cm}^{-2}$, the voltage drop V_c is then limited to 10 mV and 100 mV, respectively. Voltage limiters for voltages smaller than 6 V are frequently based on Zener diodes, where Zener tunnelling (band to band tunnelling from the valence band to the conduction band across a high doped p/n-junction) is responsible for the voltage limiting breakthrough. The limits of breakdown can easily be shifted between 2 V and 6 V by appropriate doping between $5 \times 10^{18}\,\mathrm{cm}^{-3}$ and $1 \times 10^{18}\,\mathrm{cm}^{-3}$. Heavier doping leads to backward diodes and finally to Esaki diodes where Esaki tunnelling (from conduction band to valence band) dominates in the forward direction (0 - 200 mV) leading to negative differential resistance. Both diode types are not very often used in silicon microelectronics but may find a strong upsurge with heterojunctions as explained later.

The widespread use of the metal oxide silicon field-effect transistors (MOSFETs) should not be forgotten since the vast majority of low power logic ICs are produced using such devices. This wide spread use is related to the basic unit of the CMOS circuit architecture, the inverter, only consuming significant power when switching, i.e. dynamic power. The static power dissipation is very small as it is only related to leakage currents in the transistor or circuit. The MOS inversion transistor is switched on by a gate signal creating a thin two-dimensional surface channel with highly concentrated minority carriers. The potential well for minority carriers is confined (Fig. 1.2) to the surface (inversion layer). The inversion layer thickness Z_{av} in Si (100) is given at low temperatures by

$$Z_{av}(\mathrm{nm}) \cong \frac{7}{\sqrt{n_s/\left(10^{12}\mathrm{cm}^{-2}\right)}} \tag{1.9}$$

where n_s is the sheet carrier density given by

$$n_s = \left(\frac{\varepsilon_{ox}\varepsilon_0}{q t_{ox}}\right)(V_g - V_T) \tag{1.10}$$

with t_{ox} the oxide thickness, ε_{ox} the oxide dielectric constant, V_g the gate voltage and V_T the threshold voltage. The inversion layer thickness, Z_{av} increases at room temperature at most by a factor of three, because of occupation of higher subbands.

It was in this type of MOSFET inversion layer device and not in the later developed but higher mobility GaAs MODFET that much of the semiconductor physics and low dimensional devices work in the 1960s, 1970s and 1980s was completed. Indeed, Klaus von Klitzing and colleagues discovered the quantum Hall effect using MOSFET inversion devices before the effect

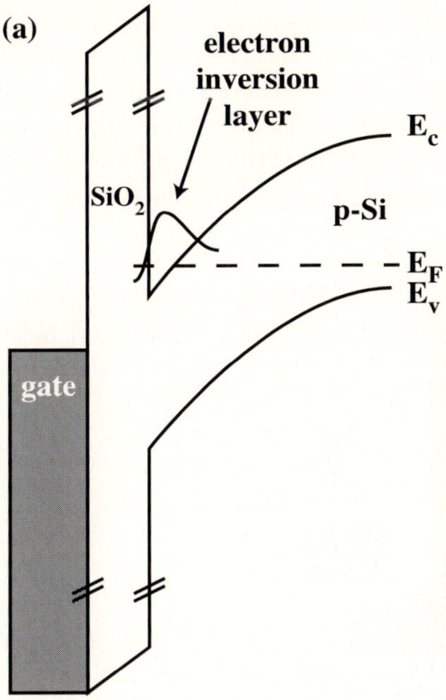

Fig. 1.2. A schematic diagram of (a) n-MOSFET and (b) p-MOSFET showing the inversion layers for the charge carrier

was observed in GaAs devices. He was awarded the Nobel Prize in 1985 for this achievement.

1.2 From Microelectronics to Nanoelectronics

Simply by shrinking the lateral dimensions into the sub-100 nm regime and the vertical dimensions into the sub-10 nm regions (e.g. the gate dielectrics thickness will be 1.5 to 3 nm) the existing microelectronics has already been converted into nanoelectronics.

This definition, however, is not the usual meaning of the term nanoelectronics as used by many researchers; it is more frequently used to describe new types of technology such as molecules, wires and other exotic structures on the nanometre scale. For the systems use of such new technologies, consideration of all the properties are required including

- materials
- devices
- integration techniques
- circuit architectures
- novel applications

Integration has been the key factor in the success story of silicon based microelectronics and should be considered as an essential step when assessing the potential of new solutions and technologies. We will therefore discuss some visible routes in a separate section. For the other levels we will structure the discussion by starting with the route of existing main stream technology and by comparing the new ideas with those either given by silicon based materials or by competitive materials.

The basic transistor type of recent digital circuits is the CMOS field effect transistor (FET) which uses inversion layers below the gate electrode (Fig. 1.3) to switch on the currents. The device operation will be explained later in the book (Chap. 5).

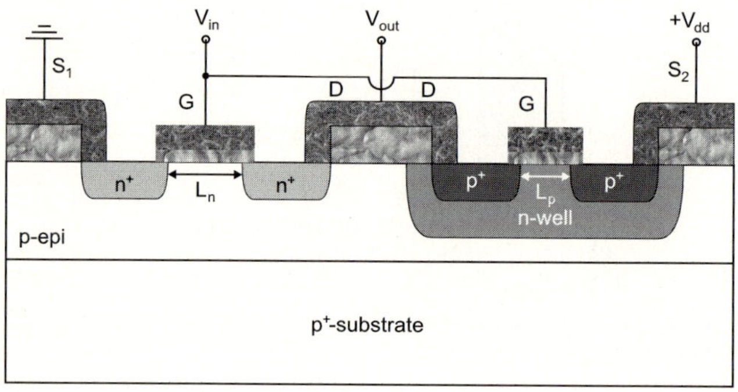

Fig. 1.3. The CMOS inverter scheme with a n–channel MOS (left side) and a p–channel MOS (right side). The input voltage, V_{in} is given to the connected gate electrodes **G**, the inverted output voltage V_{out} is taken from the common drain electrodes **D**. The source electrodes $\mathbf{S}_1$, $\mathbf{S}_2$ are on ground and at the supply voltage V_{DD}, respectively

The large scale integration (LSI) of MOSFETs reached the million transistor mark of very large scale integration (VLSI) over a decade ago and is now at the gigascale of ultra large scale integration (ULSI)). The reader interested in technological details of the breathtaking advance in complexity of ICs should consult one of the books in further reading at the end of the chapter.

The CMOS transistor size is predominantly reduced to increase the packing density in ICs but smaller gate length devices also benefit from higher speed. Shrinking of the most important lateral device dimensions – the gate length L_G – seems possible down to 20 nm or perhaps even smaller before partially or fully depleted schemes such as silicon-on-insulator (SOI) or double gate transistor schemes are required. One major concern is the power dissipation, as the density of transistors increases continuously by scaling. In general, the power dissipation P_{dis} of a CMOS circuit may be estimated from static power dissipation in standby (first term, equation 1.11) and from the dynamic switching power (second term, equation 1.11)

$$P_{dis} = V_{DD} I_{on} W 10^{-V_T/S} + C_L V_{DD}^2 f_c \frac{a}{2} \tag{1.11}$$

where V_{DD} is the supply voltage, V_T is the threshold voltage, S is the subthreshold swing (58 mV for fully depleted devices to 70 mV for partially depleted devices), W is the total width of the transistors, C_L is the total node capacitance, f_c clock frequency and a represents the transition probability of the logic gates. A full description of the CMOS architecture and power dissipation is given in Sect. 10.1. The requirement for having a reasonable current on/off ratio in a transistor (I_{on}/I_{off}) in order to minimize the standby power dictates a minimal threshold voltage

$$V_T = log \frac{I_{on}}{I_{off}} S \tag{1.12}$$

which is more than 0.4 V - 0.5 V for room temperature circuits. The power supply V_{DD} should be at least several tenths of a volt higher than V_T to cause the necessary driving current I_D which is related to the gate delay by

$$t_{\text{delay}} = C_L \frac{V_{DD}}{I_D} \tag{1.13}$$

In the short–channel devices, the threshold voltage decreases as the drain voltage V_{ds} increases due to two–dimensional electrostatic charge sharing between gate and the source/drain. Higher channel doping, retrograde vertical doping profiles and self–aligned halo implants have been shown to significantly reduce the short–channel effect around 100 nm channel length. As the gate oxide thickness is scaled down to below 2.5 nm the gate–tunnelling current increases exponentially. Alternatively, new high–k dielectric gate materials may replace the thermal oxide to yield the larger capacitance of a thinner oxide with an equivalent oxide thickness (EOT) while keeping the tunnelling current under control at the same time.

Multiple levels of interconnects stacked on top of the silicon chip dominate the design and placement of transistors in present ULSI chips. Effects of the interconnects as signal reflections, cross talk and propagation delays have become major barriers in the evolution of modern high–density, high–speed systems. With the replacement of Al alloys by Cu metallization and low k–intermetal layer dielectrics (ILD) an improvement in circuit performance by at most a factor of three has been obtained. Further improvement through new materials is difficult as few metals have significantly higher electrical conductivity and the insulators are already close to the dielectric constants of a vacuum. Optical global interconnects and the use of wireless microwave interconnects have been proposed to further improve circuit performance. The integration of wireless interconnects such as those in a cellular phone network is principally possible with existing Si technology. An efficient, switchable, silicon-compatible light source such as a laser or light emitting diode (LED) still remains a major problem for the realisation of optical interconnects.

For the ultimate CMOS scaling the short channel problems require alternative approaches by reforming the geometrical shape and/or by introducing heterostructure barriers. The geometrical approaches are mainly based on silicon on insulator (SOI), double gates and vertical transistors. Heterobarriers are either used to suppress drain induced barrier lowering (DIBL) or to increase the channel performance, e.g. by strained-silicon channels which may be created by virtual substrates with strain relaxed SiGe on Si.

The consideration of speed in a circuit environment necessitates re–evaluation of suitable architectures for computing and signal processing. Since the current drive of shrunken devices with only a few electrons will be sufficiently small, the interconnects and fan out numbers must be small as well. The best way to alleviate the global interconnect bottleneck is to adopt different architectures which can minimize interconnects. Neural networks (NN) and cellular automation (CA) belong to these architecture classes. The interested reader is referred to the references in further reading.

Quantum mechanics offers a new fundamental unit of information, the quantum bit or qubit. In contrast to the state of a bit which is specified either "0" or "1" the qubit can also exist in a complex superposition of both states. In addition, different qubits can be entangled within a short time provided the respective phases are preserved. Entanglement allows massive quantum parallelism for computation entitled quantum computing. Such technology is extremely far from market and does not provide a simple general computational machine (or Turing machine). Quantum computers if realised are likely to be used for simulation of quantum systems or solving specific non-linear problems such as factorising numbers, the travelling salesman problem or database searching.

As the feature sizes of transistors are scaled down, the number of electrons decreased and eventually reach a single electron state. In a single electron transistor (SET) a small Si island is coupled to two external electron reservoirs through tunnelling barriers. Because of the discrete nature of electric charge, q, a finite charging energy q^2/C (C capacity of the island) has to overcome by the next electron, a process which is called Coulomb blockade. For a 10 nm Si sphere within an oxide the charging energy, the selfcapacitance and the ground state are given by 72 meV, 2.2 aF and 3.8 meV, respectively.

In our device chapters we concentrate mainly on quantum effect devices for room temperature operation and on single crystal heterostructure technologies which are either already competitive with present technologies, can be integrated with present technologies to produce better performance or show the potential of near future competitiveness with present technology.

1.3 Self–ordering

The vertical nanometer structures (<20 nm) required to create substantial quantisation in Si can be easily manufactured using modern epitaxial methods

(Chap. 2). The economic manufacturing of the lateral structures of around 10^{11} devices per chip (roughly anticipated for 2016) is a much more critical issue. Self-ordering of densely packed nanostructures could be an attractive alternative, especially, if new architectures with low interconnect lengths are considered. At present, however, the lack of an appropriate architecture combined with the random nature of deposition precludes the significant development of such schemes for microelectronic circuit demonstrators.

An obvious way for self-ordering uses the Stranski–Krastanov (SK) growth mode. In this growth mode, which is explained in more detail in the Chap. 2, a high number of islands (10^9 to $10^{11}\,\mathrm{cm}^{-2}$) is nucleated on top of a wetting layer (Fig. 1.4). This growth mode is favoured under certain circumstances in strained heterostructures, e.g. in Si/Ge. Homogeneous nucleation is dominated by the statistics of supersaturation, leading to a rather broad distribution of sizes and distances. Research is devoted to heterogeneous starting conditions which may result in narrower distributions and periodical arrays. From the many approaches, we mention three ones in order to demonstrate the principles.

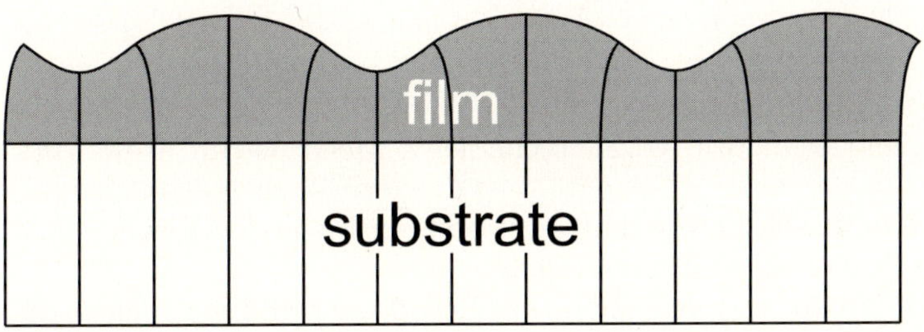

Fig. 1.4. Film surface and lattice plane distortion in a strained film grown in the Stranski–Krastanov growth mode

In stacked multilayers the nucleation in the next layer is predominantly above larger islands and the distance is more regular. The nucleation on already strained sites is energetically favourable, but many kinetic parameters such as temperature, coverage, layer spacing influence the result. Whereas the above treatment improves the periodicity of the array it leaves the absolute position of the net completely undetermined. In order to position the array, oxide windows or etched surface ridges can be used because the islands preferentially nucleate at the window or ridge edge. Surface corrugations created by misfit dislocations act as heterogeneous nucleation centers. Periodic arrays will be created if the dislocation network is periodically arranged as obtained under certain equilibrium conditions. An arrangement of this type is shown in Fig. 1.5.

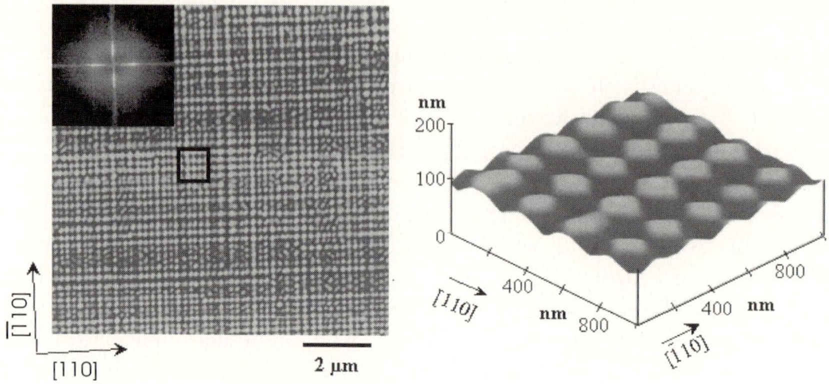

Fig. 1.5. An array of SiGe islands on Si with preferred nucleation on the misfit dislocation glide planes. With permission from C.Teichert, 2001

1.4 Further Reading

1. C.Y. Chang and S.M. Sze, ULSI Technology, McGraw-Hill, New York (1996)
2. S. Luryi, I. Xu, A. Zaslovsky, Future Trends in Microelectronics - The Road Ahead, J. Wiley, New York (1999)
3. R. Campano, Technology Roadmap for European Nanoelectronics, EC (2000)
4. K. Brunner, Rep. Prog. Phys. **65** 27 (2002)
5. International Technology Roadmap for Semiconductors, 2003 Edition
6. C. Teichert, Physics Reports, Vol.365 Number 5-6 (2002)

2. Material Science

2.1 Growth and Preparation Methods (MBE, CVD, Implantation, Annealing)

In silicon (Si) based systems quantum effect structures need small dimensions (typically less than 20 nm), because of the high effective masses. Such dimensions are difficult to obtain with lateral patterning methods. Lateral structures are usually first created in a surface resist layer and then transferred by etching, ion implantation, and or diffusion. Microelectronic manufacturing is expected to reach 25 nm lateral feature sizes in the year 2010 (Tab. 2.1).

Table 2.1. Lateral structure dimensions in processor manufacturing. Source: ITRS Roadmap 2003

Year	2003	2004	2005	2006	2007	2008	2009	2010	2013	2016
Node (nm)		90			65			45	32	22
Printed (nm)	65	53	45	40	35	32	28	25	18	13
Physical (nm)	45	37	32	28	25	22	20	18	13	9

For experimental research, however, several methods with high lateral resolutions are available, predominately electron beam lithography (EBL) and focused ion beam (FIB) methods. Standard EBL (Fig. 2.1) is now used routinely for mask preparation for optical lithography but the writing time increases significantly as the resolution is increased, since EBL is a serial process.

The vertical control of dimensions by deposition techniques can be at least ten times smaller than the lateral control using state of the art manufacturing methods. The reason is the intrinsic vertical atomic layer ordering present in most deposition methods. A good interface normally requires a single crystalline structure, which typically requires epitaxy conditions for the deposition.

Epitaxy is the oriented growth on top of a substrate. The substrate is assumed to be single crystalline and the deposited epitaxial film can be single

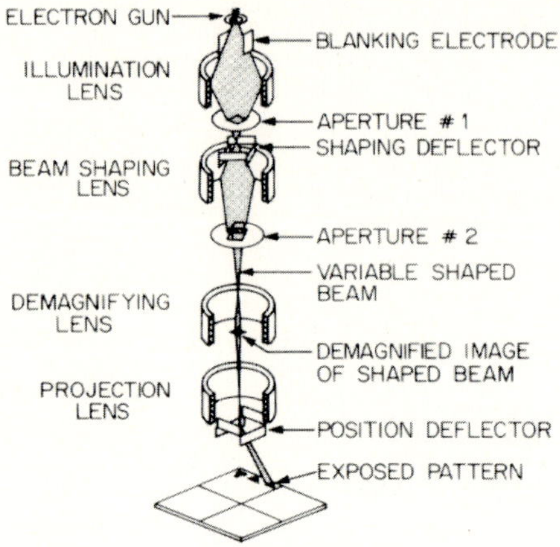

Fig. 2.1. A schematic diagram of an electron beam lithography (EBL) system. An electron beam sensitive resist layer on top of the substrate is serially irradiated by a finely focused electron beam. The resolution depends on the focus width, on the resist thickness, on back scattering of electrons by the substrate, and on the development process. The resolution of the final structure is additionally dependent on the exact pattern transfer from the resist pattern

crystal, poly-crystalline or amorphous. For the rest of this chapter we will only consider single crystal epitaxial films. In the case of covalent bonded diamond crystals the orientation of the film is the same as the substrate, e.g. a (100) oriented film is growing on a (100) substrate. The term heteroepitaxy is used when the deposited film material differs from that of the substrate. The deposited film is still homoepitaxial if the doping is different from the substrate. Epitaxial techniques, therefore, provide a number of options for producing structures designed for quantum based electronics.

There are many material combinations which are interesting for quantum electronic devices, such as semiconductor/semiconductor, metal/semiconductor or insulator/semiconductor interfaces. Examples of crystalline insulator/Si structures and of metal/Si structures include the calcium fluoride (CaF_2)/ diamond lattice (Table 2.2) and the silicide materials (Table 2.3), respectively.

Table 2.2. Lattice constant a_0 of selected insulator materials and mismatch with Si (300K)

Insulator	CaF_2	SrF_2	BaF_2
Lattice constant a_0 (nm); (300 K)	0.54629	0.57996	0.62001
Mismatch with Si (300 K) (%)	+0.60	+6.8	+14.2

For comparison: Si $a_0 = 0.543$ nm; Ge $a_0 = 0.565$ nm

Different modifications, chemical instabilities, different layer stackings, thermal expansion and lattice mismatch are considerable obstacles for insulator application in single crystalline devices, whereas silicides are widely accepted as contact materials. Chemical similarity and good lattice match are the essential factors for the ability to grow high quality covalent semiconductor/semiconductor interfaces.

Table 2.3. Lattice mismatch of selected metal / Si interfaces

Silicide	$MnSi_2$	$FeSi_2$	$CoSi_2$	$NiSi_2$	$TiSi_2$	PtSi
Lattice type	tetra-gonal	tetra-gonal	cubic	cubic	ortho-rhombic	ortho-rhombic
Lattice mismatch (%)	1.7	0.9	1.2	0.4	-	9.5

Within the group IV column of the periodic table (Table 2.4) silicon and germanium have a completely miscible alloy ($Si_{1-x}Ge_x$) with lattice mismatches ranging from 0 to 4.2% for pure Ge lattice matched to silicon. Carbon concentrations between 10^{18} cm^3 and 10^{21} cm^3 may be incorporated into Si or Ge lattice sites under metastable growth conditions.

Table 2.4. Properties of Group IV compounds (diamond or zincblende lattice)

Compound	C (diamond)	$\alpha - SiC$ (3C)	Si	Ge	$\alpha - Sn$
Lattice constant a_0 (nm)	0.3567	0.436	0.5431	0.5646	0.65
Indirect Bandgap ($E_{g,ind}$ in eV)	5.45	2.2	1.12	0.66	0
Direct Bandgap ($E_{g,dir}$ in eV)	6.5		3.2	0.80	0
Miscibility with Si	< 3 % , metastable	< 20%	-	complete	only metastable

Under equilibrium, the carbon (C) concentration is low (10^{17} cm^{-3} at temperatures below the melting point) with the carbon being mainly incorporated at interstitial sites. Some additional material properties of the SiGe system are given in Table 2.5. The completely miscible $Si_{1-x}Ge_x$ alloy follows rather closely a linear dependence (Vegard's law) (mismatch f to Si: f

Table 2.5. Properties of Si and Ge

Material	Si	Ge
Bandgap (eV)	1.12	0.66
Electron affinity χ (V)	4.05	4.0
Effective masses		
of heavy holes m_{hh}/m_0	0.49	0.28
Effective masses		
of light holes m_{lh}/m_0	0.16	0.044
Effective masses of (100) electrons		
in longitudinal direction m_l/m_0	0.98	1.64
Effective masses of (100) electrons		
in transversal direction m_t/m_0	0.19	0.082

$m_0 = 9.1091 \times 10^{-31} kg$
Note: The conduction band minimum for Ge is in the (111) direction (L-point) but for nearly all strained SiGe alloys the minimum occurs in (100) direction (X-Point). For interpolation the properties of L-electrons in Ge ($m_l = 1.64$, $m_t = 0.082$) should not be used.

$= 0.042x$) with a small quadratic deviation (Fig. 2.2). The lattice constant a_{SiGe} is given exactly by

$$a_{SiGe} = 0.5431(nm) + 0.01992x(nm) + 0.0002733x^2(nm) \tag{2.1}$$

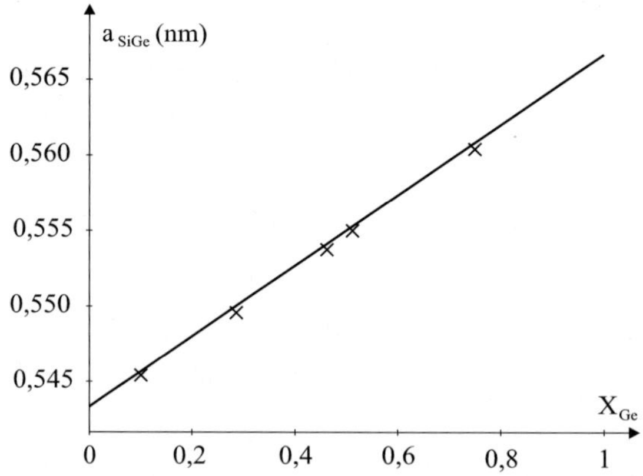

Fig. 2.2. The lattice constant, a_{SiGe}, of the SiGe alloy as a function of the Ge content x. The solid line compares Vegard's law with experimental values (crosses)

In Si based microelectronics the most important surface orientation is the (100) orientation at which the best controlled amorphous oxide/silicon structure for CMOS transistors can be realized. A cut along the (100) plane gives two dangling bonds per atom (Fig. 2.3a). By a surface reconstruction (Fig. 2.3b) the number of dangling bonds is reduced delivering lower surface energy, lower symmetry and larger cell sizes, e.g. in the case of a (100) face a (2×1) a_0 cell.

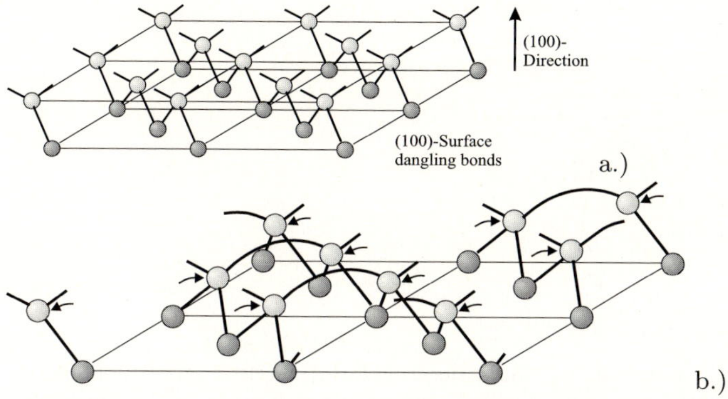

Fig. 2.3. Reconstruction of the (100) surface. A cut through the bulk (a) is compared with a (2 x 1) surface reconstruction (b). In the diamond lattice the distance between adjacent planes in (100) direction is given by $a_0/4$ (0.136 nm for Si)

Epitaxy is performed at temperatures well beyond the melting point of the deposited material, T_m (1428° C for Si) either from a metallic melt solution (liquid phase epitaxy (LPE)), from the vapour phase by a chemical reaction (vapour phase epitaxy (VPE) or chemical vapour deposition (CVD)) or from atoms and ions in a vacuum surrounding (physical vapour deposition (PVD)). Molecular beam epitaxy (MBE), a sophisticated PVD method, and advanced CVD methods are the main fabrication techniques for role for quantum device structures in research and production.

A typical growth process will now be explained using as an example silicon MBE. The principal design of a MBE system is shown in Fig. 2.4. The process is performed in a very clean ultra high vacuum (UHV) surrounding. The substrate is precleaned by a combination of ex-situ chemical treatment and in-situ thermal treatment. A thin (1 nm) chemical oxide is desorbed by a 5 minute anneal at 900 °C. Hydrogen covered substrates may be cleaned at lower temperatures (550 °C - 650 °C). The epitaxial growth is started immediately after the clean by opening the shutters above the effusion cells, which are temperature stabilized to obtain a constant molecular flux density. Chemically aggressive materials like molten silicon are evaporated by electron beam heating of the center of a piece of material. The flux composition can be changed abruptly by opening and closing the corresponding effusion cell shutters. Two doping sources (n-type and p-type) and three matrix element

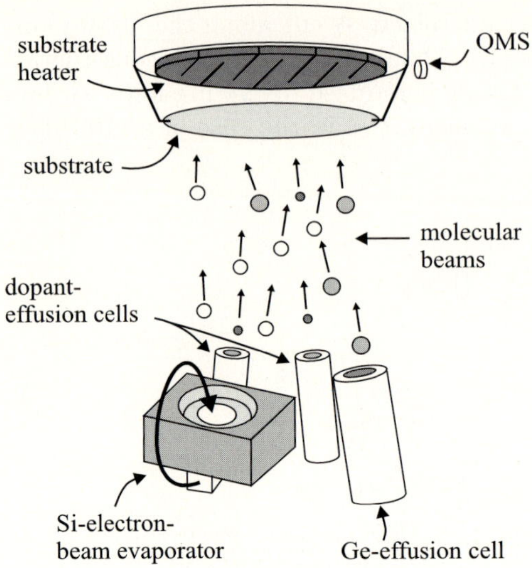

Fig. 2.4. A schematic diagram of the silicon MBE process. The substrate is mounted on top and heated from the backside by an infrared heater. The molecular beams are evaporated from separate thermal sources (effusion cells for the dopants and germanium or carbon, electron beam evaporators for silicon and sometimes also for germanium). In situ monitoring is achieved by several methods, here a quadrupole mass spectrometer (QMS) for flux measurements is shown

sources (Si, Ge and C) are the minimum source configuration with additional possibilities for other sources (metals, insulators, gases).

The growth process is monitored by in-situ analysis with electrons and optical beams. Examples are reflection high energy electron diffraction (RHEED) to observe the surface; electron induced emission spectroscopy (EIES) or quadrupole mass spectrometry (QMS) for individual flux control; pyrometry or thermoelectric voltage measurement of temperatures; and ellipsometry or reflection interferometry for film thickness monitoring. A clean surface turned out to be an essential prerequisite for reducing the growth temperature from the usual 1050 °C-1150 °C to 500 °C-700 °C. Even lower temperatures are required if the limits set by amorphous growth at 100 °C to 200 °C are to be reached. The substrate is radiation heated by a graphite meander mounted on the backside of the substrate. Several kilowatts of power are needed to operate the sources and the substrate heater. During operation the pressure raises from the base pressure of 3×10^{-11} mbar to the 10^{-10} mbar range. This is acceptable as long as hydrogen (H_2) is the dominating gas which has to be monitored by residual gas analysis (Note: The unit mbar is 100 times the international unit Pascal (Pa) $= 1\,\mathrm{Nm}^{-2}$, atmospheric pressure (AP) is roughly 10^5 Pa). The pressure defines the molecular density which is about $10^{13}\,\mathrm{m}^{-3}$ in the given pressure regime producing a mean free path for a molecule of about 100 km. This defines the impinging rate on a wall ($10^{15}\,\mathrm{m}^{-2}$) and the time for monolayer coverage of a surface (10^4 s/S, when S is the sticking coefficient of a specific gas). A technical system (Fig. 2.5) needs in addition to the growth chamber additional UHV-chambers for a load lock, wafer storage, pretreatment and analysis. A wafer transfer system transports the wafer between the load lock and the wafer storage or the wafer holder.

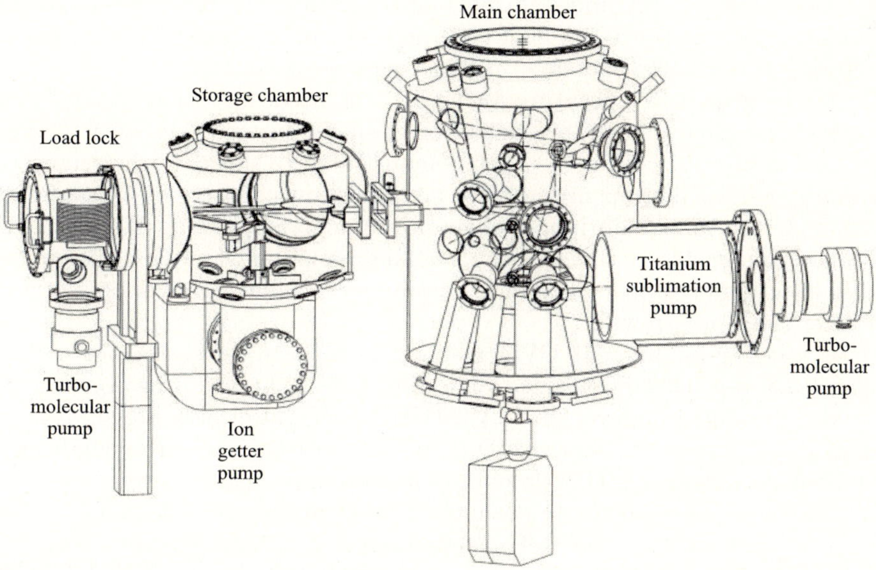

Fig. 2.5. A commercial (Leonardo) three chamber Si-MBE system with the main chamber for growth, a storage chamber with a 25 wafer magazine, and a load lock. UHV-conditions are maintained by two turbomolecular pumps, a titanium sublimation pump and an ion getter pump. The transfer system operates from the storage chamber

Standard silicon CVD (Fig. 2.6) is typically carried out at atmospheric pressure and involves the pyrolysis at an elevated temperature of the precursor gas of silane or silicon halide ($SiH_{4-z}Cl_z$ with $1 \leq z \leq 4$). Radio frequency coils are used to heat the system to temperatures ranging from 900 ^{o}C to greater than 1100 ^{o}C to volatilise nominally contaminating species such as water, oxygen or carbon.

While such high temperature can be tolerated for the homoepitaxial blanket growth of silicon onto a silicon wafer without dopants, the addition of either doping or germanium into the growth system requires significantly lower temperatures. Autodoping occurs at temperatures above 1000 ^{o}C which involves the diffusion of dopants from the substrate into the epitaxial film creating unwanted anisotropic distortions in the epitaxial layer. The reduction of the system operating pressure serves to eliminate a slowly floating boundary layer of gas immediately above the substrate, allowing the more rapid transport of evaporated dopant away from the substrate, thus reducing the auto-doping effect. The reduction of the operating temperature also reduces both the rate of dopant evaporation into the gas stream and solid state diffusion. For strained $Si_{1-x}Ge_x$ layers there are two main problems with high temperature. The first is the roughing or development of surface undulations from high temperature growth and the second is diffusion of the germanium. The activation energy and diffusivity of Ge from a $Si_{0.7}Ge_{0.3}$ layer into silicon have been measured to be 4.7 eV and 0.04 m^2/s respectively suggesting diffusion of some nanometers even at a temperature of 1000 ^oC for 1 minute.

As the growth temperature for CVD is reduced, lower background pressures are required to maintain an oxide free silicon surface to grow on. Oxygen content in $Si_{1-x}Ge_x$ films has been demonstrated to substantially reduce the minority carrier lifetime in the films, an important property for bipolar transistors. Chemical equilibrium data for the maintenance of an oxide free silicon surfaces demonstrates that it is the partial pressure of water which is the limiting effect and requires ultra high vacuum background chamber pressures for low temperature growth.

A number of different CVD reactors have been developed for the low temperature growth of strained $Si_{1-x}Ge_x$ films. These can be conveniently divided into ultra-high vacuum CVD (UHV-CVD) at growth pressures of less than 10 Pa and low pressure CVD (LPCVD) with pressures ranging from 10 to 1000 Pa. Other systems do exist but most have been research tools and have not been developed into production tools. Source gases for CVD include SiH_4, Si_2H_6, SiH_2Cl_2 and GeH_4 while doping is achieved using AsH_3, PH_3 and B_2H_6. The majority of commercial LPCVD reactors are single wafer tools while the IBM UHV-CVD system is a batch tool allowing growth of 25 wafers or more at a time.

Whatever is the method of deposition of the epitaxial layer, the result is an atom sticking to the surface. These adsorbed atoms are called adatoms and are a precursor state before the atoms are incorporation into the lattice. The adsorption energy, E_{ad} is lower than the binding energy, E of an atom in the crystal, usually $\frac{1}{2}E$ to $\frac{2}{3}E$. When an atom is moving toward the surface (Fig. 2.7) the balance of attractive and short ranged (atomic radius) repulsive atomic forces create a potential well at the equilibrium position for the adatom. In order to stick at this adatom position, the atom or molecule must transfer its energy, momentum and moment of torque to the solid. If this does not happens rapidly enough, as is often the case with molecules with a moment of torque, we refer to a sticking coefficient S smaller than unity.

Even if the adatom is adsorbed, the adatom may later escape by a desorption step caused by thermal vibrations. If there is an equilibrium between the solid and the vapour phases, the adsorption and desorption events are balanced. For growth the vapour pressure has to be higher than the equilibrium

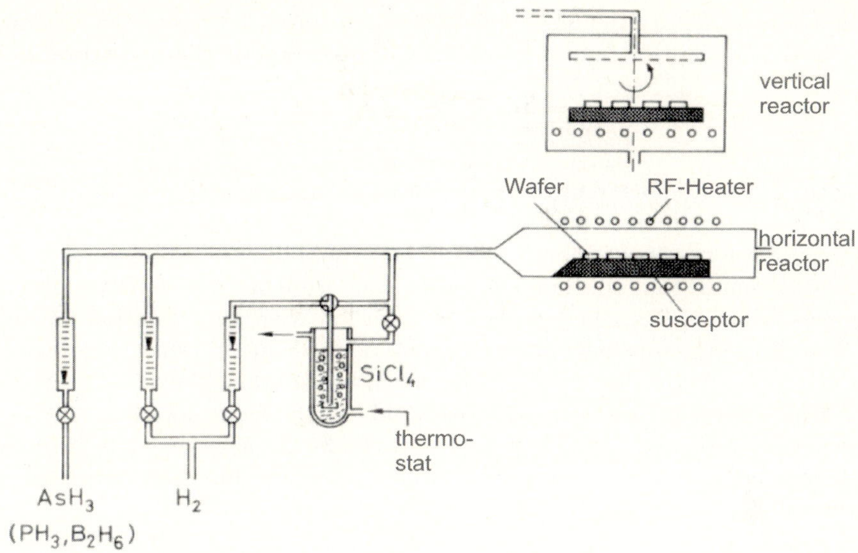

Fig. 2.6. A schematic diagram of a CVD epitaxy tool

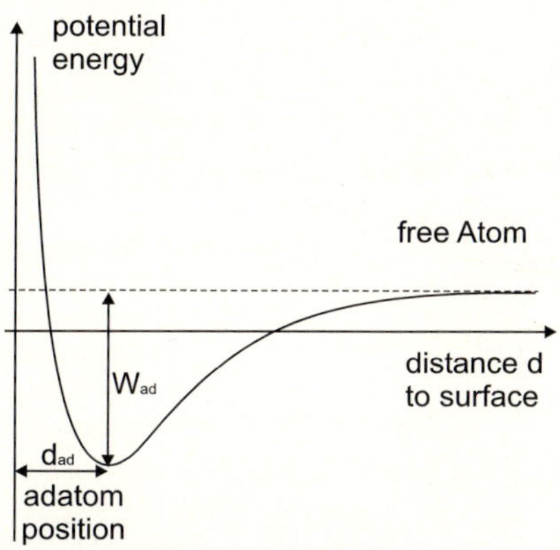

Fig. 2.7. The potential energy as function of distance to the surface for the adsorption of adatoms

pressure (supersaturation). The flux F impinging on a surface is connected to the vapour pressure, p by

$$F = p\sqrt{\frac{N_A}{2\pi M k_B T}} \tag{2.2}$$

where $N_A = 6.022 \times 10^{29}\,\mathrm{m}^{-3}$ is Avogadro's number and M is the molecular weight.

In a simple picture the desorbing flux F_{des} is proportional to the adatom density n_s and the Boltzmann probability of an energetic thermal vibration.

$$F_{des} = n_s \omega_q \exp\left\{-\frac{E_{ad}}{k_B T}\right\} \tag{2.3}$$

where ω_q is the frequency of thermal vibrations usually assumed as 10^{12} Hz to 10^{13} Hz (see Sect. 3.6).

A regular network of surface positions is available for the adatoms. Indeed, we have to assume that adatoms can easily jump from one position to another, a process which is described as surface diffusion with an activation barrier U_S. The diffusing atom may desorb or may be incorporated into the crystal. The energy gain for the incorporation from an adatom place is E_s where $E_s + E_{ad} = E$ (Fig. 2.8). In a 70 years old "Gedankenexperiment" Kossel showed that repeated joining of atoms to kinks on surface steps delivers continuous crystal growth with every atom gaining the binding energy when incorporated at the kink.

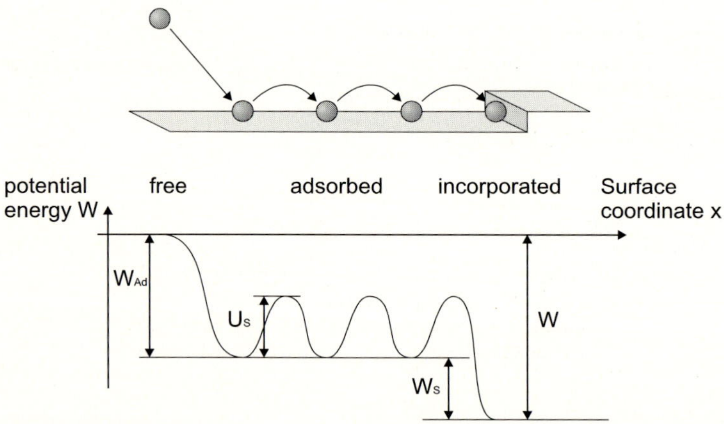

Fig. 2.8. A schematic diagram of the incorporation of adatoms. The process involves the adsorption of the adatom (with energy gain E_{ad}), the surface diffusion (with energy barrier U_S) and the incorporation of the adatom in the crystal at a surface step (with binding energy E)

The equilibrium concentration of adatoms n_{S0} is then given by the balance between the adatom position and any energetically favourable step positions.

$$n_{S0} = N_S \exp\left\{-\frac{E_S}{k_B T}\right\} \tag{2.4}$$

where N_S is the surface density of atoms. The diffusion coefficient D_S is given by

$$D_S = \frac{1}{N_S} \nu \exp\left\{-\frac{U_S}{k_B T}\right\} \tag{2.5}$$

where the mean distance between neighbouring adatom places is given by $\sqrt{\frac{1}{N_S}}$. Before an adatom can be adsorbed it walks on average for the desorption time, τ_{Des} which defines a diffusion length λ_S on the surface of

$$\lambda_S = \sqrt{D_S\ \tau_{Des}} = \sqrt{\frac{1}{N_S}} \exp\left\{\frac{E_{ad} - U_S}{2k_B T}\right\} \tag{2.6}$$

Note that under growth conditions, the incorporation of the adatom in the crystal is a competing process to the desorption process and therefore the majority of adatoms will walk smaller distances than given by λ_S.

Let us now consider a growth experiment by supplying fluxes F of the different matrix and doping elements to a surface. For one element we define a linear supersaturation σ by

$$\sigma = \frac{F}{F_0} - 1 \tag{2.7}$$

(F_0 equilibrium flux of the element given by $F_0 = n_{S0}/\tau_{Des}$).

On the surface the adatom concentration n_S will also increase above the equilibrium value n_{S0}. We define a surface supersaturation σ_S by

$$\sigma_S = \frac{n_S}{n_{S0}} - 1 \tag{2.8}$$

Depending on the composition, three different basic growth modes are possible (Fig. 2.9).

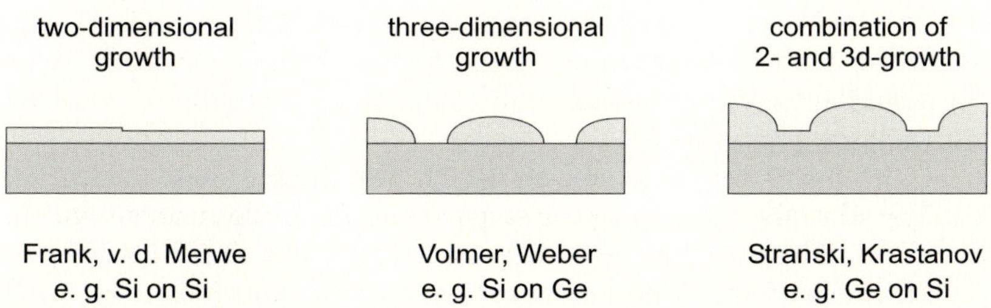

Fig. 2.9. The basic growth modes of two-dimensional (or Frank–v. d. Merwe) growth (2D), three-dimensional (or Volmer–Weber) growth (3D) and a mixture from 2D to 3D growth, the Stranski–Krastanov mode

A qualitative understanding is possible by using the droplet model (Fig. 2.10) where the force balance at the rim of a droplet is considered. The surface tension Σ (specific surface energy) is composition dependent, the interface energy Σ_i depends also on the strain and dislocation structure.

$\Sigma_S = \Sigma_i + \Sigma_f \times \cos(\Theta)$
$3D\ growth: \Sigma_S - \Sigma_i < \Sigma_f$
$2D\ growth: \Sigma_S - \Sigma_i > \Sigma_f$

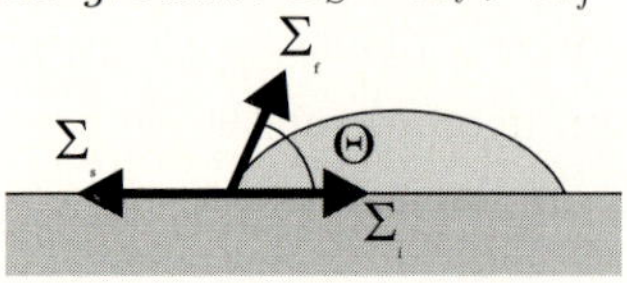

Fig. 2.10. The droplet model of three-dimensional island growth. The force between the surface tensions of Σ_S and Σ_f balanced by the interface energy Σ_i leads to an island inclination angle Θ when $\Sigma_S - \Sigma_i < \Sigma_f$

The most simple case is homoepitaxy, the growth of for example a Si epilayer on a Si substrate. The minimum surface energy is obtained by two-dimensional, flat growth. The surface morphology involves atomic steps of usually monolayer (h) or bilayer (2h) thickness. The low temperature regime (below 100°C for Si), where amorphous or highly defective growth proceeds, or the high temperature regime (above 1250°C for Si), where surface roughening occurs, are not taken into account, because these regimes are not presently used for the growth of epilayers which are used in quantum electronic structures. On silicon where the dislocation density is negligible, there are two sources which create surface steps. At higher temperatures (roughly above $T_m/2$, T_m melting point, Si $\approx$ 1700 K) steps from the misorientation, i, of the substrate are present. Even with nominally oriented substrates (e.g. (100)) a small misorientation ($i < 0.5°$, arc $i \cong \tan i < 0.0085$) is technically unavoidable which leads to terraces of width > 15 nm separated by monoatomic steps. At lower temperatures (roughly below $T_m/2$) adatoms are not fast enough to move to the steps and nucleate into two-dimensional islands (Fig. 2.11). When all adatoms reach the already existing misorientation steps (in substrates which are not as perfect as Si the dislocation steps also act in a similar way) the monoatomic steps move laterally forward by the adatom capture (step flow).

When at the lower temperatures in which two-dimensional nucleation takes place, the adatoms can join to the steps at the rim of the nucleus within smaller distances. The nuclei grow and coalesce to form a single monolayer, so that the 2D nucleation is a periodic process. A critical nucleus (Fig. 2.12) is defined by the size in which the growth by the capture of adatoms is more probable than the decay of the nucleus.

Therefore with high supersaturation which results in high adatom density the critical size of the nucleus is smaller. For the extremely high supersaturation which occurs during Si-MBE, two joining adatoms probably already create a critical nucleus. The basic picture is somewhat blurred by the loss of symmetry from the surface reconstruction which results in highly anisotropic diffusion and two step types with different kink densities (Fig. 2.13). A detailed discussion is beyond the scope of this book. In either case the minimum

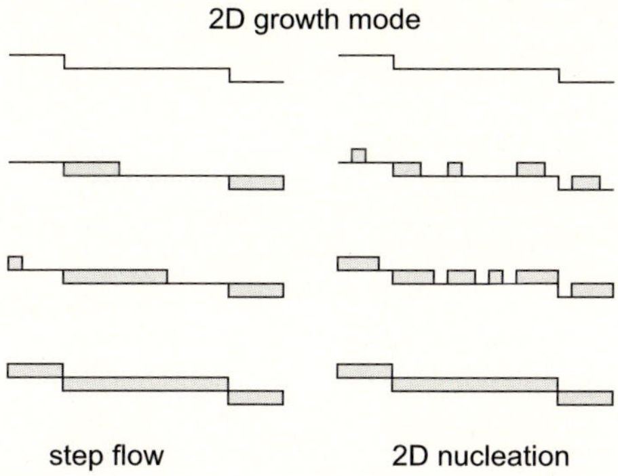

Fig. 2.11. The two-dimensional growth by step flow or 2D island nucleation (side view)

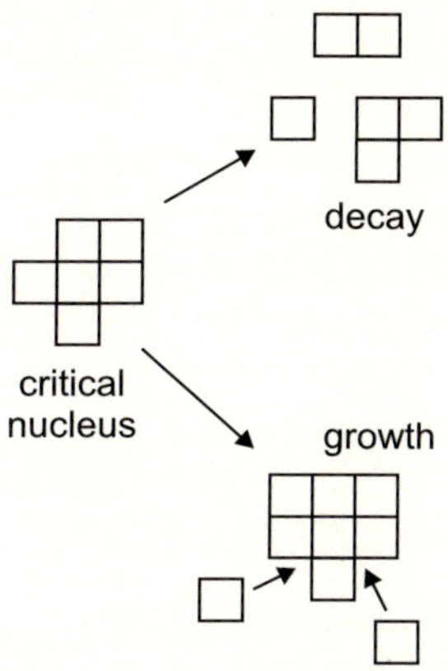

Fig. 2.12. The two-dimensional nucleation (top view). The size of the nucleus fluctuations and a critical nucleus size is achieved when several adatoms have joined. The size and binding energy of this nucleus is high enough that a decay of the nucleus is less probable than the further growth by adatom capture

step density is defined by the misorientation. In the step flow regime the number of steps is constant whereas in the 2D nucleation regime the step density oscillates above the minimum step density. With very sensitive surface monitoring methods like electron diffraction (RHEED) these oscillations in the 2D nucleation regime can be observed through intensity variations.

We will treat as an example the simplest case of step flow within the framework of a theory developed by Burton, Cabrera, and Frank (BCF-theory). Let us consider a regular array of misorientation steps (Fig. 2.14). The terrace width L is defined by the misorientation i.

$$\frac{h}{L} = \tan i \approx \arg i \tag{2.9}$$

Fig. 2.13. A scanning tunnelling microscopy (STM) image of a Si (100) surface depicting two different step types (S_A, S_B) along with separation terraces with (2×1) and (1×2) reconstructions

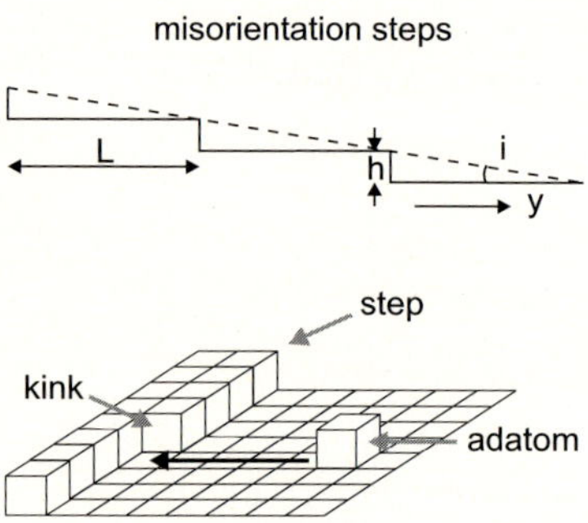

Fig. 2.14. A regular array of misorientation steps. The misorientation (inclination i) leads to terraces of width L separated by steps with height h. The steps move by the capture of adatoms on sites with kinks

The BCF-theory is a surface diffusion theory with specific conditions for particle conservation at the steps. Generally particle conservation is described by the continuity equation,

$$\frac{\mathrm{d}n_S}{\mathrm{d}t} + \nabla \cdot \boldsymbol{S} = G_S - R_S \tag{2.10}$$

$$\boldsymbol{S} = -\nabla n_S \tag{2.11}$$

where $\boldsymbol{S}$ is the surface flux vector, G_S, R_S the generation and recombination rates, respectively. The trick in the BCF-theory is the choice of the boundary conditions of (2.10). Only the terraces are considered where the steps are outside. With this choice and the assumption that the adatoms are only captured at steps, the recombination term in the differential equation contains only the desorption term. The adatom incorporation, therefore, will be treated by the boundary conditions.

In the one-dimensional (coordinate y- perpendicular to the steps along the surface) and stationary form ($\frac{dn_S}{dt} = 0$), the equation reads

$$\frac{\mathrm{d}^2 n_S}{\mathrm{d}y^2}\lambda_S^2 - n_S + F\tau_{Des} = 0 \tag{2.12}$$

If we use the assumption that the step acts as a perfect sink (Fig. 2.15) for adatoms, the boundary conditions may be written as $n_S = n_{S0}$, at $y = \pm\frac{L}{2}$ (the y-axis origin is given between two steps to obtain symmetrical solutions). The solution for the local surface supersaturation is given by

$$\sigma_S = \frac{n_S}{n_{S0}} - 1 = \sigma\left[1 - \frac{\cosh\frac{y}{\lambda_S}}{\cosh\frac{L}{2\lambda_S}}\right] \tag{2.13}$$

with $\sigma = \frac{F}{F_0} - 1$, $\cosh(u) = \frac{1}{2}[\exp(u) + \exp(-u)]$.

The adatom concentration has its maximum half way between two steps. The local concentration gradient drives a diffusion flux $|\mathbf{S}|$ which is highest at the steps.

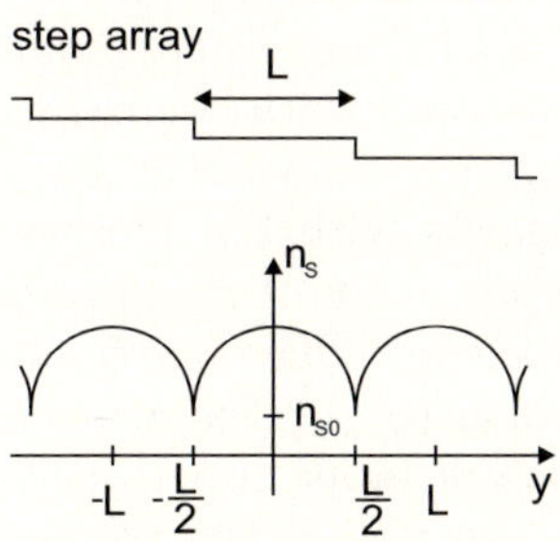

Fig. 2.15. Local concentration of adatoms on a step array. BCF-theory with the assumption of steps as perfect sinks for adatoms.

The simple BCF-theory of step flow describes the homoepitaxial growth of MBE-silicon fairly well in the temperature regime between roughly 550 °C and 900 °C. The lower temperature bound is caused by the onset of two-dimensional nucleation which is well documented by the appearance of RHEED oscillations. The transition temperature depends on the misorientation i (with terrace length L) and growth rate R (for supersaturation σ). The upper temperature bound is caused by surface defects and the surface roughening which allows adatoms to also be incorporated outside the steps and nuclei. This temperature value is uncertain, because MBE experiments are usually well below 900 °C and in CVD experiments the surface kinetics are overlapped by mass transport in the vapour phase, by chemical reactions and by adsorption of hydrogen and reaction products.

In the typical Si-MBE temperature regime of step flow growth (typically 550 °C-750°C) a further simplification of the BCF theory can be made. Si-desorption is very weak in this temperature regime and can be neglected, which is mathematically described by the inequality $\lambda_S \gg L$. The differential equation (2.12) then reads

$$D_S \frac{\mathrm{d}^2 n_S}{\mathrm{d}y^2} + F = 0 \tag{2.14}$$

with the solution for the step array

$$n_S - n_{S0} = \frac{F}{2D_S}\left[\left(\frac{L}{2}\right)^2 - y^2\right]. \tag{2.15}$$

We will now give a simple example, where we choose a temperature $T = 900\,\mathrm{K}$ (627°C), $F = 7 \times 10^{14}\mathrm{cm}^{-2}\mathrm{s}^{-}1$ (1 monolayer (ML) per second), $L = 30$ nm ($i = \frac{1}{4}^{\circ}$), $E_S = 2.0\,\mathrm{eV}$, $U_S = 0.6\,\mathrm{eV}$, $\omega_q = 10^{13}$ Hz. Then we calculate a low value for the equilibrium adatom density $n_{S0} = 1.84{\times}10^3\,\mathrm{cm}^{-2}$, a rather high surface diffusion coefficient $D_S = 4.8{\times}10^{-6}\,\mathrm{cm}^{-2}\mathrm{Hz}$ describing the good surface mobility of adatoms and a maximum adatom density of n_{Smax} between two steps ($y = 0$) of $n_{Smax} = 1.5{\times}10^8\,\mathrm{cm}^{-2}$. The equilibrium adatom density is already low and decreases steeply with decreasing temperature (70 K decrease yields an order of magnitude decrease in n_{S0}). The diffusion coefficient also decreases but more slowly with the temperature (a 200 K temperature decrease is needed for an order of magnitude decrease in D_S corresponding to the lower activation energy). The maximum adatom density increases with decreasing temperature as ($\frac{1}{D_S}$). This increase in adatom density favours two-dimensional nucleation at lower growth temperatures. To prove the inequality $\lambda_S \gg L$ we calculate $\tau_{Des} = 60\,\mathrm{s}$ ($E_{ad} = 2.55\,\mathrm{eV}$ is assumed) and $\lambda_S = 17\,\mu\mathrm{m}$ which is 500 times higher than the step distance L. The diffusion length increases with decreasing temperature with an activation energy of $\frac{1}{2}(E_{ad} - U_S)$.

The simple assumption that a step acts as a perfect sink for adatoms can be replaced by more sophisticated models. In one of these models the adatoms at the upper terrace have to overcome an energy barrier (Schwoebel barrier) to be captured. This model predicts step bunching where a step array is separated in regions with lower step densities and ripples with higher step densities. In Si, step bunching is probably also influenced by diffusion anisotropy and step energies.

The BCF-theory is not applicable to segregating dopants, because the steps then lose their sink properties, or to strained heteroepitaxial layers, because then the diffusion of adatoms is not solely controlled by the concentration gradients but also by chemical and strain gradients.

In an epitaxial growth process the technologically controlled parameters are the substrate orientation, the material flux and the growth temperature. From the BCF theory we learned that the adatoms behave like a two-dimensional atom gas with much higher diffusivity in the plane of the surface than in the bulk. The laws of surface physics govern the movement of an atom to its final position in the crystal.

At the moderate temperatures used for the epitaxial growth of quantum device structures, each atom is effectively fixed in its position in the bulk. The bulk diffusivity is determined by the diffusion of lattice defects (vacancies, interstitials) and the positional interchange between an atom and a defect. The equilibrium concentration and the mobility of these defects decreases with temperature, so the bulk diffusivity, D of substitutional dopants reduces with temperature rather steeply (typical activation energies E_A around 4 eV).

$$D = D_0 \exp\left(\frac{-E_A}{k_B T}\right). \tag{2.16}$$

For example with $D_0 = 0.1\,\mathrm{m^2s^{-1}}$, $E_A = 4\,\mathrm{eV}$, $T = 900\,\mathrm{K}$, time $t = 3600$ s one obtains a bulk diffusion coefficient D $= 4 \times 10^{-26}\,\mathrm{m^2s^{-1}}$ which is 16 orders of magnitude lower than the calculated surface diffusivity. The diffusion length $2(Dt)^{\frac{1}{2}}$ amounts to 2.4×10^{-11} m.

The processing of many quantum device structures may need high thermal budget processing including the annealling of any implants, thermal oxidation and silicide formation. Most of these processing steps require temperatures above $800\,^\circ C$ where bulk diffusion cannot be neglected. The equilibrium diffusion may be masked at this temperature regime by what is known as transient enhanced diffusion (TED). Transient enhanced diffusion is caused by a nonequlibrium concentration of point defects (vacancies, interstitials) with which an atom can exchange its position. Implantation, oxidation, nitride formation and silicide formation, especially at lower temperatures, severely disturb the point defect equilibrium. A very precise method to measure TED is given by observing the out diffusion of δ-doped layers after point defect injection.

2.2 Segregation and Diffusion of Dopants and Alloy Materials

The replacement of a Si atom in the crystal lattice by a dopant atom (substitutional impurity) delivers the extrinsic semiconductor properties required for device operation. Group III materials and Group V materials as dopants result in hole conduction (p-type) and electron conduction (n-type) for Group IV semiconductors, respectively. The most important p-type dopant in silicon is boron (B), important n-type dopants are phosphorus (P), arsenic (As) and antimony (Sb). A selection of Group III to Group V elements is shown in Table 2.6.

Important parameters for dopant atoms are the covalent radius (the radius in a covalent bond which is roughly the atomic radius without the outer

Table 2.6. Part of the Periodic Table of elements showing a selection of dopant and alloy elements for silicon. Lower indices show the atomic mass number and upper indices the atomic number of each element

Main Group		
III	IV	V
$B^{5}_{10.8}$	$C^{6}_{12.0}$	$N^{7}_{14.0}$
$Al^{13}_{27.0}$	$Si^{14}_{28.1}$	$P^{15}_{31.0}$
$Ga^{31}_{69.7}$	$Ge^{32}_{72.6}$	$As^{33}_{74.9}$
$In^{49}_{114.8}$	$Sn^{50}_{118.7}$	$Sb^{51}_{121.7}$

shell electrons), the chemical behaviour, the solid solubility and the diffusion constant. The atomic radius of nitrogen is too small for a substitutional position in Si, it therefore fails as a donor in Si. Boron and phosphorus are smaller than Si (covalent radius of Si: 0.117 nm), therefore thick and heavily doped layers are under tensile strain. Arsenic has a near identical covalent radius to Si, therefore heavily n-doped layers (e.g. buried layers for collector contacts) are preferably As doped. Antimony (as Ga, In) is larger than Si, which results in compressively strained layers. The maximum solid solubility is obtained 100 °C to 150 °C below the melting point. It ranges from about $10^{21}\,\mathrm{cm}^{-3}$ for B and As to several $10^{18}\,\mathrm{cm}^{-3}$ for Ga. At the low temperatures used for the fabrication of many quantum effect devices the solid solubilities are considerably lower, but metastable high solubilities can be obtained easily. From the four main dopants B, P, As, Sb the former two (B, P) diffuse faster than the later ones (As, Sb). At lower process temperatures ($< 850\,°\mathrm{C}$) transient enhanced diffusion (TED) created by nonequilibrium point defect concentrations has to be considered.

During growth at lower temperatures the phenomenon of surface segregation turned out to be the most important transport mechanism of dopants. To explain this phenomenon, consider a two atom system with matrix atoms (e.g. Si) and dopant atoms (e.g. Sb). The impinging matrix atoms are incorporated at steps (either from misorientation or from nucleation) and the steps move forward due to this mechanism. Surface segregation occurs when the second atom (dopant) is not incorporated into the first but manages to continue to reside on the growing surface. This can happen by either climbing across the steps or by an atomic exchange (Fig. 2.16) from a subsurface position to a surface (adatom) position. The driving force for this exchange stems from an energy gain when comparing the pair dopant atom/matrix adatom (initial state) with matrix atom/dopant adatom (final state).

It is immediately clear, that a BCF-type of diffusion theory with its adatom capture at steps cannot describe a segregating dopant. In the case of a strongly segregating species we can completely neglect the matrix steps and assume a locally homogeneous adatom density of dopants, n_S (note: in BCF

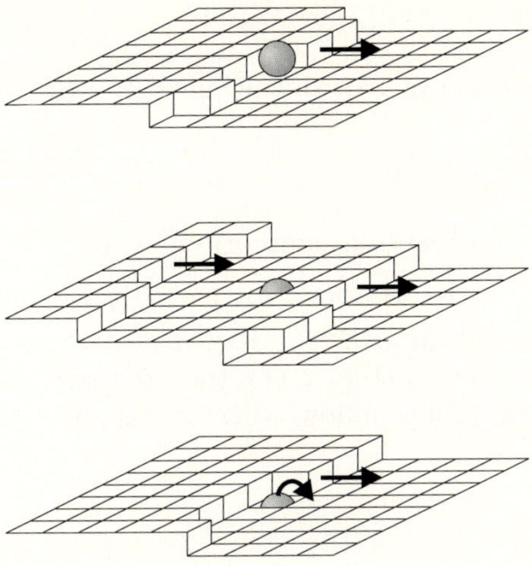

Fig. 2.16. The surface segregation of a dopant atom (grey) on a growing matrix surface. For simplicity, step flow growth of the matrix is assumed and an atomic position exchange is shown

theory the term n_S was used for the matrix adatom density). The continuity equation ($\nabla \mathbf{S} = 0$)

$$\frac{\mathrm{d}n_S}{\mathrm{d}t} = F - \frac{n_S}{\tau_{Des}} - N_D \times R \tag{2.17}$$

where $\nabla \cdot \boldsymbol{S} = 0$, because of the locally homogeneous concentration, n_S, the particle gain is defined by the dopant flux F, the particle loss is described by desorption with a mean desorption time τ_{Des} and by partial incorporation into the bulk (N_D is the bulk donor concentration of the dopant (N_A is the equivalent bulk acceptor concentration), R growth rate). Equation 2.17 contains three technology dependent quantities F (dopant flux), R (growth rate dependent on matrix flux) and τ_{Des} (desorption time dependent on growth temperature). There are also the two variables n_S and N_D. In a first approximation (valid at least at low concentrations) a direct relation between n_S and N_D is assumed.

$$\Delta_S = \frac{n_S}{N_D} \tag{2.18}$$

The segregation width Δ_S is a measure of the relationship between adatom density n_S and incorporated bulk density N_D (Note: Sometimes the adatom density n_S is compared with the bulk dopant atoms in a monolayer $N_D \times h$ which gives a dimensionless segregation ratio). Then the differential equation reads

$$\frac{\mathrm{d}n_S}{\mathrm{d}t} + n_S \left(\frac{1}{\tau_{Des}} + \frac{1}{\tau_{inc}} \right) = F \tag{2.19}$$

with $\tau_{inc} = \frac{\Delta_S}{R}$ incorporation time

or in the equivalent form for N_D as function of growth thickness $Z = Rt$

$$\frac{\mathrm{d}N_D}{\mathrm{d}t} + N_D \frac{1}{R}\left(\frac{1}{\tau_{Des}} + \frac{1}{\tau_{inc}}\right) = \frac{F}{R\Delta_S} \tag{2.20}$$

The characteristic time constant for obtaining equilibrium is given by $\left(\frac{1}{\tau_{Des}} + \frac{1}{\tau_{inc}}\right)^{-1}$. In the case of negligible desorption ($\tau_{Des} \rightarrow \infty$) this time constant is completely described by segregation with τ_{inc}. To give an example, let the flux F switch on and off after a time t_0 (Fig. 2.17), then the adatom concentration, n_S and the bulk doping profile follow with an exponential decay.

Switch on ($\tau_{Des} \rightarrow \infty$):

$$n_S = F \times \tau_{inc}\left[1 - \exp\left(-\frac{t}{\tau_{inc}}\right)\right] \tag{2.21a}$$

$$N_D = \frac{F}{R}\left[1 - \exp\left(-\frac{Z}{\Delta_S}\right)\right] \tag{2.21b}$$

Switch off:

$$n_S = F \times \tau_{inc}\left[1 - \exp\left(-\frac{t_0}{\tau_{inc}}\right)\right]\exp\left(-\frac{t - t_0}{\tau_{inc}}\right) \tag{2.22a}$$

$$N_D = \frac{F}{R}\left[1 - \exp\left(-\frac{Z_0}{\Delta_S}\right)\right]\exp\left(-\frac{Z - Z_0}{\Delta_S}\right) \tag{2.22b}$$

So, segregation converts an abrupt flux profile into a smeared doping profile. At low epitaxy temperatures, this mechanism has to be considered for all nanometer structures. The segregation is especially strong for Sb in Si and may be studied easily with these elements. The fundamental behaviour (Fig. 2.18) is that above a transition temperature T^* (which is 550 °C-600 °C for Sb/Si), the segregation width follows an equilibrium law with an activation energy which is given by the gain of a position exchange. Below the transition temperature, segregation is kinetically limited as one can easily understand when you consider that in a typical growth experiment roughly every second a monolayer is grown. When the adatom cannot manage to climb onto the next layer within one second it is buried and practically fixed, because bulk diffusion is much slower than surface processes.

Several strategies have been developed to overcome the limitations set by segregation. A number of these strategies are listed in Table 2.7.
Common strategies include the growth and co-evaporation at low temperatures where the segregation width is limited by kinetic effects along with the pre-build up and flash off of an adatom concentration before and after growth of the doped layer. Other techniques are the solid phase recrystallization of a deposited amorphous layer, implantation of dopants (Fig. 2.19) and the use of surfactants. Surfactants are accessorily added surface atoms which modify

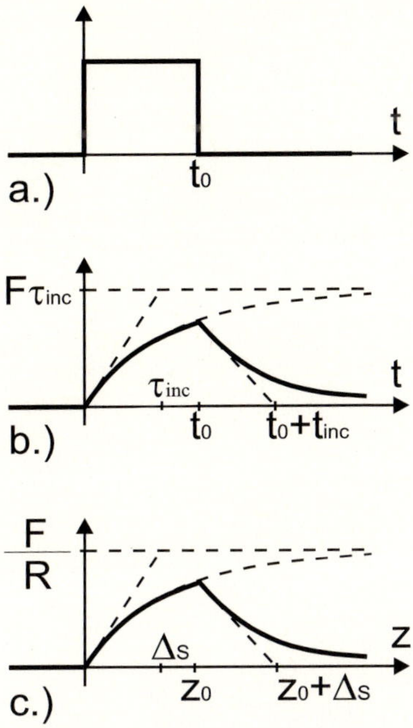

Fig. 2.17. The smearing of the profile by surface segregation. A rectangular flux profile is assumed between $t = 0$ and $t = t_0$ (a.). The adatom density n_S(t) (b.) and the bulk profile N_D(Z) (c.) demonstrate an exponential decay depending on the value of the segregation width Δ_S and the corresponding incorporation time $\tau_{inc} = \Delta_S/R$

Table 2.7. Doping strategies to overcome segregation. Typical temperatures are given for the Si/Sb system

strategy	typical temperatures (Sb/Si)
coevaporation	325 °C - 450 °C
pre-build up / flash off	450 °C - 800 °C
solid phase epitaxy (SPE)	100 °C - 550 °C
doping by secondary implantation (DSI)	550 °C - 600 °C
direct implantation	500 °C - 700 °C
surfactant	450 °C - 600 °C

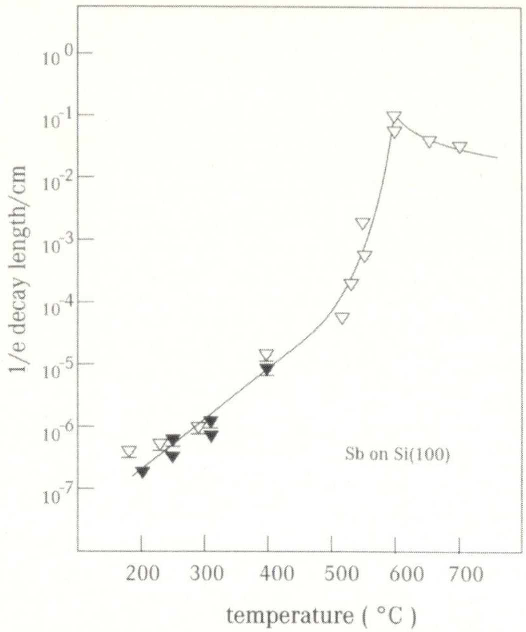

Fig. 2.18. The segregation width of Sb in Si. Kinetic limitations reduce the segregation width strongly below a transition temperature T^*. The transition temperature T^* depends on the growth rate R

the surface energy, in our example a third kind of atom: matrix, dopant, surfactant. Examples of successful surfactants are heavily segregating elements which suppress the segregation of the intended dopant or alloy. Examples of surfactants in Si are As, Sb, Bi, Ga, Sn and atomic Hydrogen (H).

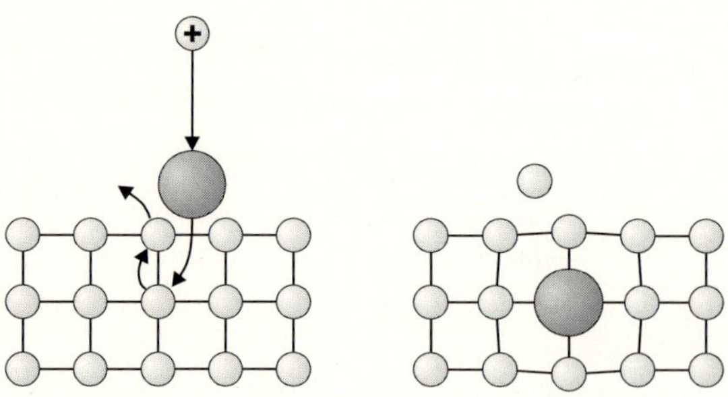

Fig. 2.19. Doping by secondary ion implantation (DSI). A segregation dopant adatom (grey) is pushed into the lattice by an accelerated matrix ion (typical ion energies are several hundred electron volts)

2.3 Lattice Mismatch and its Implication on Critical Thickness and Interface Structure

In general, a lattice mismatch will be between a film material (such as an epitaxial layer) and a substrate. We consider the thick substrate as the reference crystal and define the lattice mismatch f for cubic crystals as:

$$f = \frac{a_f - a_0}{a_0} \tag{2.23}$$

where a_f and a_0 are the lattice constants of the film and the substrate, respectively.

Nature has several answers for the growth of single crystalline mismatched films:

- Film accomodation by strain (elastic accomodation; pseudomorphic growth)
- Film accomodation by misfit dislocations at the interface (plastic relaxation)
- Morphological relaxation by surface undulations
- Cracks

As a starting point, let us consider a strained film on a rigid substrate (Fig. 2.20).

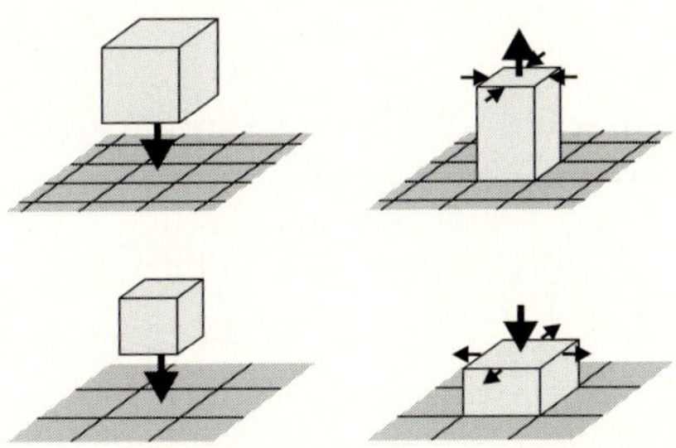

Fig. 2.20. Elastic accomodation of a film cell to the substrate mesh

A larger film cell fits to the substrate cell by being compressed in the plane of the interface, and becomes a smaller film through tension. In the vertical direction, the opposite strain results following the laws of elasticity. A cubic cell will be transformed into a tetragonal one by the biaxial stress in the interface (Fig. 2.21) (stress $\sigma_x = \sigma_y = \sigma$, $\sigma_z = 0$).

The strain components, $\epsilon_{x,y,z}$ follow from isotropic elasticity theory

$$\epsilon_x = \epsilon_y = \epsilon = \frac{1-\nu}{E}\sigma \tag{2.24a}$$

$$\epsilon_z = -\frac{2\nu}{E}\sigma = -\frac{2\nu}{1-\nu}\epsilon \tag{2.24b}$$

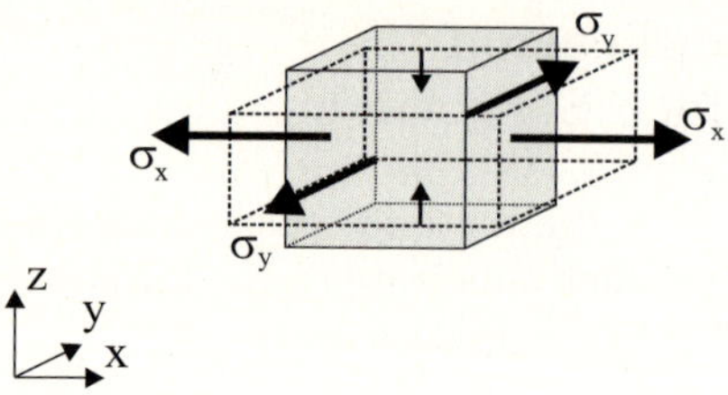

Fig. 2.21. Biaxial stress leading to a tetragonal deformation

with E the elastic modulus and ν Poisson's number. Both elastic modules are connected with the shear modulus μ by

$$\mu = \frac{E}{2(1+\nu)} \tag{2.25}$$

For example, elastic relaxation of a film with lattice mismatch f and strain ϵ cancel each other if there is elastic accommodation of the strain in the film.

$$\epsilon + f = 0 \quad \text{elastic accommodation} \tag{2.26}$$

The energy E_{hom} of a homogeneously strained layer is proportional to the square of the strain and linearly dependent on film thickness t.

$$E_{hom} = 2\mu\frac{1+\nu}{1-\nu}\epsilon^2 t \tag{2.27}$$

A finite substrate is bent by the stress. If equal elastic constants for the film and the substrate are assumed, the curvature ($1/\rho$ with ρ radius of curvature) is a measure of the strain ϵ.

$$\frac{1}{\rho} = 6\epsilon\frac{t}{t_S}^2 \qquad t_s \text{ substrate thickness} \tag{2.28}$$

With increasing thickness (and increasing elastic energy) other accommodation mechanisms start to be favourable. Let us first consider accomodation of strain by misfit dislocations at the interface (Fig. 2.22). When the film cell is larger than the substrate cell, some atomic planes will end at the interface without continuation into the film. This kind of atomic line defect is called a dislocation. The essential properties of a dislocation are shown in (Fig. 2.23). The dislocation line has a direction **l**, the deformation field around the dislocation is characterized by the Burgers vector **b**.

The Burgers vector is easily found by surrounding the dislocation in a closed cycle (1, 2, 3, ..., 8 in Fig. 2.23) and then projecting the path into an ideal crystal (1', 2', 3'...8' in Fig. 2.23). In the ideal crystal the path can be closed only with an additional vector $\boldsymbol{b}$, the Burgers vector. The Burgers vector of a single dislocation is conserved whereas the direction may change. The plane defined by the line direction and the Burgers vector ($\boldsymbol{b} \times \boldsymbol{l}$) is

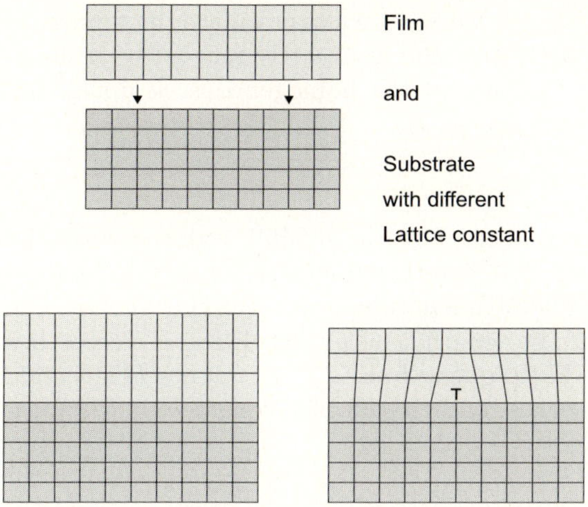

Fig. 2.22. Lattice mismatched films. Left side: a pseudomorphic film by elastic accommodation. Right side: a strain relaxed layer with a misfit dislocation in the centre

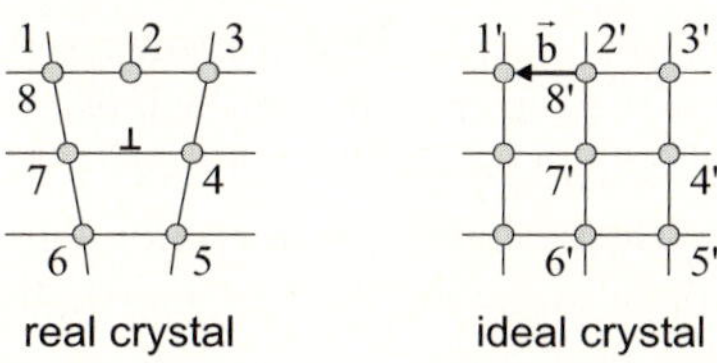

Fig. 2.23. Elements of a dislocation. Line direction **l**, Burgers vector **b**, glide plane

called the glide plane. Dislocations can easily move within this plane because only small atomic displacements are necessary for gliding. Movements outside the glide plane (called climbing) require generation or annihilation of point defects (vacancies, interstitials).

The energy E_{ds} of a single dislocation is given by,

$$E_{ds} = A\left[\ln\frac{r_a}{r_i}\right] \qquad \text{with } A = \frac{\mu b^2}{4\pi(1-\nu)} \tag{2.29}$$

where μ is the shear modulus, $\boldsymbol{b}$ is the Burgers vector, ν Poisson's number and r_a and r_i are the outer and inner cut off radii, respectively. The inner cut off radius is of the order of the Burgers vector length, the outer cut off radius is either determined by the nearby surface for single dislocations ($\mathrm{r}_a = \mathrm{t}$) or by the distance p to neighbouring dislocations in dense networks ($\mathrm{r}_a = p/2$). The areal energy, E_d, of a orthogonal network (usual network on 100 surfaces) of dislocations is given by

$$E_d = \frac{2}{p}E_{ds}. \tag{2.30}$$

Following the treatment of van der Merwe, we can calculate the minimum energy configuration of a film partly strained and partly relaxed by dislocations. The total energy E_{tot} is the sum of the homogeneous strain energy E_{hom} and the dislocation networks energy E_d

$$E_{tot} = E_{hom} + E_d. \tag{2.31}$$

To find the minimum one has to know how the strain is reduced when the number of dislocations $(1/p)$ per unit length is increased. Each dislocation displaces the atomic net by the effective length b' of the Burgers vector. The effective length is smaller than the length b when the Burgers vector lies outside the interface, e.g. for the frequent $\boldsymbol{b} = \frac{1}{2}\langle 111\rangle$ follows (111) glide plane dislocation in diamond lattices, and the effective length $b' = b/2$ for a (100) interface.

$$f + \epsilon = \frac{b'}{p} \quad \text{partially strained film.} \tag{2.32}$$

Taking (2.31) and (2.32) one can find the minimum energy configuration. An equivalent technique to the minimisation of energy is to consider the forces on dislocations as first suggested by Matthews-Blakeslee. Different numerical values stem from different choices of the cut off radii r_a and r_i and the effective lengths b'. By this procedure one can also find a thickness - called the critical thickness h_c - at which the first dislocations are energetically favourable. The relationship for this equilibrium critical thickness is given by the intrinsic equation

$$h_c \times f - \ln\left(\frac{h_c}{r_i}\right)\frac{b^2}{b'}\frac{1}{8\pi(1+\nu)} = 0. \tag{2.33}$$

In brittle materials (as Si and Ge are at temperatures below 750 °C) often kinetic limitations (dislocation nucleation, dislocation movement) causes the appearance of the first dislocations at their equilibrium thickness h_c. The material grows pseudomorphic to higher thickness (metastable regime). The experimentally found critical thickness depends on the growth temperature (Fig. 2.24). In the SiGe system very often a curve is fitted to the experimental results as found by People/Bean. This fit was found for 550 °C growth temperature.

The different values of critical thicknesses often confuse the reader. We give here the index m for the equilibrium critical thickness h_{cm} with a certain choice of constants (essentially the choice Matthews-Blakeslee made) and the index p for the curve fit given by People/Bean (h_{cp} valid for about 550 °C growth temperature).

$$\left(\frac{h_{cm}}{b}\right) f - 5.78 \times 10^{-2} \ln\left(\frac{h_{cm}}{b}\right) = 0 \quad \text{(equilibrium)} \tag{2.34}$$

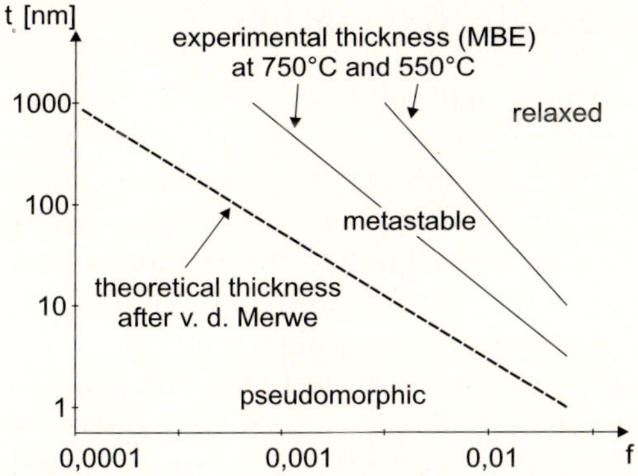

Fig. 2.24. The critical thickness h_c as function of lattice mismatch f. Shown are the regions of equilibrium pseudomorphic growth, metastable growth (also pseudomorphic up to a critical thickness which depends on growth temperature), and a (partly) relaxed regime

$$\left(\frac{h_{cp}}{b}\right) f^2 - \frac{1}{200} \ln \left(\frac{h_{cp}}{b}\right) = 0 \quad (550°\mathrm{C\ fit\ curve}). \tag{2.35}$$

Beyond the critical thickness a network of misfit dislocations is created with increasing density up to a limiting dislocation spacing p_0 (Fig. 2.25).

$$p_0 = \frac{b'}{f} \tag{2.36}$$

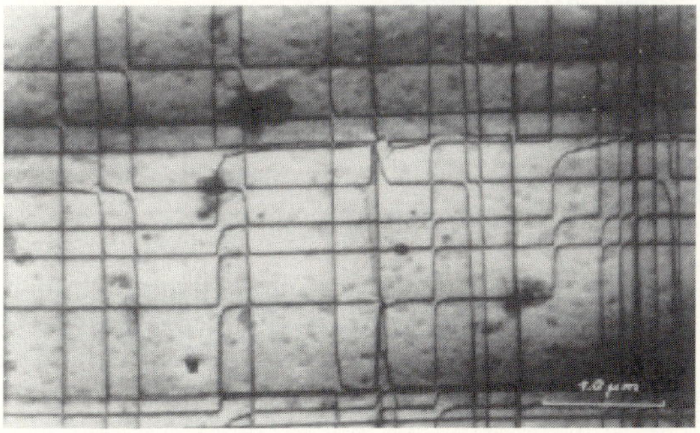

Fig. 2.25. A plan view electron microscope image of the two dimensional network of misfit dislocations in a sample

2.4 Virtual Substrates and Strain Relaxation

A number of the quantum device concepts that will be reviewed in the later chapters require to be grown on relaxed $Si_{1-y}Ge_y$ substrates to produce strain in heterolayers grown on top of these substrates so that the correct band structure can be obtained. There are many applications where a relaxed $Si_{1-y}Ge_y$ substrate is therefore desirable. Unfortunately the liquidus solidus curves of the Si and Ge material system prevents the formation of $Si_{1-y}Ge_y$ crystals with any value of y being pulled from the molten constituent elements. Some small diameter $Si_{1-y}Ge_y$ wafers have recently become available with low values of y (below 0.10) but these wafers have used the constituent elements along with molten metals to allow SiGe crystals to be extracted. A significant problem is the contamination of the wafers with a number of non-Group IV elements. Therefore as no bulk SiGe substrates are available, strain relaxation buffers must be grown on bulk silicon substrates. These are frequently called virtual substrates (or sometimes pseudo-substrates) as they produce the relaxed $Si_{1-y}Ge_y$ substrate on top of a silicon substrate.

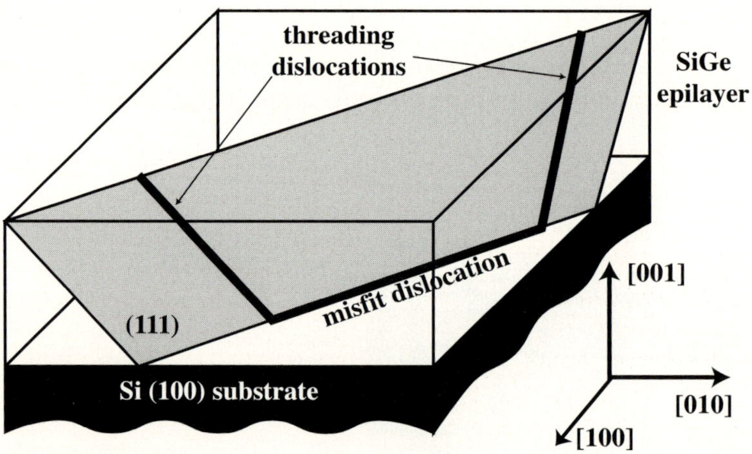

Fig. 2.26. A schematic diagram of the relaxation of a compressively strained-$Si_{1-x}Ge_y$ heterolayer by the formation of a misfit dislocation along the interface of the bulk-Si substrate and the $Si_{1-x}Ge_y$ heterolayer along with the two threading segments of the dislocation which thread to the surface

If a thick $Si_{1-y}Ge_y$ layer is grown directly on top of a silicon wafer but well above the critical thickness ((2.33), (2.34) and (2.35)) the layer will relax. In particular as shown in Fig. 2.26, a misfit dislocation with two segments which thread upwards at 60° on the (111) plane to the surface are produced. It should be noted that relaxation can only occur through the formation of dislocations or defects. If there are no dislocations or defects then there can be no lattice mismatch between the two layers and the heterolayer cannot, therefore, be relaxed. The problem with this technique is that a large number of dislocations are created which interact and penetrate through the epilayer resulting in threading dislocation densities of over $10^9\,\mathrm{cm}^{-2}$ and frequently $10^{12}\,\mathrm{cm}^{-2}$ for the useful Ge contents for devices. This density is significantly higher than that which can be tolerated by any electronic or optoelectronic device as the dislocations can either trap or interact with the electrons and holes. The problem with this particular technique is that misfit segments nucleate at the silicon / $\mathrm{Si}_{1-x}\mathrm{Ge}_x$ interface and glide along the $\{111\}$ planes resulting in line vectors along the $[110]$ and $[1\bar{1}0]$ directions (Fig. 2.26). As all the dislocations nucleate at this one interface, there are substantial interactions between the two networks of dislocations which impedes the glide of the threading segments to the edges of the wafer. Therefore shorter misfit segments result which produces a higher density of threading dislocations since the end of the misfit segment has to be connected to a free surface via a threading dislocation. The only other option is the formation of a loop dislocation where the two threading segments join.

One solution to the high dislocation density is to find a method of extending the misfit segment lengths. The ideal scenario is where the two threads have glided to the edge of the wafer so that they cannot interfere with the active epitaxial layers grown above the buffer. Three conditions are required to extend the misfit segment lengths: (a) The growth temperature or any subsequent anneal temperature requires to be high enough to enable a high misfit dislocation velocity ($\propto$ constant $\times \exp\left[-\left(\frac{2.25\,(\mathrm{eV})}{k_B T}\right)\right]$), much higher than the growth rate. (b) Secondly the density of pinning centres or interactions with other dislocation threads which result in pinning require to be low enough to promote long misfit segments. (c) Third, misfit nucleation is required for strain relaxation but the activation energy requires to be higher than that for dislocation glide.

A number of different approaches have been used to achieve the above criteria. One of the first and still to a large extent the most successful is to slowly grade the Ge content from 0% up to the required Ge composition. This technique works very well for Ge contents up to approximately 30% using conventional CVD or MBE growth techniques provided the grading rate is below 10% Ge per μm. Figure 2.27 shows a scanning optical micrograph of the surface of a wafer grown at 560 °C with a grading rate of 52% per μm. The corresponding transmission electron microscope (TEM) image is shown in Fig. 2.28. The surface is quite clearly pitted even though no threading segments can be observed to thread through the surface of the wafer in the cross sectional TEM image. The TEM image also demonstrates that some of the 60 °C dislocations thread down into the substrate as well as up to the surface.

If the grading is reduced to 5% per μm with the growth temperature increased to 800 °C then the surface is substantially improved as demonstrated in Figure 2.29. Comparison of the TEM picture in Fig. 2.30 with that of the

Fig. 2.27. A scanning optical microscope image of the surface of a virtual substrate grown at 560 ^{o}C using LPCVD with a Ge grading rate of 52% per μm up to $Si_{0.74}Ge_{0.26}$

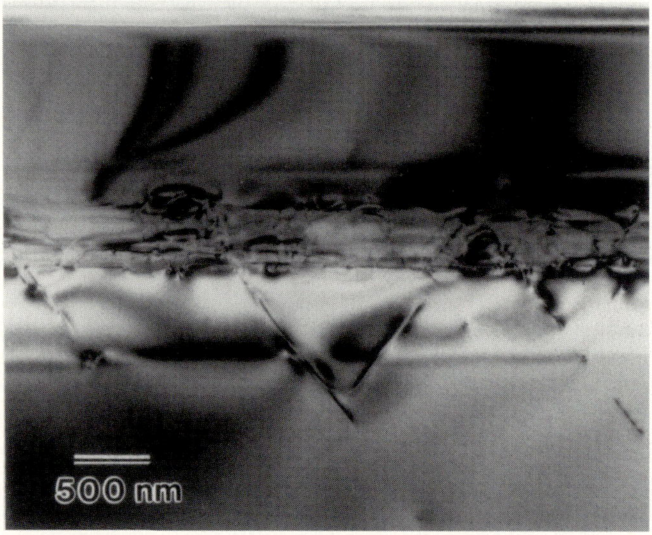

Fig. 2.28. A transmission electron microscope image of the surface of a virtual substrate grown at 560 ^{o}C using LPCVD with a Ge grading rate of 52% per μm up to $Si_{0.74}Ge_{0.26}$

Fig. 2.29. A scanning optical microscope image of the surface of a virtual substrate grown at 800 °C using LPCVD with a Ge grading rate of 5% per µm

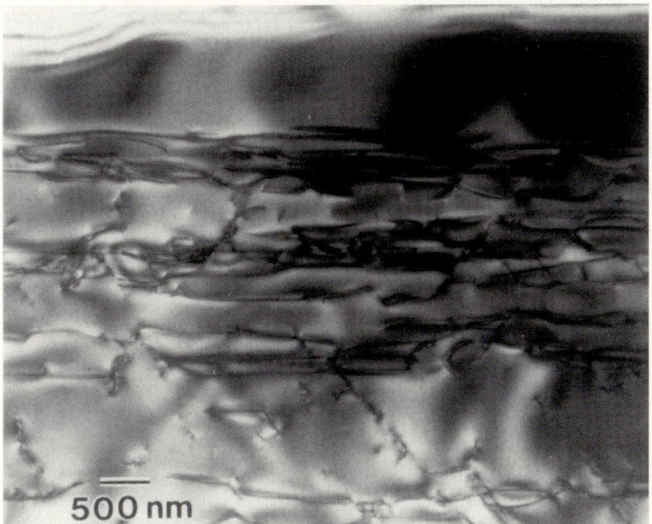

Fig. 2.30. A transmission electron microscope image of the surface of a virtual substrate grown at 800 °C using LPCVD with a Ge grading rate of 5% per µm

higher grading rate (Fig. 2.28) demonstrates how the dislocation segments and visible threads are spread out in the vertical direction resulting in fewer threading segments vertically. Close inspection demonstrates the reduced interaction between the dislocations with the lower grading rate. While these interactions are difficult to image using cross sectional TEM, the surface quality of the wafer strongly depends on the number of interactions (Figs. 2.27 and 2.29). Low temperature electrical measurements of modulation-doped strained-Si quantum wells grown on top of the buffers also demonstrate the difference in virtual substrate quality. The 52% per μm buffer had a 1.5 K Hall mobility of 128 000 cm^2/Vs while the 5% per μm substrate had a Hall mobility of 258 000 cm^2/Vs. The active heterolayers in this case were grown at 560 °C after the high temperature buffer had been completed.

At low growth temperatures (irrespective of the grading rate), a small number of heterogeneous dislocation nucleation sources are activated resulting in large dislocation pileups containing a particular Burgers vector. These pileups give rise to a large spread of mosaic crystal tilts relative to the silicon substrate. The size of these dislocation pileups may be correlated with the surface roughness measured by atomic force microscopy (AFM). At high growth temperatures, more dislocation nucleation sources are activated and the pileup size and hence the spread in mosaic crystal tilts is reduced. In both cases the presence of these dislocation pileups controls the surface morphology although through different mechanisms. Large pileups of 60 ° dislocations result in a tilting of the surface at low growth temperatures because of the surface disruption caused by the tilt component of the dislocation's Burgers vector. At higher growth temperatures this effect is reduced and as the pileup size is reduced the surface morphology is dominated by strain driven roughening due to the residual strain field around the dislocation pileups and the increased surface mobility afforded by higher temperature growth due to different surface energies of the heteromaterials. The origins of these roughening mechanisms are found to occur during the very earliest stages of relaxation suggesting that it is this part of the growth which must be controlled to remove these effects.

Table 2.8 provides a summary of different virtual substrate growth temperatures along with grading rates and thicknesses of the constant composition buffer above the grading region. 1.5 K mobilities have also been provided where n-type modulation-doped structures have been grown on top with a strained-Si quantum well. The sample with the lowest electron mobility has the thinnest top constant composition buffer. This sample actually has a lower mobility than growth of a single constant composition buffer without any grading which has a dislocation density above 10^{11} cm^{-2}. Therefore in the thin constant composition buffer sample, the misfit segments are frequently within 100 nm of the electrons in the quantum well and the mobility is reduced due to strong interactions with the dislocations. Increasing the thickness of this buffer to over 0.6 μm can remove most of this interaction, thereby significantly increasing the mobility. This is probably due to the misfit dislocations trapping charge and it is the interaction of this charge through remote Coulomb scattering which is detrimental to the mobility.

Table 2.8. A comparison of the grading rate, growth temperature and thickness of the constant composition layer of different strain relaxation buffer grown by LPCVD, solid source MBE and LEPECVD. Root mean squared (rms) roughness values have been measured by AFM while mobility measurements are for n-MODFETs except the final row which is a p-type Ge channel MODFET with mobility measured at 4.2 K.

growth system	growth temp. (^{o}C)	final Ge comp.	grade rate (% Ge/μm)	constant comp. thickness (μm)	rms roughness (nm)	1.5 K mobility (cm^2/Vs)
LPCVD	800	28 %	45	0.33	20-25	-
LPCVD	800	27 %	42	0.33	23-25	-
LPCVD	560	29 %	36	0.09	2-3	2,390
LPCVD	560	26 %	52	0.87	3-4	137,000
LPCVD	800	23 %	7	0.68	2	133,000
LPCVD	800	24 %	5	0.88	2	258,000
MBE	450	30 %	single step	0.5		13,100
MBE	750	25 %	30	0.58		88,500
MBE	750	29 %	8.5	0.6		173,000
LEPECVD	720	35 %	5.8	1.8	<1	150,000
LEPECVD	720	70 %	10	1.5	4.5	87,000 (p)

The results in table 2.8 demonstrate that relatively high electron mobilities can still be achieved even with aggressive grading rates. The thickness of the constant composition buffer is far more important to distance the electrons from the misfit segments. The best mobilities at low temperatures can only be achieved by grading at less than about 10% per μm and for these n-types modulation-doped samples the Ge content requires to be between about 20 and 30 %. As virtual substrates are grown with the same techniques above 30 % Ge content, the surface roughness of the buffers is increased and the grading rate becomes more important for high mobility.

Low energy plasma enhanced CVD has been developed specifically to grow thick relaxed buffer layers at high growth rates and low Ge grading per μm to reduce the surface roughness and threading dislocation density especially for Ge contents above 30 %. Growth rates up to 7 nm per second have been demonstrated and the material has demonstrated very high mobilities both at room and low-temperatures. A significant problem with these thick virtual substrates is that the thermal conductivity of SiGe layers is significantly lower than bulk silicon or bulk germanium ($Si_{0.8}Ge_{0.2}$ has a thermal conductivity of 5.1 $Wm^{-1}K^{-1}$ compared to Si of 140 $Wm^{-1}K^{-1}$ and 70 $Wm^{-1}K^{-1}$ for Ge). This is a serious issue for a number of electronic and optoelectronic device applications where the ability to remove the power which is dissipated as heat is paramount. Therefore there has been a significant interest in developing thin virtual substrate techniques both to reduce the thermal problems but also to reduce the fabrication time and hence cost of virtual substrates.

A number of alternative techniques for thin virtual substrates have also been researched or are presently being developed. The first of these used silicon-on-insulator (SOI) substrates as a base where the buried oxide is used as a compliant layer. A pseudomorphic $Si_{1-x}Ge_x$ layer is grown at low temperature above the top silicon layer of the SOI substrate. Once the $Si_{1-x}Ge_x$ is deposited, the whole wafer is annealled at high temperatures so that the $Si_{1-x}Ge_x$ layer relaxes on the compliant oxide layer.

A second technique uses point-defects (vacancies, interstitials) produced either by growth at very low temperature (150°C-400°C) or by ion bombardment. If there are a large number of point defects close together then the point defects can condense and form prismatic dislocation loops which supports the relaxation without threading segments through the surface of the heterolayer. There can also be interaction of the point defects with the dislocations. Such interactions allow dislocation climb as well as the previously discussed glide along the [110] directions which can lead to the annihilation of defects. High Ge content up to 50%, high degree of relaxation up to 100% and cross-hatch free surfaces have been demonstrated with sub 100nm virtual substrates, while supersaturation of point-defects and micro twins formation are remaining problems.

A similar techniques for virtual substrates was demonstrated using hydrogen ion implantation a process which has similarities to the SmartCut^{TM} silicon-on-insulator (SOI) technique. The hydrogen implantation into a pseudomorphic $Si_{1-x}Ge_x$ layer is used to produce a thin defect band slightly below the $Si_{1-x}Ge_x$ / Si interface. During subsequent annealling, platelets and cavities form giving rise to strongly enhanced strain relaxation. Initial experiments demonstrated a reduced threading dislocation density compared to relaxed constant composition layers. It has been suggested that the defect band induces stain relaxation via preferred nucleation of dislocation loops in the defect band which extend to the interface to form misfit segments.

The latest technology that has emerged especially for strained-Si CMOS is strained-Si on insulator which is similar to SOI. For MOSFETs with channel lengths below about 30 nm partially or fully depleted SOI devices may be required to allow MOSFETs to be switched off with a low enough off-current for circuit applications. CMOS microprocessors using SOI wafers are already in production as the reduced capacitance from the buried oxide improves the circuit performance by up to 30 % compared to a bulk silicon substrate. As strained-Si CMOS gets closer to production SOI versions of the technology are likely to be developed. There have been a number of demonstrations of strained-Si directly on the oxide which remains stable up to anneal temperatures of 950 ^oC. Both wafer bonding and oxide implantation (SIMOX) techniques have been demonstrated. Strained-Si CMOS devices have been produced on both types of SOI wafers with enhanced performance over standard bulk Si and SOI substrates.

From the materials side of silicon and silicon-germanium technology, virtual substrate development is one of the most active research areas at present as many different device technologies require a relaxed $Si_{1-y}Ge_y$ substrate. It is highly likely that there will be new developments in this area in the future as the buffers as such an important part of quantum devices.

2.5 Further Reading

1. J. Murota, B. Tillack, M. Caymax, J. Sturm, Y. Yasuda, S. Zaima , Proc. First Internat. SiGe Technology and Device Meeting, Appl. Surf. Sci., Vol. 224 (2004)
2. R. Waser, Nanoelectronics and Information Technology, Wiley-VCH, Weinheim (2003)
3. C.K. Maiti, N.B. Chakrabarti, S.K. Ray, Strained Silicon Heterostructures Materials and Devices, The Institution of Electrical Engineers, London (2001)
4. E. Kapser and K. Lyutovich, Properties of Silicon Germanium and SiGe:Carbon, EMIS Datareviews, Vol.24, INSPEC, London (2000)
5. M.A. Herman and H. Sitter, Molecular Beam Epitaxy: Fundamentals and Current Status, 2nd Edition, Springer-Verlag, Berlin (1996)
6. J.F.A. Nijs, Advanced Silicon and Semiconducting Silicon-Alloy Based Materials and Devices, IOP publishing, Bristol (1994)
7. F. Schaeffler, Semicond. Sci. Technol. 12, 1515 (1997)
8. E. Kasper, K. Lyutovich, Solid-State Electronics 48, 1257 (2004)

3. Resumé of Semiconductor Physics

This chapter presents a brief summary of the physics required to understand the operation of the devices presented in later chapters. It is not the intent here to provide a complete volume on the whole of semiconductor physics and it is assumed that the reader already has a basis in solid state theory and quantum mechanics. The reader seeking further information should consult one of the many text books covering the areas in this chapter in more detail.

3.1 Quantum Mechanics

3.1.1 The Wave Behaviour of Particles

It was originally de Broglie who suggested that particles such as electrons could have wave behaviour and defined the de Broglie wavelength, λ for a particle of momentum, p to be

$$\lambda = \frac{h}{p} \tag{3.1}$$

where h is Planck's constant. Therefore, an electron in a vacuum at position $\mathbf{r}$ without any external forces could be described by a state or wavefunction, $\psi(\mathbf{r})$ which is a form of a travelling wave such that

$$\psi(\mathbf{r}) = \exp\left[\mathrm{i}\left(\mathbf{k} \cdot \mathbf{r} - \omega t\right)\right] \tag{3.2}$$

This describes the motion of the wave at time, t with angular frequency, ω and wavevector, $\mathbf{k}$ such that

$$k = |\mathbf{k}| = \frac{2\pi}{\lambda} \tag{3.3}$$

The complete description of the wave behaviour of electrons is described by the Schrödinger equation. For this chapter it will be the single-particle, time independent Schrödinger equation which is required and is defined in the general case as

$$\hat{H}\psi(\mathbf{r}) = E\psi(\mathbf{r}) \tag{3.4}$$

where $\hat{H}$ is the Hamiltonian operator of the system. From (3.1), the momentum in quantum mechanics can be written as

$$\mathbf{p} = \frac{h}{\lambda} = h\frac{\mathbf{k}}{2\pi} = \hbar\mathbf{k} \tag{3.5}$$

where $\hbar = \frac{h}{2\pi}$. From this the momentum operator, $\hat{\mathbf{p}}$ can be defined such that

$$\hat{\mathbf{p}}\psi(\mathbf{r}) = -\mathrm{i}\hbar\nabla\psi(\mathbf{r}) \tag{3.6}$$

Since classically kinetic energy of a particle with velocity, v and mass, m is given by

$$\frac{1}{2}m\mathbf{v}^2 = \frac{(m\mathbf{v})^2}{2m} = \frac{\mathbf{p}^2}{2m} \tag{3.7}$$

the quantum mechanical analogy defines the operator for kinetic energy as

$$\frac{\hat{\mathbf{p}}^2}{2m}\psi(\mathbf{r}) = -\frac{\hbar^2}{2m}\nabla^2\psi(\mathbf{r}) \tag{3.8}$$

and combining this with the potential energy in the system, $V(\mathbf{r})$ gives the single-particle, time independent Schrödinger equation as

$$-\frac{\hbar^2}{2m}\nabla^2\psi(\mathbf{r}) + V(\mathbf{r})\psi(\mathbf{r}) = E\psi(\mathbf{r}) \tag{3.9}$$

It is the solutions to this equation which will allow the physics behind the electrons in semiconductors to be understood.

The other major piece of quantum mechanics required for this chapter is that due to Planck who postulated that atoms in the walls of a cavity emitted radiation in quanta. A few years later, Einstein concluded that electromagnetic radiation of frequency, ν behaved as if it consisted of a collection of energy quanta with energy $h\nu$. These quanta are now called photons and behave with Bose-Einstein statisics such that the energy of each photon is

$$E = h\nu = \hbar\omega = \frac{hc}{\lambda} \tag{3.10}$$

where ω is the angular frequency which equals $2\pi\nu$.

3.1.2 The Potential Barrier and Quantum Mechanical Tunnelling

Tunnelling is a purely quantum mechanical effect with no similar effect in classical mechanics. If a particle is incident on a potential barrier of height V_0 then classically if it has energy lower than V_0 it is reflected. The particle is only transmitted when it has higher energy than V_0 allowing the barrier

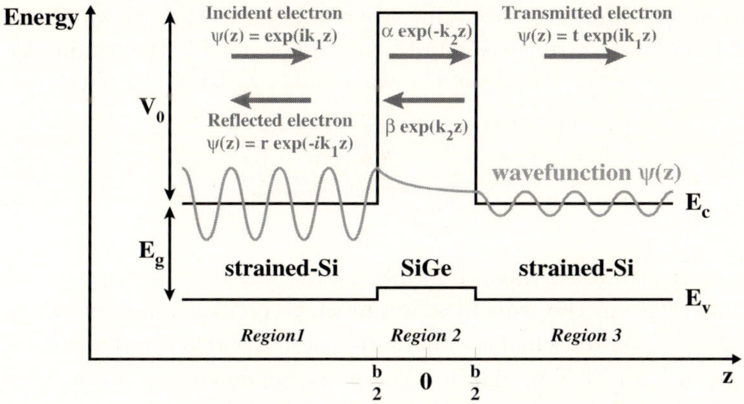

Fig. 3.1. A schematic diagram of a single barrier produced in Si/SiGe heterostructures using relaxed SiGe and strained-Si. The wavefunctions in the three regions are also drawn along with the appropriate equations

to be surmounted. In the quantum mechanical case, a particle of energy less than V_0 may tunnel through the barrier.

Figure 3.1 shows the band structure and trial wavefunctions for the case of a single potential barrier of height V_0 and width b created using strained-Si and relaxed SiGe. We will only consider the one dimensional case of an incident electron of kinetic energy E and mass m^* in the conduction band incident on the barrier, and ignore the transverse components in the xy-plane. [The effective mass, m^*, in a semiconductor will be defined fully in Sect. 3.2.1.] The system is described by the one-dimensional, time independent Schrödinger equation

$$-\frac{\hbar^2}{2m^*}\frac{\partial^2\psi(z)}{\partial z^2}+V(z)\psi(z)=E\psi(z) \tag{3.11}$$

For electrons of incident energy less than the barrier height, $E < V_0$

Region 1 $\quad z < -\frac{b}{2} \quad V(z) = 0$

$$\psi(z)=\exp(\mathrm{i}k_1 z)+\tilde{r}\exp(-\mathrm{i}k_1 z) \qquad k_1=\sqrt{\frac{2m^*E}{\hbar^2}} \tag{3.12}$$

representing a normalised incident plane wave solution in addition to a reflected component of amplitude $\tilde{r}$ and wavenumber k_1.

Region 2 $\quad -\frac{b}{2} < z < \frac{b}{2} \quad V(z) = V_0$

$$\psi(z)=\alpha\exp(-k_2 z)+\beta\exp(k_2 z) \qquad k_2=\sqrt{\frac{2m^*(V_0-E)}{\hbar^2}} \tag{3.13}$$

which describes an exponentially decaying wavefunction in the barrier where α is the amplitude of the wave travelling to the right and β is the amplitude of the wave travelling to the left. k_2 is the wavenumber inside the barrier

Region 3 $\quad z > \frac{b}{2} \quad V(z) = 0$

$$\psi(z) = \tilde{t}\exp(\mathrm{i}k_1 z) \qquad k_1 = \sqrt{\frac{2m^* E}{\hbar^2}} \tag{3.14}$$

which describes a transmitted wave of amplitude $\tilde{t}$.

The value and slope of the wavefunction must be continuous everywhere and so $\psi(z)$ and $\frac{d\psi(z)}{dz}$ are matched at the boundaries. To calculate the current in the system we require to find the flux transmission coefficient $\tilde{T}$. A wave $\tilde{F}\exp(ikz)$ carries a current density $\left(\hbar k/m^*\right)\left|\tilde{F}\right|^2$ and so the transmission coefficient is

$$\tilde{T} = \frac{\hbar k_1/m^*}{\hbar k_1/m^*}\left|\tilde{t}\right|^2 = \left|\tilde{t}\right|^2 \tag{3.15}$$

One may also define the reflection coefficient, $\tilde{R}$ in a similar manner

$$\tilde{R} = \frac{\hbar k_1/m^*}{\hbar k_1/m^*}\left|\tilde{r}\right|^2 = \left|\tilde{r}\right|^2 \tag{3.16}$$

and the conservation of particles produces

$$\tilde{T} + \tilde{R} = 1 \tag{3.17}$$

One may solve for $\tilde{t}$ and hence $\tilde{T}(E)$ using the information above and find $(V_0 > E)$

$$\tilde{t} = \frac{2k_1k_2\exp(-\mathrm{i}k_1 b)}{2k_1k_2\cosh(k_2 b) - \mathrm{i}\left(k_1^2 - k_2^2\right)\sinh(k_2 b)} \tag{3.18}$$

and

$$\begin{aligned}\tilde{T} &= \frac{4k_1^2k_2^2}{4k_1^2k_2^2 + \left(k_1^2 + k_2^2\right)^2\sinh^2(k_2 b)} \\ &= \frac{1}{\left[1 + \frac{V_0^2}{4E(V_0 - E)}\sinh^2(k_2 b)\right]}\end{aligned} \tag{3.19}$$

In the approximation $k_2 b \gg 1$ (i.e. for a high or wide barrier)

$$\tilde{T} \approx \frac{16E(V_0 - E)}{V_0^2}\exp(-2k_2 b) \approx \frac{16E}{V_0}\exp(-2k_2 b) \tag{3.20}$$

The important result is that the transmission coefficient decays exponentially as the width of the barrier is increased.

3.1.3 Quantum Wells

Quantum wells play an enormous role in microelectronics and are found in many different shapes and forms. For many cases explicit solutions exist to Schrödinger's equation and we start by solving the case of a simple one dimensional quantum well of depth V_0 and width w (Fig. 3.2(a)). We require to calculate the energy levels E_n of a particle in the well of mass m^*. Inside the well, the potential $V(x)$ is 0 and so the Schrödinger equation takes the form

$$\frac{d^2\psi}{dx^2} + \alpha^2\psi = 0 \quad \text{for} - \frac{w}{2} \leq x \leq \frac{w}{2} \text{ where} \qquad \alpha^2 = \frac{2m^*E}{\hbar^2} \tag{3.21}$$

and outside the well where the potential is V_0

$$\frac{d^2\psi}{dx^2} - \beta^2\psi = 0 \quad \text{for } x < -\frac{w}{2} \text{ and } x > \frac{w}{2} \qquad \beta^2 = \frac{2m^*(V_0 - E)}{\hbar^2} \tag{3.22}$$

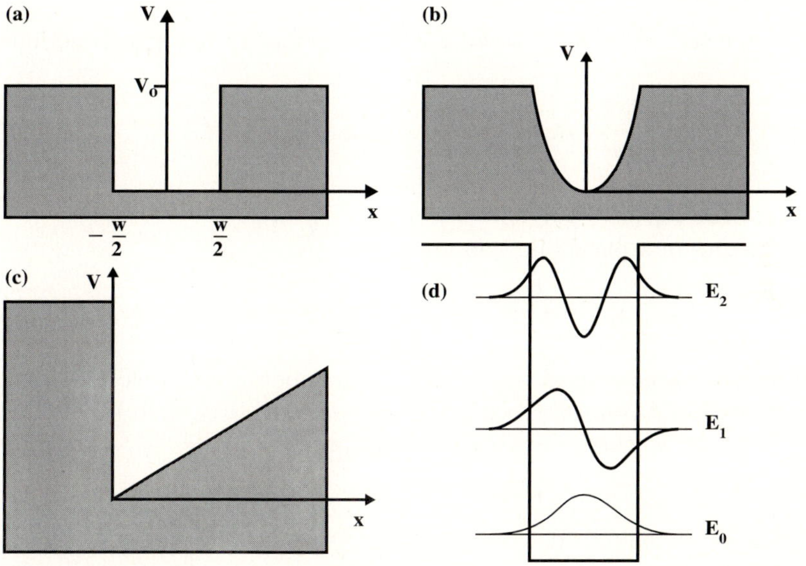

Fig. 3.2. The three major types of quantum wells found in microelectronics: (a) finite square quantum well, (b) parabolic quantum well and (c) triangular quantum well. (d) shows the wavefunctions of the first three subbands in a finite quantum well

At an infinite distance from the well, the probablity of finding the particle, $|\psi|^2$ vanishes and so the boundary conditions are:

$$\psi = A\exp\left[-\mathrm{i}\beta x\right] \qquad \text{for } x < -\tfrac{w}{2} \tag{3.23}$$

$$\psi = A\exp\left[\mathrm{i}\beta x\right] \qquad \text{for } x > \tfrac{w}{2} \tag{3.24}$$

$$\psi = C\exp\left[\mathrm{i}\alpha x\right] + D\exp\left[-\alpha x\right] \qquad \text{for } \tfrac{w}{2} \leq x \leq \tfrac{w}{2} \tag{3.25}$$

where A, B, C and D are integration constants which are bound by the continuity requirements of ψ and $\frac{d\psi}{dx}$ at $x = \pm\frac{w}{2}$. The normalised eigenfunctions may also be written in the form

$$\psi = \begin{cases} \sqrt{\frac{2}{w}}\sin\left[\frac{n\pi}{w}x\right] & \text{with } n = \text{even integer} \\ \sqrt{\frac{2}{w}}\cos\left[\frac{n\pi}{w}x\right] & \text{with } n = \text{odd integer} \end{cases} \tag{3.26}$$

For an infinitely deep quantum well with $V_0 \gg E_n$, the solutions are

$$E_n = \frac{(n+1)^2\,\pi^2\hbar^2}{2m^*w^2} \qquad \text{for } n = 0, 1, 2, 3, \ldots \tag{3.27}$$

while for a finite well

$$E_n = \frac{\hbar^2}{2m^*w^2}\left[\pi\,(n+1) - \arcsin\sqrt{E_n/V_0}\right]^2 \qquad \text{for } n = 0, 1, 2, 3, \ldots \tag{3.28}$$

For a parabolic well (Fig. 3.2(b)) with $V(x) = \frac{1}{2}m^*\omega^2x^2$ the eigenfunctions are

$$\psi_n\,(x) \sim \exp\left[-\frac{x^2}{2a^2}\right] H_n\left(\frac{x}{a}\right) \qquad \text{with } n = 0, 1, 2, 3 \ldots \tag{3.29}$$

where H_n are the Hermite polynomials and $a = \sqrt{\hbar/m^*\omega}$
The eigenenergies of a parabolic well are given by

$$E_n = \hbar\omega\left(n + \frac{1}{2}\right) \qquad \text{with } n = 0, 1, 2, 3, \ldots \tag{3.30}$$

For many microelectronic devices, the confining potential for electrons can be approximated as a triangular well. A triangular well (Fig. 3.2(c)) confined by an electric field, F corresponding to the potential

$$V(x) = qFx \qquad \text{for } x > 0 \text{ and} \tag{3.31}$$

$$V(x) = \infty \qquad \text{for } x < 0 \tag{3.32}$$

and the solutions are given by

$$\psi_n\,(x) = \mathrm{Ai}\left[\left(\frac{2m^*qF}{\hbar^2}\right)^{1/3}\left(x - \frac{E_n}{qF}\right)\right] \tag{3.33}$$

where $\mathrm{Ai}\,[x]$ is the Airy's function. The energies are given by

$$E_n \approx \left(\frac{\hbar^2}{2m^*}\right)^{1/3}\left[\frac{3\pi qF}{2}\left(n + \frac{3}{4}\right)\right]^{2/3} \tag{3.34}$$

3.1.4 The Hydrogen Atom

Before considering the electronic band structure of crystals, it is worthwhile considering the electronic structure of a simple one electron hydrogen atom. This is the simplest atom to investigate as it is a two particle system with a positive nucleus of mass, m_1 at a position, $\mathbf{r}_1$ and an electron of mass, m_2 at position $\mathbf{r}_2$. The standard method of solving the problem is to transform into the centre of mass coordinate, $\mathbf{R}$ and the difference coordinates, $\mathbf{r}$ defined as

$$\mathbf{R} = \frac{m_1\mathbf{r}_1 + m_2\mathbf{r}_2}{m_1 + m_2} \tag{3.35}$$

and

$$\mathbf{r} = \mathbf{r}_1 - \mathbf{r}_2 \tag{3.36}$$

along with the reduced mass, m_μ

$$\frac{1}{m_\mu} = \frac{1}{m_1} + \frac{1}{m_2} \tag{3.37}$$

The solution can be found in any good quantum mechanics text book and is the result of being able to separate the equation into two parts by letting

$$\psi(\mathbf{r}_1, \mathbf{r}_2) = f(\mathbf{R})\,\psi(\mathbf{r}) \tag{3.38}$$

where

$$-\frac{\hbar^2}{2M}\nabla^2 f(\mathbf{R}) = E_K f(\mathbf{R}) \tag{3.39}$$

$$\left(-\frac{\hbar^2}{2m_\mu}\nabla^2 - \frac{q^2}{4\pi\varepsilon_0\mathbf{r}}\right)\psi(\mathbf{r}) = E\psi(\mathbf{r}) \tag{3.40}$$

and the total energy, E_T is given by

$$E_T = E_K + E \tag{3.41}$$

The eigenenergy and eigenfunction of the centre of mass part (see (3.39)) are simply those of a free particle of mass, M with energy

$$E_K = \frac{\hbar^2\mathbf{K}^2}{2M} \tag{3.42}$$

and

$$f(\mathbf{R}) = \frac{\exp[\mathrm{i}\mathbf{K}\cdot\mathbf{R}]}{\sqrt{V}} \tag{3.43}$$

so that the total solution is given by

$$\psi(\mathbf{r}_1, \mathbf{r}_2) = \frac{\exp[\mathrm{i}\mathbf{K}\cdot\mathbf{R}]}{\sqrt{V}}\psi(\mathbf{r}) \tag{3.44}$$

where V is the volume of the space. The solutions of these equations are extremely involved and the reader is refered to a number of quantum mechanical texts. For the crystal and electronic structure, it is the shape and energy of the solutions that are important. The solutions are labelled from the lowest energy upwards as s-, p-, d-, f-,.... For the bandstructure of silicon, only the s- and p- orbitals need to be considered and the shapes of these wavefunctions are shown in Fig. 3.3. Each orbital state (s-, p-, d-, f-,...) can only be occupied by a fixed number of electrons. The s-state can only have two electrons and is spherical (Fig. 3.3(a)). The p-states have three different possible orientations and each can be occupied by 2 electrons. Therefore the p-states can have a total occupation of six electrons. The d-states can have up to ten electrons occupying them. The orbitals states also form shells. The inner shell is just an s-state (1s). The second shell can have both s and p-states (2s 2p). The third shell can have s-, p- and d-states (3s 3p 3d) and the fourth can have s-, p-, d- and f- (4s 4p 4d 4d). It is this structure which defines the periodic table. The number of electrons in each filled or non-filled orbital state is normally denoted by a superscript after the shell number and orbital letter. Therefore Si with 14 electrons has the electronic configuration of $1s^2 2s^2 2p^6 3s^2 3p^2$ and Ge with 32 electrons has the configuration $1s^2 2s^2 2p^6 3s^2 3p^6 3d^{10} 4s^2 4p^2$. It is only the electrons in the outer shell or valence electrons which will be important for electronic properties since the inner shells are strongly bound to the nucleus through the Coulomb force.

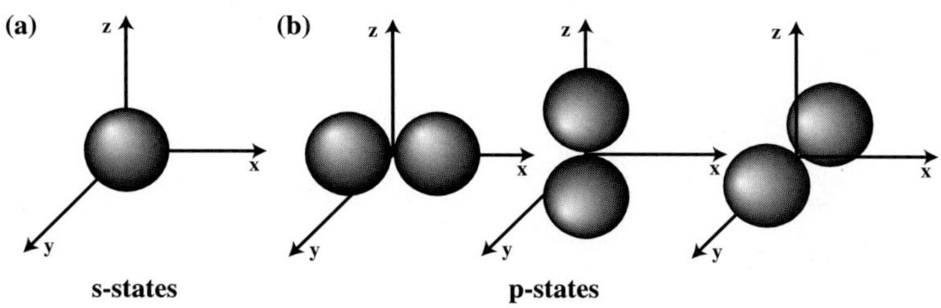

Fig. 3.3. The probability densities for the s- and p-states of an electron in a hydrogen atom

While these wavefunctions are derived for the hydrogen atom, by replacing the masses with those for other larger atoms, these solutions are taken as an approximation to the energy levels for the larger atoms. It is the interaction of these hydrogen-like orbitals which form the bonds between atoms that creates molecules and crystals. The Pauli Exclusion Principle states that for fermions such as electrons, only one electron can be in each quantum state or energy level and so with the spin degeneracy there are two electrons in each energy level (one with spin up and the other with spin down). Two electrons are also required to form a covalent bond between atoms to form molecules or crystals. Through the wave behaviour of the electrons, the wavefunctions for bonds can have a number of positive or negative interference effects when the wavefunctions are added. The positive are termed bonds and the negative are termed antibonds.

In the next section, the electronic structure of a single atom will be expanded to that for a crystal structure using a number of different techniques. This will allow the electronic properties of semiconductors to be calculated.

3.2 The Band Structure of Semiconductors

3.2.1 The Free Electron Picture and the Effective Mass

The Schrödinger equation for a free particle in 3D is

$$-\frac{\hbar^2}{2m}\nabla^2\psi(\mathbf{r}) = E\psi(\mathbf{r}) \tag{3.45}$$

In orthogonal coordinates, the simplest crystal that the electrons can be confined to is a cube of edge, L. If we define the origin at one corner of the cube, the wavefunction in the cube must be a standing wave of the form

$$\psi(\mathbf{r}) = A\sin\left(\frac{\pi n_x x}{L}\right)\sin\left(\frac{\pi n_y y}{L}\right)\sin\left(\frac{\pi n_z z}{L}\right) \tag{3.46}$$

where n_x, n_y and n_z are positive integers. It is convenient to introduce wavefunctions which satisfy periodic boundary conditions in $x-$, $y-$ and $z-$directions with period L. Thus

$$\psi(x+L, y, z) = \psi(x, y, z) \tag{3.47}$$

$$\psi(x, y+L, z) = \psi(x, y, z) \tag{3.48}$$

$$\psi(x, y, z+L) = \psi(x, y, z) \tag{3.49}$$

Wavefunctions which satisfy the free particle Schrödinger equation and the periodic boundary equations are of the form of a travelling plane wave such as

$$\psi(\mathbf{r}) = \exp[\mathrm{i}\mathbf{k}\cdot\mathbf{r}] \tag{3.50}$$

provided that the components of the wavevector $\mathbf{k}$ satisfy

$$k_x, k_y, k_z = 0;\ \pm\frac{2\pi}{L};\ \pm\frac{4\pi}{L};\ \ldots \tag{3.51}$$

The components of $\mathbf{k}$ are the quantum numbers for the problem along with the quantum number for electron spin in the system, m_s.

On substituting (3.50) into (3.45), the energy of the electrons for a wavevector $\mathbf{k}$ is

$$E = \frac{\hbar^2 \mathbf{k}^2}{2m} \tag{3.52}$$

remembering that m is the mass of the electron in vacuum. In reality the electron does not behave as if it is in a vacuum and the free electron picture basically replaces the mass of the electron in vacuum by an "effective mass", m^* which accounts for the interactions that the electron feels in the crystal by using the mass as an empirical fitting parameter. This technique works quite effectively for simple parabolic bands and for small applied electric fields. In real semiconductors, the bands can be quite complicated and the effective mass may be anisotropic. This difference in the mass of the electron should not be too surprising as the transport is by waves through the crystal rather than particles in a vacuum in the free electron picture. The next section will derive the effective mass by comparison of the quantum mechanical picture with classical mechanics which will allow a more general definition of the effective mass.

If the momentum operator $\hat{\mathbf{p}} = \frac{\hbar}{i}\nabla$ is used on the wavefunction ψ then

$$\hat{\mathbf{p}}\psi\left(\mathbf{r}\right) = \frac{\hbar}{\mathrm{i}}\nabla\psi\left(\mathbf{r}\right) = \hbar\mathbf{k}\psi\left(\mathbf{r}\right) \tag{3.53}$$

The plane wave function of the electron is therefore an eigenfunction of the linear momentum with the eigenvalue $\hbar\mathbf{k}$. Relating this crystal momentum with the classical momentum produces

$$m^*\mathbf{v} = \hbar\mathbf{k} \tag{3.54}$$

Now since the group velocity of a wave is related to the angular frequency of the wave, ω and from Planck's quantisation of energy $E = \hbar\omega$

$$\mathbf{v} = \frac{\partial\omega}{\partial\mathbf{k}} = \frac{1}{\hbar}\frac{\partial E}{\partial\mathbf{k}} = \frac{\hbar}{2m^*}\frac{\partial}{\partial\mathbf{k}}\left\{\mathbf{k}^2\right\} = \frac{\hbar\mathbf{k}}{m^*} \tag{3.55}$$

While (3.54) suggests by comparison that the semi-classical particle velocity of the electron equals the crystal momentum divided by the effective mass, (3.55) shows how the group velocity of the electron wave is equal to the crystal momentum divided by the effective mass. If this is now put into Newton's law relating the electric field, $\mathbf{F}$ to the velocity of an electron of charge q then

$$q\mathbf{F} = -m^*\frac{\partial\mathbf{v}}{\partial t} = -m^*\frac{\partial\mathbf{v}}{\partial\mathbf{k}}\frac{\partial\mathbf{k}}{\partial t} = \frac{m^*}{\hbar^2}\frac{\partial^2 E}{\partial\mathbf{k}^2}\cdot q\mathbf{F} \tag{3.56}$$

Now (3.56) suggests that the effective mass can be defined by

$$\frac{1}{m^*} = \frac{1}{\hbar^2}\frac{\partial^2 E}{\partial\mathbf{k}^2} \tag{3.57}$$

and therefore may be obtained from the energy dispersion relation or band structure of the isotropic material. More generally the effective mass is anisotropic in many crystals and has therefore tensor properties.

$$\frac{1}{m_{ij}^*} = \frac{\partial^2 E}{\partial p_i \partial p_j} = \frac{1}{\hbar^2}\frac{\partial^2 E(k)}{\partial k_i \partial k_j} \tag{3.58}$$

There are more complete descriptions of the effective mass such as that defined in **k.p**-theory (Sect. 3.2.6) which calculates an effective mass which takes into account mixing from different bands which are interacting. This will be briefly introduced in section 3.2.6 but is well beyond the scope of the present text. The major point is that the effective mass is a useful construct that naively determines the amount of interaction a conduction electron in a crystal has with the lattice and other electrons in the system.

3.2.2 The Crystal Structure

All major semiconductors which are used in manufacturing such as Si, Ge or GaAs are crystalline materials. Each has a regular structure of the constituent atoms along with a lattice constant.

By defining the basis vectors **a**, **b** and **c** describing a crystal solid such that the crystal structure remains invariant under translation through any vector that is an integer multiple of the basis vectors, the direct lattice sites **r** can be defined as:

$$\mathbf{r} = i\mathbf{a} + j\mathbf{b} + k\mathbf{c} \tag{3.59}$$

(for i, j, k = integers).

At room temperature and pressure, the group IV elements of Si and Ge both form the diamond lattice structure. Each of the four electrons in the outer shell form sp^3 hybridised orbitals with covalent bonds between the atoms being formed by one electron from each atom with opposite spin. The sp^3 hybridised orbitals produce the tetrahedral structure which creates the face centred cubic diamond lattice (Fig. 3.4). The structure results in the $\{111\}$ planes being the easiest cleavage planes for both Si and Ge.

For a given set of direct basis vectors, a set of reciprocal lattice vectors $\mathbf{a}^*$, $\mathbf{b}^*$, $\mathbf{c}^*$ can be defined such that

$$\mathbf{a}^* \equiv 2\pi \frac{\mathbf{b} \times \mathbf{c}}{\mathbf{a} * \mathbf{b} \times \mathbf{c}} \tag{3.60}$$

$$\mathbf{b}^* \equiv 2\pi \frac{\mathbf{c} \times \mathbf{a}}{\mathbf{a} * \mathbf{b} \times \mathbf{c}} \tag{3.61}$$

$$\mathbf{c}^* \equiv 2\pi \frac{\mathbf{a} \times \mathbf{b}}{\mathbf{a} * \mathbf{b} \times \mathbf{c}} \tag{3.62}$$

so that the reciprocal lattice point is given by

$$\mathbf{r}^* = l\mathbf{a}^* + m\mathbf{b}^* + n\mathbf{c}^* \tag{3.63}$$

(for l, m, n = integers)

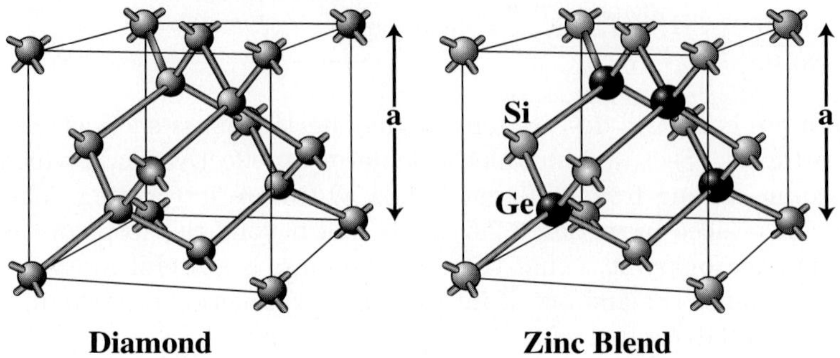

Fig. 3.4. (a) The diamond crystal structure of elemental Si and Ge. The zinc blend structure is that found for compound semiconductors such as GaAs and ordered $Si_{0.5}Ge_{0.5}$

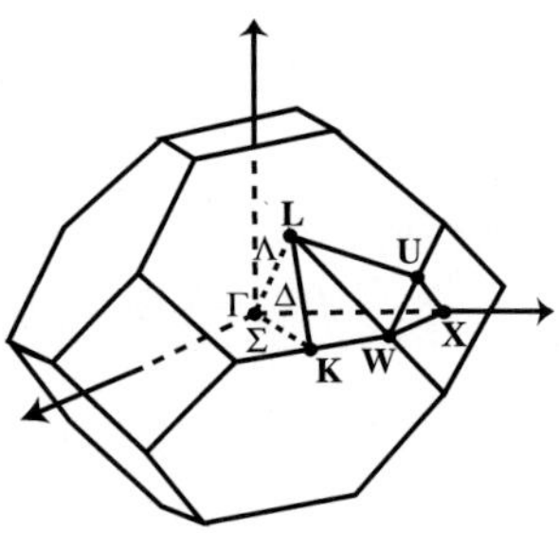

Fig. 3.5. The body centred cubic reciprocal lattice along with the Brillouin zone boundaries and the major symmetry points marked on the using the standard group theory symbols

The unit cell of the reciprocal lattice can be formed by constructing the Wigner-Seitz cell. The Wigner-Seitz cell is primitive and displays the symmetry of the crystal system. To obtain this cell one must start at any of the lattice points and the origin and draw vectors to all the neighbouring lattice points. Planes perpendicular to and passing through the midpoints of these vectors are constructed. The Wigner-Seitz cell is then the smallest volume about the origin bounded by these planes.

For a face centred cubic lattice, the reciprocal lattice which forms the Wigner-Seitz cell is a body centred cubic lattice. Figure 3.5 shows the body centred cubic reciprocal lattice along with the corresponding Brillouin zone boundaries with the major symmetry points marked by the standard group theory symbols. Once the crystal structure of a material is known, this frequently allows the calculation of the energy band structure from Schrödinger's equation to be simplified when the symmetry of the system is taken into account.

3.2.3 Bloch's Theorem and Bloch Functions

For a periodic potential, the electronic band structure and the wavefunctions are derived from the Hamiltonian which must satisfy the symmetry of the crystal. For an electron in a periodic potential, V

$$V(\mathbf{r}) = V(\mathbf{r} + \triangle\mathbf{r}) \tag{3.64}$$

where $\triangle\mathbf{r} = n_1\mathbf{a} + n_2\mathbf{b} + n_3\mathbf{c}$ and $\mathbf{a}$, $\mathbf{b}$ and $\mathbf{c}$ are the lattice basis states and n_1, n_2 and n_3 are integers. This translation is therefore just a translation by a lattice vector or multiples thereof. The electron wavefunction, as before, must satisfy the Schrödinger equation

$$\hat{H}\psi(\mathbf{r}) = \left[-\frac{\hbar^2}{2m^*}\nabla^2 + V(\mathbf{r})\right]\psi(\mathbf{r}) = E(\mathbf{k})\,\psi(\mathbf{r}) \tag{3.65}$$

The Hamiltonian must be invariant under translation by the lattice vector $\mathbf{r} \to \mathbf{r} + \triangle\mathbf{r}$. If $\psi(\mathbf{r})$ is a solution to (3.65) then so must be $\psi(\mathbf{r} + \triangle\mathbf{r})$. Thus $\psi(\mathbf{r} + \triangle r)$ can only differ from $\psi(\mathbf{r})$ by at most a constant. This constant must have a unity magnitude otherwise the wavefunction will grow to infinity if the translation $\triangle\mathbf{r}$ is repeated indefinitely. Therefore the general solution to Schrödinger's equation must be of the form

$$\psi_{n\mathbf{k}}(\mathbf{r}) = \exp[\mathrm{i}\mathbf{k}\cdot\mathbf{r}]\,u_{n\mathbf{k}}(\mathbf{r}) \tag{3.66}$$

where $u_{n\mathbf{k}}$ is defined as the periodic function

$$u_{n\mathbf{k}}(\mathbf{r} + \triangle\mathbf{r}) = u_{n\mathbf{k}}(\mathbf{r}) \tag{3.67}$$

This result is Bloch's theorem and the wavefunctions, $\psi_{n\mathbf{k}}(\mathbf{r})$ defined in (3.66) are called Bloch functions. It is important to note that Bloch's theory is a result of the crystal lattice periodicity and not related to quantum mechanical effects.

3.2.4 The Kronig-Penney Model

The simplest periodic array to consider is that of a set of square well potentials. In modern semiconductor systems this can be grown epitaxially and is called a superlattice. The system is also appropriate for considering a 1D model of electrons travelling through a crystal in that the electron experiences a periodic potential due to the crystal structure. Each electron will be accelerated as it approaches the positive nucleus of each atom and then a decceleration as it moves past before experiencing the acceleration from the next atom. The starting point is to approximate the crystal by assuming that each loosely bound valence electron in the outer shell of an atom is confined by the positive nuclear charge to a square potential well

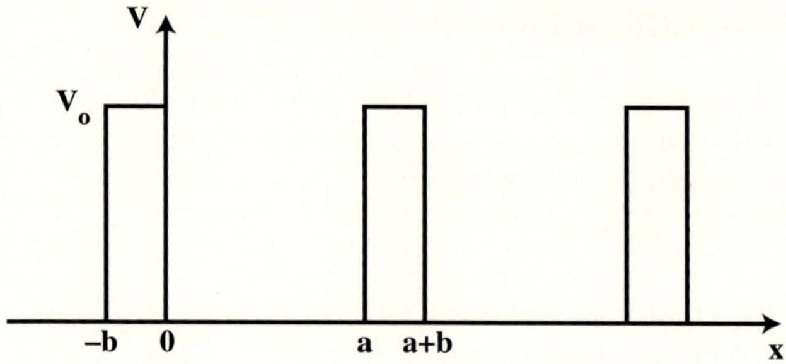

Fig. 3.6. The periodic potential which is used in the Kronig-Penney model

The solutions for a single quantum well have already been calculated in Sect. 3.1.3. For the present case the solutions can be divided into the quantum well and the barrier (Fig. 3.6) so that we have respectively

$$\psi(x) = A\exp[\mathrm{i}\alpha x] + B\exp[-\mathrm{i}\alpha x] \qquad \text{for } 0 < x < a \tag{3.68}$$

$$\psi(x) = C\exp[\mathrm{i}\beta x] + D\exp[-\mathrm{i}\beta x] \qquad \text{for } -b < x < 0 \tag{3.69}$$

with

$$\alpha^2 = \frac{2m_0 E}{\hbar^2} \quad ; \quad \beta^2 = \frac{2m_0(E - V_0)}{\hbar^2} \tag{3.70}$$

The solutions require to be periodic and therefore we require the solutions to have a Bloch form. Thus the solution in the region $a < x < a + b$ must be related to the solution in the barrier (equation 3.69) in the region $-b < x < 0$ by the Bloch theorem so that

$$\psi(a < x < a + b) = \psi(-b < x < 0)\exp[\mathrm{i}k_x(a + b)] \tag{3.71}$$

which naturely defines the wavevector, k_x as an index for the solution. As before for the quantum well, the constants A, B, C and D must be chosen so that ψ and $\frac{d\psi}{dx}$ are continuous at $x = 0$ and $x = a$. At $x = 0$

$$A + B = C + D \tag{3.72}$$

$$i\alpha(A - B) = \mathrm{i}\beta(C - D) \tag{3.73}$$

At $x = a$ by applying equation 3.71 for $\psi(a)$ under the barrier in terms of $\psi(-b)$

$$A\exp[\mathrm{i}\alpha a] + B\exp[-\mathrm{i}\alpha a] = \\ (C\exp[-\mathrm{i}\beta b] + D\exp[\mathrm{i}\beta b])\exp[\mathrm{i}k(a + b)] \tag{3.74}$$

$$\mathrm{i}\alpha(A\exp[\mathrm{i}\alpha a] - B\exp[-\mathrm{i}\alpha a]) = \\ \mathrm{i}\beta(C\exp[-\mathrm{i}\beta b] - D\exp[\mathrm{i}\beta b])\exp[\mathrm{i}k_x(a + b)] \tag{3.75}$$

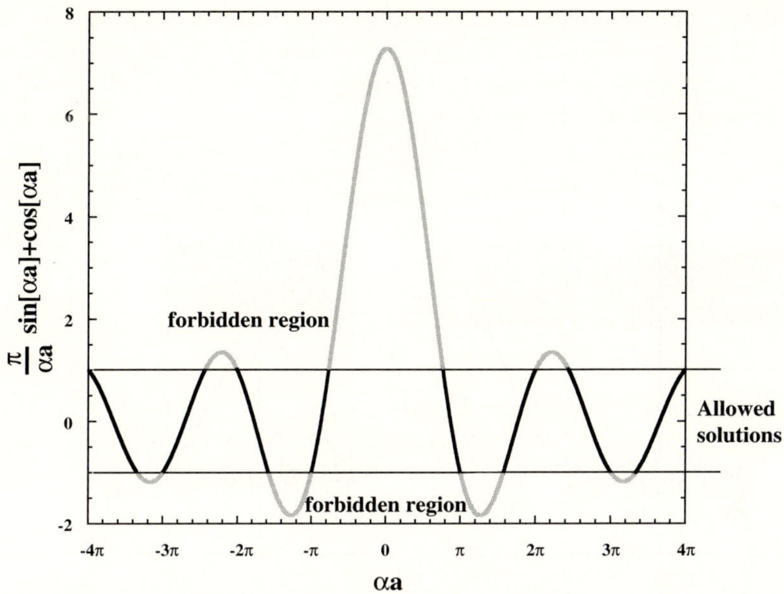

Fig. 3.7. A plot of the left-hand side function from (3.77) with $\frac{\beta^2 ba}{2} = \pi$. The allowed values of energy of an electron in the system is related to the values of αa which lie between ± 1 on the vertical axis since the right-hand side of (3.77) is a cosine function. Outside this range, no Bloch or travelling wave solutions can exist. Therefore a number of forbidden energy gaps are created

The solution to set of equations (3.72) to (3.76) can be found only if the coefficients to A, B, C and D can be made to vanish. For $E < V_0$, $\beta^* = i\beta$, it can be shown with much algebra that this is when

$$\left(\frac{\beta^{*2} - \alpha^2}{2\alpha\beta^*}\right) \sinh\left[\beta^* b\right] \sin\left[\alpha a\right] + \cosh\left[\beta^* b\right] \cos\left[\alpha a\right] = \cos\left[k_x \left(a + b\right)\right] \quad (3.76)$$

A much simpler result can be found if the periodic potential is replaced by a set of δ-functions so that (3.76) is being taken in the limit of $b = 0$ and $V_0 \rightarrow \infty$. In this limit $\beta^* \gg \alpha$ and $\beta^* b \ll 1$ which allows (3.76) to be simplified to

$$\left(\frac{\beta^{*2} ba}{2}\right) \frac{\sin\left[\alpha a\right]}{\alpha a} + \cos\left[\alpha a\right] = \cos\left[ka\right] \quad (3.77)$$

This equation does not have solutions for all α. As an example, if we set $\frac{\beta^{*2} ba}{2} = \pi$, then Fig. 3.7 shows the solutions plotted. The only allowed values of energy are for which the left-hand side of (3.7) exists for real k. Since the right-hand side is a cosine function, which can only have values ± 1, any solutions outside ± 1 are not allowed and energy gaps form. The energy as a function of k is plotted in Fig. 3.8. The energy is parabolic when averaged over the whole dispersion curve but energy gaps open up at values of $k = \frac{n\pi}{a}$ where n is an integer. While the curves can be plotted over all k, since the structure is periodic, the solutions are typically folded back into $\pm\frac{n\pi}{a}$ which is called the reduced zone scheme (Fig. 3.8(b)).

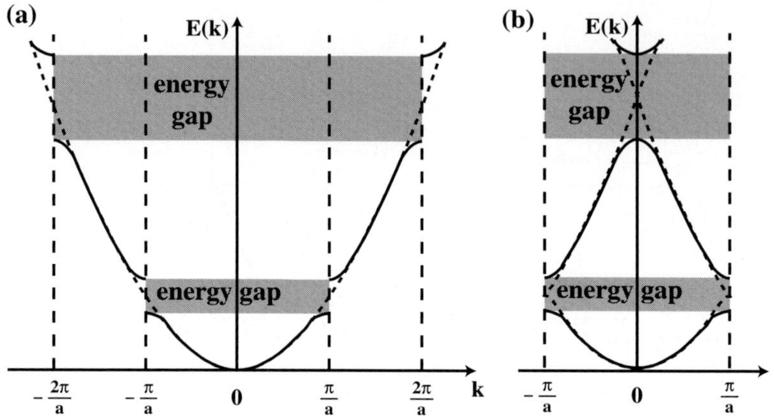

Fig. 3.8. (a) A plot of energy against wavenumber in the extended zone scheme for the Kronig-Penney potential demonstrating the energy gaps at $k_x = \frac{n\pi}{a}$ with n and integer. (b) A plot of the same potential as (a) but in the reduced zone scheme

These energy gaps open up in any 1D system such as a superlattice and are useful in designing a number of quantum devices since electrons can only have a certain energy and are forbidden to have energies corresponding to those in the band gaps. The origin of these gaps is from the periodicity of the crystal structure and the wave behaviour of the electrons. In the next sections, the band structure for bulk 3D crystals such as Si and Ge will be calculated where again band gaps will form due to the periodicity of the material.

3.2.5 The Tight Binding Model

There are two different starting points for calculating the band structure of materials. Either the starting point is from isolated atoms being put together into a crystal with the individual wavefunctions of the atoms overlapping to form bands (e.g. the Kronig-Penney model in 1D, Sect. 3.2.4) or you start from a large bulk crystal and use the periodicity and wave properties of electrons through Bloch's theory to find the band structure. In this section, the tight binding method will be shown to be the former by constructing the band structure from the energy structure of the constituent atoms.

We have already found the molecular orbitals for the hydrogen atom, Sect. 3.1.4. It will become important for the tight binding method but also for the optical properties of quantum wells to understand how atoms bond together. Figure 3.9 shows the different types of bonds that form for s-states and p-states. In each case either a bonding or an antibonding state can form which is related to the parity of the wavefunctions and whether they are added or subtracted. This again is related to the wave behaviour of electrons from quantum mechanics.

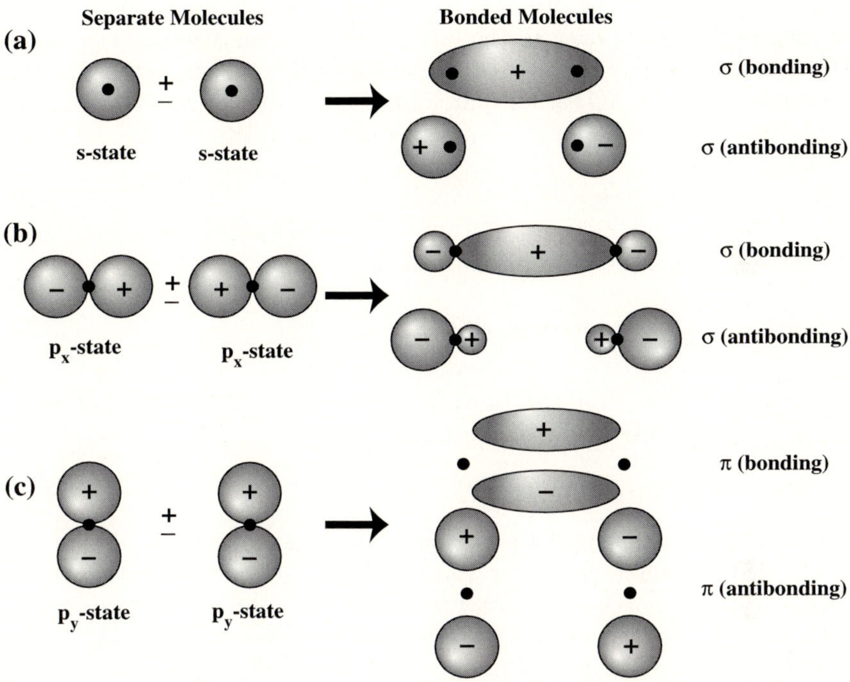

Fig. 3.9. A plot showing how the separate molecular orbitals of atoms form bonding and antibonding states for (a) s-orbitals forming σ-bonds, (b) for p_x-orbitals forming σ-bonds and for (c) p_y-orbitals forming π-bonds

We will now derive the important equations for the tight binding method. Let the eigenstate of an isolated atom be $\phi(\mathbf{r})$ with eigenenergy, E_0 and assume that it is a normalised and non-degenerate state such as an s-state. The basic assumption is that the overlap of this atomic state with the neighbouring atoms is small which is called the tight binding approximation. We also require the assumption that the extra potential energy felt by the electrons in the crystal is small compared to the atomic potential energy. The Hamiltonian of the system is split into that for the atomic potential, $\hat{H}_{atomic}$ and that for the crystal, $\hat{H}_{crystal}$. Since we have assumed that the crystal potential is much weaker, the effect of the crystal is treated as a perturbation in the system.

If we now define the atoms as sitting on the lattice points, $\mathbf{r}_m$ then when an electron is close to the atom at $\mathbf{r}_m = 0$ its eigenfunction is approximately $\phi(\mathbf{r})$. Similarly when the electron is near the atom at the lattice point $\mathbf{r}_m$, its wavefunction is approximately $\phi(\mathbf{r} - \mathbf{r}_m)$. Thus the wavefunction for one electron in the crystal is

$$\psi_k(\mathbf{r}) = \sum_m A_{km} \phi(\mathbf{r} - \mathbf{r}_m) \tag{3.78}$$

The summation must be over all the lattice points in the system. This wavefunction is a linear combination of atomic orbitals (LCAO) and the technique is sometimes named LCAO rather than tight binding. This wavefunction must be of a Bloch form and so for N atoms in the crystal, let

$$A_{km} = \frac{1}{\sqrt{N}} \exp[\mathrm{i}(\mathbf{k}.\mathbf{r}_m)] \tag{3.79}$$

and the wavefunction for tight binding becomes

$$\psi_k(\mathbf{r}) = \frac{1}{\sqrt{N}} \sum_m \exp[\mathrm{i}(\mathbf{k}.\mathbf{r}_{m'})] \phi(\mathbf{r} - \mathbf{r}_m) \tag{3.80}$$

It is easy to show that this wavefunction satisfies Bloch's theorem by replacing $\mathbf{r}$ with $\mathbf{r} + \Delta\mathbf{r}$ and showing that

$$\psi(\mathbf{r} + \Delta\mathbf{r}) = \exp[\mathrm{i}(\mathbf{k}.\Delta\mathbf{r})] \psi(\mathbf{r}) \tag{3.81}$$

While (3.80) can be used to calculate the band structure for different materials, this is a whole field unto itself and for the purposes of this text it is important only to understand how the method is derived and used along with the results from the technique. Figure 3.10 shows how the bands in a particular crystal system form from the molecular orbitals as a function of the spacing of the atoms. At infinite seperation the atoms have the hydrogen like molecular orbitals and as the atoms are brought together in a crystal and bonds are formed between molecules, bands of allowed energies for the electrons to occupy form. On paper one can bring the nuclei together but this is not a useful concept as it cannot be done in real crystals. The important point for the crystal is when the distance between atoms corresponds to the lattice spacing of the crystal. In the particular case shown in Fig. 3.10, two bands with a band gap between them are formed. This particular diagram is appropriate for the case of Si.

The actual band states that are formed for Si and for Ge are shown in Fig. 3.11. As will be derived later in a rigourous manner, the Fermi energy separates the filled from the unfilled states and for pure crystals lies inside a band gap. It is this band gap that defines a semiconductor. Insulators have very large band gaps and it is impossible to excite carriers across it at room temperature. Hence electrons have no free states to occupy and no electronic transport can take place. In metals the bands all overlap and electrons at

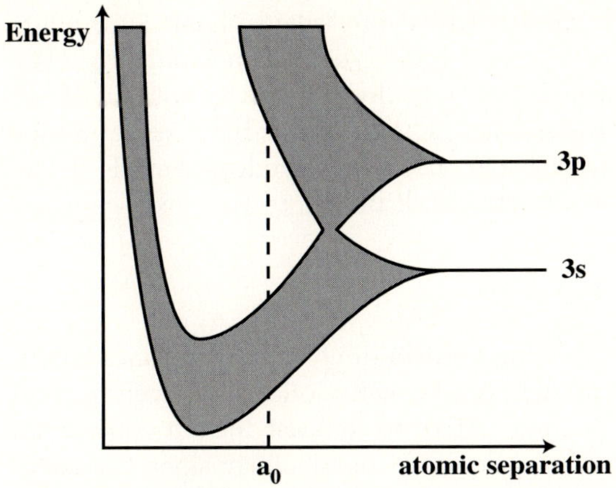

Fig. 3.10. The formation of the band structure as isolated atoms with s- and p-states are brought together. For the case of silicon the s- and p-levels form sp^3 hybridised orbitals separated by an energy gap

a finite temperature can always move to a free state in another band and so the electrons can easily move around and hence they have good electrical transport properties. Semimetals such as graphite also exist where the filled and unfilled bands touch and so are insulators at 0 K but metals at all finite temperatures. Semiconductors lie between semimetals and insulators in that the band gaps are small enough that at a high enough temperature, electrons can be excited across the band gap into free states allowing electronic transport.

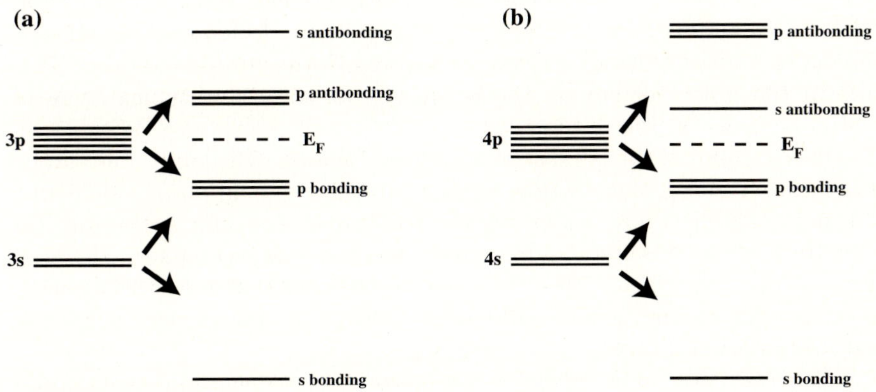

Fig. 3.11. The formation of the band structures of (a) Si and (b) Ge from the valence electrons in the outer shell of each atom

The top of the filled electron band is called the valence band (E_v) while the bottom of the unfilled states is termed the conduction band (E_c). For Si both of these have p-state characteristics while for Ge, the bottom of the conduction band has s-state characteristics. The complete band structure for materials is normally calculated using **k.p** theory or pseudopotentials so the complete band structure of semiconductors will be left to the next section.

3.2.6 Pseudopotentials and k.p Theory

Most band structures that are used for semiconductors are the ones calculated using the pseudopotential model. While such modelling is well beyond the level of the present text, it is informative to at least have a conceptual idea of how such band structures are calculated particularly since the pseudopotential band structures will be the ones typically used to explain the properties of almost all bulk semiconductors.

The pseudopotential relies on splitting the potential of an atom into two separate parts: one for the electrons closest to the core of the atom and a second for the valence electrons in the atom. Therefore the pseudopotential reproduces the valence states as the lowest eigenstates of the problem and neglects the core states. From chemistry and the concept of bonding with valence electrons (see Sect. 3.2.5) this is not too difficult to justify. The main reason for this division from the band structrue calculation point of view is that the wavefunctions of the electrons near the core of the atom will have strong spacial oscillations which makes it very difficult to solve Schrödinger's equation for the system. The pseudopotential technique relys on the splitting of the potential for the atom into a part for the valence electrons and a soft potential part for the core. This allows Schrödinger's equation to be solved by splitting it into the parts of the core and the valence electrons. A summation of Bloch-like plane waves are used for the valence electron wavefunctions which are modified by making them orthogonal to the core-level states. This results in the potential having the required soft core. The typical type of potential is shown in Fig. 3.12.

In the **k.p** method the band structure over the entire Brillouin zone can be extrapolated from the zone centre energy gaps and optical matrix elements. The **k.p** method is therefore extremely convenient for interpreting optical spectra from semiconductors and heterostructures. The technique is also useful for strained materials as the strain can be calculated as a perturbation to the system. The technique allows the band dispersion and effective masses to be calculated around high symmetry points.

The **k.p** method can be drived from the one-electron Schrödinger equation (3.9) by applying Bloch's theorem. Using a Bloch wavefunction of the form

$$\psi_{n\mathbf{k}}(\mathbf{r}) = \exp\left[\mathrm{i}\left(\mathbf{k}.\mathbf{r}\right)\right] u_{n\mathbf{k}}(\mathbf{r}) \tag{3.82}$$

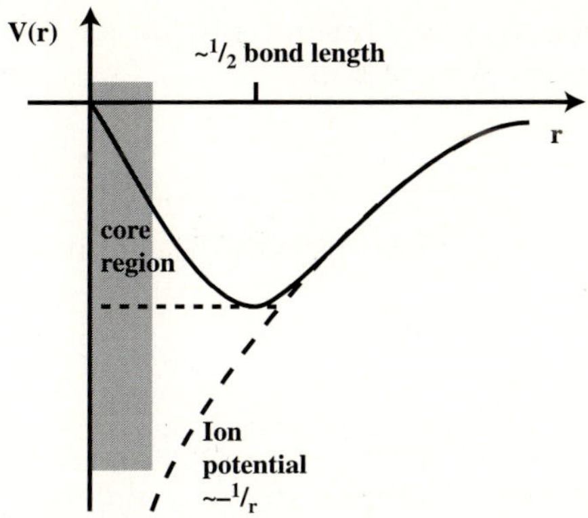

Fig. 3.12. The typical pseudopotenial for a Si atom in real space. The solid curve in which $V(r) \rightarrow = 0$ in the core region is called a ''soft core" pseudopotential. The dashed curve in which $V(r) \rightarrow$ constant is a ''hard core" pseudopotential

where n is the band index, k lies within the first Brillouin zone and $u_{n\mathbf{k}}$ is a function with the periodicity of the lattice. Substituting (3.82) into (3.9) produces

$$\left(\frac{p^2}{2m_0}+\frac{\hbar\mathbf{k}.\mathbf{p}}{m_0}+\frac{\hbar^2k^2}{2m_0}+V(r)\right)u_{n\mathbf{k}}(\mathbf{r})=E_{n\mathbf{k}}u_{n\mathbf{k}}(\mathbf{r}) \tag{3.83}$$

where m_0 is the free electron mass. At $\mathbf{k}=(0,0,0)$ this reduces to

$$\left(\frac{p^2}{2m_0}+V(r)\right)u_{n0}(r)=E_{n0}u_{n0}(r) \tag{3.84}$$

While (3.83) could be solved for any k-point, the advantage of solving at (0,0,0) is that the functions of $u_{n\mathbf{k}}$ are then periodic. Therefore once E_{n0} and u_{n0} are known, the terms $\hbar\mathbf{k}.\mathbf{p}/m_0$ and $\hbar^2k^2/(2m_0)$ are pertubations and can be calulated using degenerate or non-degenerate perturbation theory. It should be noted that since the perturbations are proportional to k and k^2, this method works best for small values of k.

As an example, let us calculate using non-degenerate perturbation theory the first terms at least to second order for a band structure which has an extremum at the energy, E_{n0}.

$$u_{nk}(r)=u_{n0}(r)+\frac{\hbar}{m_0}\sum_{n'\neq n}\frac{\langle u_{n0}(r)|\,k.p\,|u_{n'0}(r)\rangle}{E_{n0}-E_{n'0}}u_{n'0}(r) \tag{3.85}$$

and

$$E_{nk}=E_{n0}+\frac{\hbar^2k^2}{2m_0}+\frac{\hbar^2}{m_0^2}\sum_{n'\neq n}\frac{|\langle u_{n0}(r)|\,k.p\,|u_{n'0}(r)\rangle|^2}{E_{n0}-E_{n'0}} \tag{3.86}$$

The linear term in k vanishes because E_{n0} has assumed to be an extremum. If the energy is rewritten for small values of k as

$$E_{nk} = E_{n0} + \frac{\hbar^2 k^2}{2m^*} \tag{3.87}$$

then the effective mass, m^* is brought into the method but with the definition

$$\frac{1}{m^*} = \frac{1}{m_0} + \frac{2}{m_0^* k^2} \sum_{n' \neq n} \frac{\left| \langle u_{n0}(r) | k.p | u_{n'0}(r) \rangle \right|^2}{E_{n0} - E_{n'0}} \tag{3.88}$$

This provides a clearer picture of the concept of the effective mass as it defines, m^* in terms of the coupling between electronic states in different bands through the **k.p** term. The change in value from the free electron mass is therefore defined in terms of a perturbation to the electron from the electronic states in the system.

3.2.7 Bandstructures of Real Materials

In this section the band structures of Si, Ge and GaAs as calculated by pseudopotentials will be reviewed. The simplest case to consider first is GaAs where the top of the valence band is at the zone centre or Γ (Fig. 3.13). Therefore at $k = 0$ electrons and holes can either be excited by the absorption of a photon or a photon may be emitted by the recombination of electron-hole pairs. Therefore GaAs is used in optoelectronics particularly for light emitting diodes (LEDs) and lasers.

The Si band structure differs from GaAs in a number of ways. Firstly the maximum of the valence band is at $k = 0$ but the minimum of the conduction band is at approximately 0.85 of the way to the X-point. Si is therefore an indirect bandgap material and as such for an electron and hole pair to be generated or recombined, a change of k and therefore crystal momentum is required. Such a process requires a phonon which is the quantum of heat or a lattice vibration and will be reviewed in Sect. 3.6. The other major difference is that the bottom of the conduction band has six equivalent points as shown in Fig. 3.15 (b) due to the symmetry of the crystal structure. GaAs has only one due to the bottom of the conduction band being at $k = 0$. These six equivalent points are called valleys.

Ge has a different band structure (Fig. 3.16). Again the top of the valence band is at $k = 0$ but this time the bottom of the conduction band is at the L-point. Due to the symmetry this has 8 equivalent directions in reciprocal space at the Brillouin zone boundary as shown in Fig. 3.15. Therefore Ge has eight half equivalent valleys.

The top of the valence band has much finer structure which is shown in Fig. 3.17. As will be shown in the next section (3.3.2), the valence band is filled with electrons but some of these can be excited out of the valence band leaving holes behind. Three seperate bands are found close to the Γ

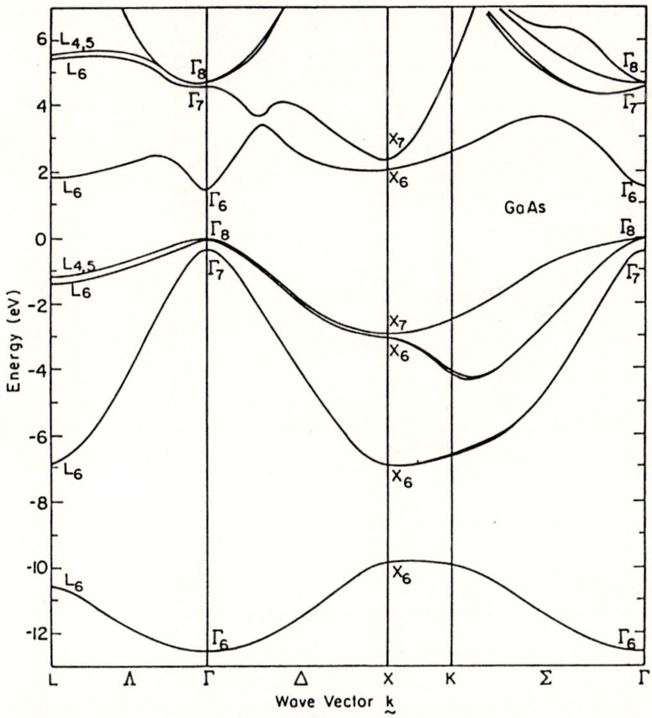

Fig. 3.13. The band structure of GaAs calculated using pseudopotentials (after Cohen and Cheklovsky)

point where $k = 0$, two of which are degenerate in energy. These are called the heavy hole and light hole bands and each has a different effective mass through the different band curvatures. The third band is termed the spin orbit split-off band and is lower in energy than the other two.

3.3 The Concentration of Carriers in a Semiconductor

3.3.1 The Density of States

The density of states, $g(E)$ is defined as the density of allowed energy states per unit volume and, as will be demonstrated later, is an important number for calculating the number of electrons or holes in a semiconductor. A second way of writing this definition is such that $g(E)dE$ is the number of solutions to Schrödinger's equation in the energy interval between E and $E + dE$. The density of states has different forms depending on the number of dimensions in the system. If we consider a phase space in d dimensions then statistical mechanics states that a hyper-volume V in the phase space contains N dis-

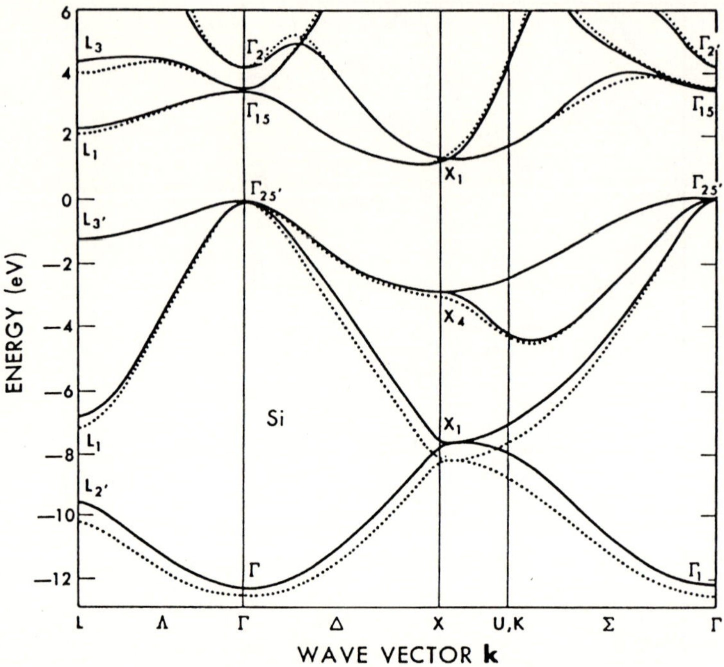

Fig. 3.14. The band structure of Si calculated using a simple local pseudopotential (dotted line) and using a non-local pseudopotential (solid line) which more accurately reproduces the lowest valence band (after Cohen and Cheklovsky).

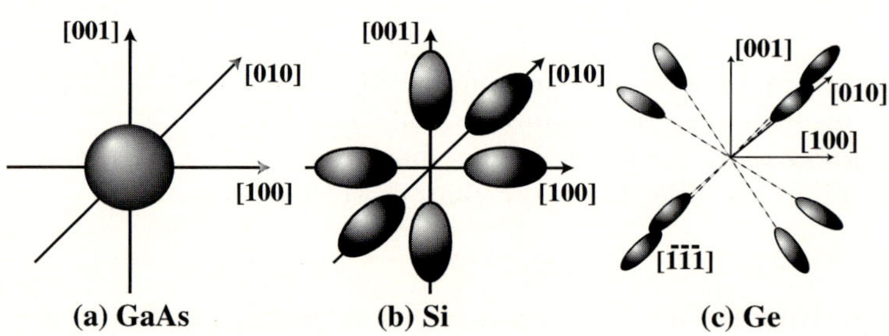

Fig. 3.15. The shape of the conduction band valleys in (a) GaAs, (b) Si and (c) Ge

tinct states where for particles with spin degeneracy g_s and valley degeneracy g_v

$$N = \frac{g_s g_v V^d}{(2\pi\hbar)^d} \tag{3.89}$$

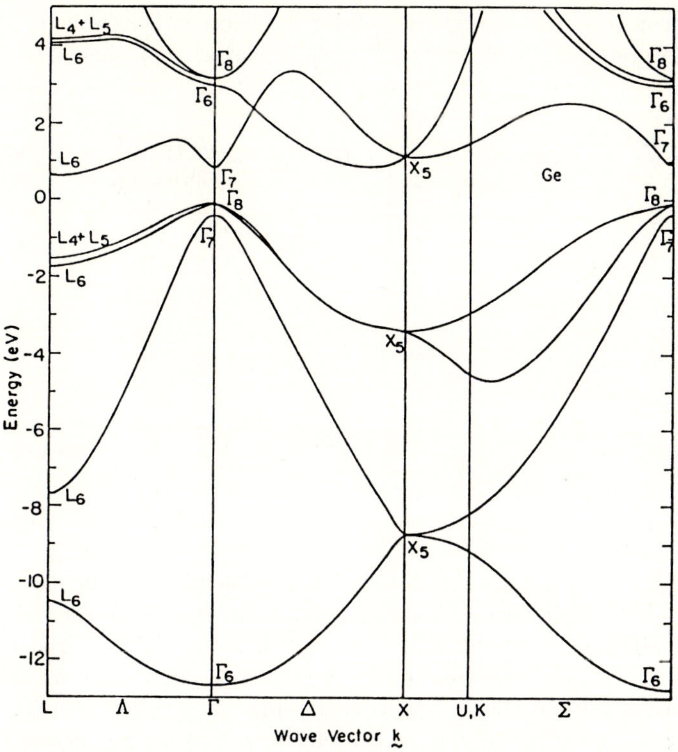

Fig. 3.16. The band structure of Ge as calculated by pseudopotentials (after Cohen and Cheklovsky)

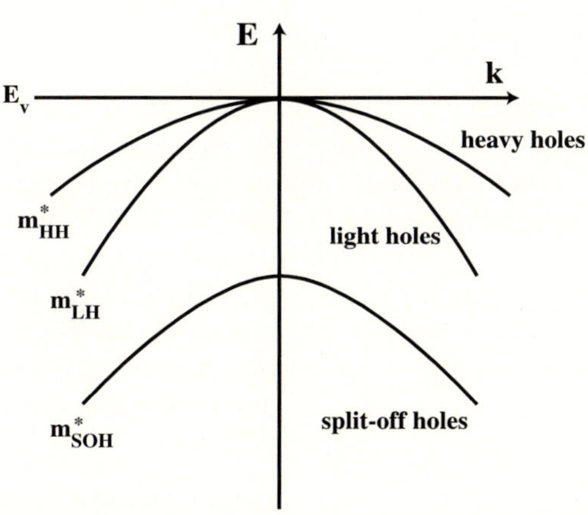

Fig. 3.17. A schematic diagram of the top of the valence band. The light hole (LH) and heavy hole (HH) bands are degenerate at $k = 0$ while the spin orbit split-off band is lower in energy. Each band has a different effective mass due to the different band curvatures

For a free electron gas, the energy dispersion relation is isotropic $E = p^2/2m^*$ and therefore counting the states in a spherical energy shell $\mathrm{d}E = (2E/m^*)^{1/2}\mathrm{d}p$

$$1D: \quad \mathrm{d}N = \frac{g_s g_v L\mathrm{d}p}{2\pi\hbar} = \left[\frac{g_s g_v \sqrt{m^*}}{2\pi\hbar\sqrt{2E}}\right] L\mathrm{d}E \quad g_{1D} = \frac{g_s g_v \sqrt{m^*}}{2\pi\hbar\sqrt{2E}} \tag{3.90}$$

$$2D: \quad \mathrm{d}N = \frac{g_s g_v L^2 2\pi p\mathrm{d}p}{(2\pi\hbar)^2} = \left[\frac{g_s g_v m^*}{2\pi\hbar^2}\right] L^2\mathrm{d}E \quad g_{2D} = \frac{g_s g_v m^*}{2\pi\hbar^2} \tag{3.91}$$

$$3D: \quad \mathrm{d}N = \frac{g_s g_v L^3 4\pi p^2\mathrm{d}p}{(2\pi\hbar)^3} = \left[\frac{g_s g_v (2m^*)^{3/2}}{4\pi^2\hbar^3}\right] L^3\mathrm{d}E$$

$$g_{3D} = \frac{g_s g_v (2m^*)^{3/2}\sqrt{E}}{4\pi^2\hbar^3} \tag{3.92}$$

For both Si and GaAs, the spin degeneracy $g_s = 2$ while for bulk three dimensional Si with the six equivalent valleys, $g_v = 6$. For two-dimensional quantum wells in silicon such as in the MOSFET (see Sect. 5.3), the quantisation in the quantum well (Sect. 3.1.3) can split the valley degeneracy since there will be different subband states with either the longitudinal or transverse m^*. Therefore $g_v = 2$ is frequently used in 2D layers such as the inversion layer of a MOSFET (Sect. 5). Also if strain is present then the number of equivalent valleys can be reduced to either $g_v = 4$ for compressive strain or $g_v = 2$ for tensile strained layers (see Chap. 4). GaAs with a single valley has $g_v = 1$.

3.3.2 Equilibrium Carrier Statistics and Doping

As already described, a pure semiconductor material at 0 K is an insulator. The reason semiconductors are the ideal material for electronics is that by introducting certain impurities, the conductivity of the material may be controllably varied by over 8 orders of magnitude as in Fig. 3.18. For Si or Ge, by introducting a group V impurity such as As into the host lattice, an electron from the As has five valence electrons in the outer shell (Fig. 3.19). Four are required for bonding to the host lattice leaving a single free electron. This is the idea of doping: by introducting impurities a number of extra electrons or holes may be introduced into the system which provide electrical conduction (Fig. 3.19).

Carrier Statistics and Intrinsic Material. At equilibrium, the available states in a conductor are filled up according to the Fermi-Dirac distribution function which is

$$f(E) = \frac{1}{1 + \exp\left[\frac{E - E_F}{k_B T}\right]} \tag{3.93}$$

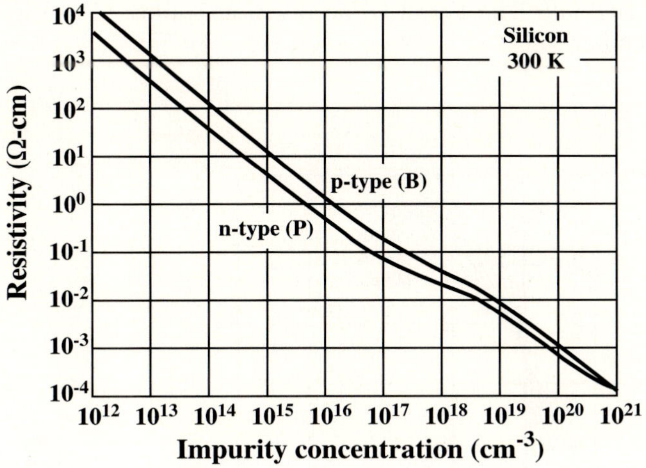

Fig. 3.18. The resistivity of n and p-doped Si as a function of doping density (after Sze)

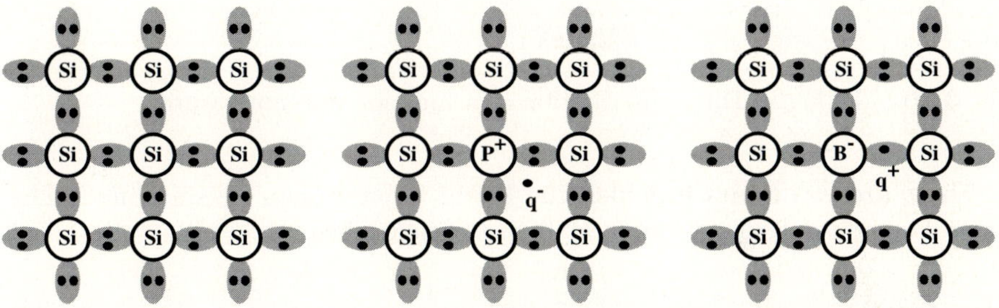

Fig. 3.19. Electrical conduction in semiconductors using the $s-p^3$ bonding picture. (a) Intrinsic silicon has no free electrons and is an insulator. (b) P doped material has a free electron for each P atom and hence is n-type. (c) B doped material has an electron missing from a bond and hence is p-type

where k_B is the Boltzmann constant, T is the absolute temperature and E_F is the Fermi energy. The Fermi energy, E_F is defined as the energy which separates the filled from the unfilled states and is strictly speaking only defined at T =0 K. At T=0 K, the chemical potential = E_F but at finite temperatures the two values may differ substantially depending on the system and any applied voltages. Here we will ignore the differences and call both the Fermi energy although one should remember that at finite temperatures you are really dealing with the chemical potential. Away from equilibrium, such as when a bias is applied across a sample, the system has no common Fermi energy but a local Fermi level is defined which can vary spacially in the system. The Fermi energy is the energy at which the probability of occupation of a state is precisely one-half. The Fermi-Dirac function is plotted in Fig. 3.20 as a function of temperature.

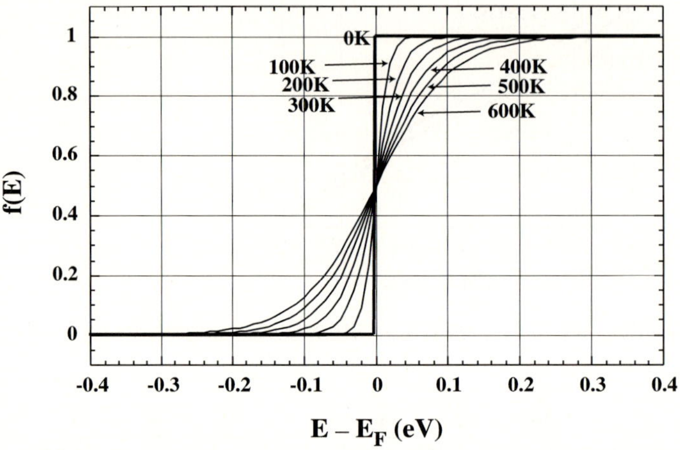

Fig. 3.20. The Fermi-Dirac distribution as a function of temperature

There are two limits in which the Fermi function can be simplified. The first is the non-degenerate limit at high temperatures

$$[E - E_F] >> k_B T \qquad f(E) \approx \exp\left[\frac{-(E - E_F)}{k_B T}\right] \tag{3.94}$$

while for the degenerate limit at low temperatures

$$[E - E_F] << k_B T \qquad f(E) \approx \Theta\left(E_F - E\right) \tag{3.95}$$

where θ is the Heaviside step function which is defined as 0 for $E_F < E$ and 1 for $E_F > E$.

To calculate the number of carriers in any semiconductor system, the density of states multiplied by the Fermi-Dirac distribution is integrated over appropriate limits, i.e.

$$n = \int_0^\infty g(E) f(E) \mathrm{d}E \tag{3.96}$$

For an intrinsic semiconductor without any dopants or impurities compared to the number of thermally generated carriers there are few electrons in the conduction band even though there are a large number of allowed states, hence the probability of an electron populating one of these states is small. All the electrons fill the allowed states in the valence band as shown in Fig. 3.21(a). The Fermi level is therefore at the centre of the bandgap and $E_F \simeq E_i$ (Fig. 3.21(c)) and the number of electrons in the conduction band equals the number of holes in the valence band (Fig. 3.21(d)).

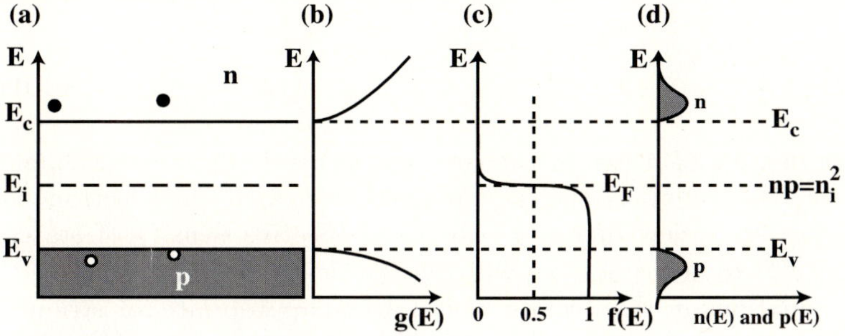

Fig. 3.21. A schematic digram of an intrinsic semiconductor (a) The electron and hole distribution in the conduction and valence bands (b) the density of states (c) the Fermi-Dirac distribution with the Fermi energy marked and (d) the number of electrons and holes in the structure

To calculate the electron density in a piece of bulk, intrinsic silicon, equations 3.92 and 3.94 must be substituted into (3.96) with a spin degeneracy of 2 and a valley degeneracy of g_v giving

$$n = 4\pi g_v \left[\frac{2m_e^*}{h^2}\right]^{3/2} \int_0^\infty E^{1/2} \exp\left\{-\frac{E-E_F}{k_BT}\right\} \mathrm{d}E \tag{3.97}$$

To solve this integral, a substitution of $x = E/k_BT$ into (3.97) produces

$$n = 4\pi g_v \left[\frac{2m_e^* k_BT}{h^2}\right]^{3/2} \exp\left\{\frac{E_F}{k_BT}\right\} \int_0^\infty x^{1/2} exp\left[-x\right] \mathrm{d}E \tag{3.98}$$

where $\int_0^\infty x^{1/2} \exp\left[-x\right]\mathrm{d}E$ is a standard integral which equals $\sqrt{\pi}/2$. Therefore

$$n = 2g_v \left[\frac{2\pi m_e^* k_BT}{h^2}\right]^{3/2} \exp\left\{\frac{E_F}{k_BT}\right\} \tag{3.99}$$

Up to now we have been integrating from 0 to infinity but electrons can only occupy the conduction band. Therefore it is more common to redefine the electron density with respect to the bottom of the conduction band rather than $E = 0$. This results in (3.99) becoming

$$n = 2g_v \left[\frac{2\pi m_e^* k_B T}{h^2}\right]^{3/2} \exp\left\{-\frac{E_c - E_F}{k_B T}\right\} \tag{3.100}$$
$$= N_C \exp\left\{-\frac{E_c - E_F}{k_B T}\right\}$$

where

$$N_C = 2g_v \left(\frac{2\pi m_e^* k_B T}{h^2}\right)^{3/2} \tag{3.101}$$

is the effective density of states in the conduction band. The reader should note that often the valley degeneracy is combined with the electron effective mass, m_e^* for a new quantity called the effective density of states, $m_{de}^{*\ 3/2} = g_v m_e^{*3/2}$. In this book we will use the definition given in (3.101) and previous equations in this section as it is more appropriate for strained materials where the degeneracy is determined by the strain and dimensionality (i.e. quantisation) of the system. In a similar fashion the hole density in the valence band may be obtained

$$p = N_V \exp\left\{-\frac{E_F - E_V}{k_B T}\right\} \tag{3.102}$$

with the effective density of states in the valence band of

$$N_V = 2\left(\frac{2\pi m_p^* k_B T}{h^2}\right)^{3/2} \tag{3.103}$$

The effective masses in the above equations are the density of states effective masses which nievely take an average value of the masses in all the bands used for conduction. The electron value is defined as

$$m_e^* = (m_1^* m_2^* m_3^*)^{\frac{1}{3}} \tag{3.104}$$

where m_1^*, m_2^* and m_3^* are the effective masses along the three principal axes of the ellipsoidal energy surface. In silicon (Fig. 3.15) this corresponds to

$$m_e^* = \left(m_l^* m_t^{*2}\right)^{\frac{1}{3}} \tag{3.105}$$

where m_l^* is the longitudinal mass ($= 0.916m_0$) and m_t^* is the transverse mass ($= 0.19m_0$). For holes in Si, Ge or GaAs the density of states effective mass for the valence band is given by

$$m_p^* = \left(m_{LH}^{*\frac{3}{2}} + m_{HH}^{*\frac{3}{2}}\right)^{\frac{2}{3}} \tag{3.106}$$

where m_{HH}^* is the heavy hole mass and m_{LH}^* is the light hole mass.

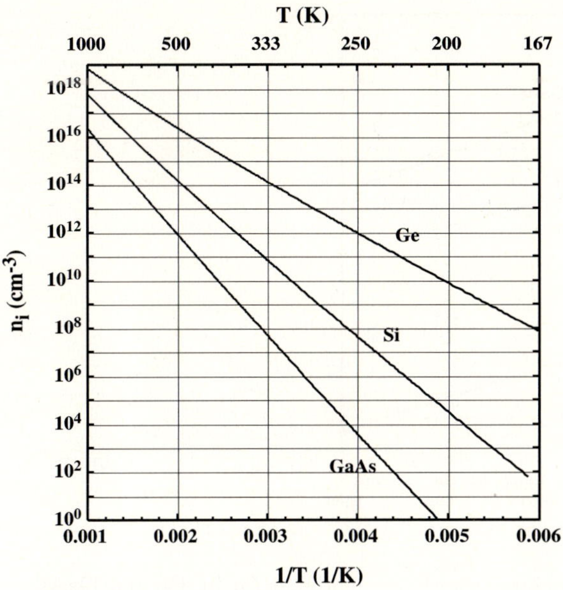

Fig. 3.22. The intrinsic carrier density as a function of temperature for Si, Ge and GaAs

As explained before and shown in Fig. 3.21(d), the number of electrons in the conduction band equals the number of holes in the valence band and this number is called the intrinsic carrier density, $n_i = n = p$. The Fermi level can be found by equating (3.100) and (3.102) to give

$$E_i = \frac{E_C + E_V}{2} + \frac{k_B T}{2} \ln\left(\frac{N_V}{N_C}\right) \tag{3.107}$$

Since at room temperature the temperature-dependent second term is much smaller than the bandgap, the intrinsic Fermi level is close to the centre of the bandgap. The intrinsic carrier density is obtained using (3.100), (3.102) and (3.107) and is given by

$$np = n_i^2 = N_C N_V \exp\left(-\frac{E_C - E_V}{k_B T}\right) \tag{3.108}$$

and since $E_g = E_c - E_v$, the intrinsic carrier density can be written in terms of the bandgap as

$$n_i = \sqrt{np} = \sqrt{N_C N_V} \exp\left[\frac{-E_g}{2k_B T}\right] \tag{3.109}$$

Figure 3.22 shows the temperature dependence of the intrinsic carrier concentration for the three main semiconductors. At high temperatures, the

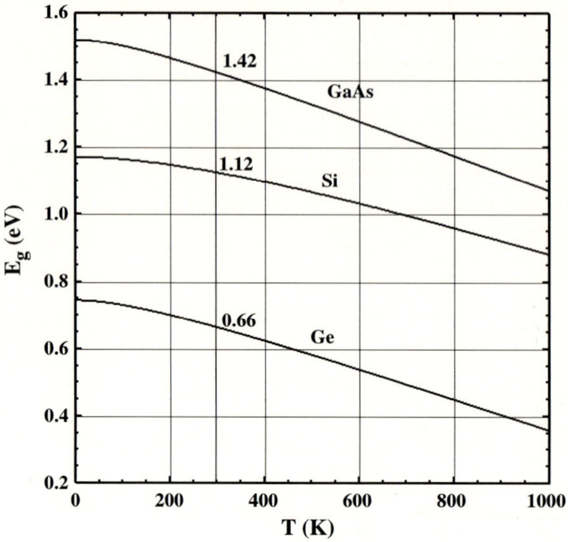

Fig. 3.23. The temperature dependence of the bandgaps for Si, Ge and GaAs

intrinsic carrier concentration can be comparable or larger than the background impurity levels in semiconductors and therefore at a high enough temperature, even doped semiconductors will become intrinsic.

It should also be clear from this section that the size of the band gap or the band gap energy, E_g plays an important role in determining the carrier concentration in the semiconductor. The temperature dependence of the band gaps of Si, Ge and GaAs are shown in Fig. 3.23.

3.3.3 Doping: The Extrinsic Semiconductor

The intrinsic carrier density in Si is $1.07 \times 10^{10}\,\mathrm{cm}^{-3}$ at room temperature which is extremely small compared to the $5.0 \times 10^{22}\,cm^{-3}$ atoms in a bulk silicon crystal. The highest purity silicon crystals produced have background impurity densities of about $10^{11}\,\mathrm{cm}^{-3}$ which corresponds to 1 impurity in every 10^{12} silicon atoms and is the most pure material available in any field. Therefore intrinsic silicon hardly exists at room temperature since it requires lower impurity densities than those available with present technology. Many of the impurities have energies which lie in the bandgap of the semiconductor and form shallow levels which can be easily excited through thermal processes to form electrons or holes in the conduction and valence band respectively. These dopants are typically from group III and group V of the Periodic Table for p-type and n-type dopants respectively. Figure 3.24 shows the energies of impurities in both silicon and germanium. In addition to shallow energy levels, deep levels also exist from a number of impurities, typically from refractory metals. These deep levels are difficult to ionise and therefore may act as traps for carriers in the system, thereby reducing the number of carriers available for electronic transport.

Luttinger and Kohn demonstrated that dopant atoms may be modelled using the modified hydrogen atom to describe the electronic structure of dopants in semiconductors. The ionisation energy of each dopant is calculated by replacing the free electron mass with the effective mass and introducing the dielectric constant, ε_r of the semiconductor so that the energy of the impurity in the bandgap is given by

$$E_D = \frac{m^* q^4}{32\pi^2 \varepsilon_0^2 \varepsilon_r^2 \hbar^2} \tag{3.110}$$

For an intrinsic semiconductor $n = p = n_i$ but once dopant impurities have been added to a semiconductor, the Fermi level will be adjusted to preserve charge neutrality. If N_D donors are added to a system then to preserve charge neutrality the number of electrons in the conduction band n and the number of holes in the valence band are related by

$$n = N_D^+ + p \tag{3.111}$$

where N_D^+ is the number of ionised donors in the system given by

$$N_D^+ = N_D \left[1 - f\left(E_D\right)\right] = N_D \left[1 - \frac{1}{1 + \frac{1}{2}\exp\left(\frac{E_D - E_F}{k_B T}\right)}\right] \tag{3.112}$$

The $\frac{1}{2}$ in the denominator of $f\left(E_D\right)$ originates from the spin degeneracy of the system to allow both up and down spin states. Similarly, if N_A acceptors are added to a semiconductor, the number of ionized acceptors is given by

$$N_A^- = N_A \left[f\left(E_A\right)\right] = N_A \left[\frac{1}{1 + \frac{1}{2}\exp\left(\frac{E_A - E_F}{k_B T}\right)}\right] \tag{3.113}$$

For the standard doping species at room temperature in the main semiconductors, there is very little difference between N_D and N_D^+. In many cases it is assumed that all the dopants are ionised.

The Fermi level is calculated using (3.111) and substituting in the values of (3.100) and (3.102) which gives

$$N_C \exp\left[-\frac{E_c - E_F}{k_B T}\right] = \frac{N_D}{1 + 2\exp\left[-\frac{E_D - E_F}{k_B T}\right]} + N_V \exp\left[-\frac{E_F - E_v}{k_B T}\right] \tag{3.114}$$

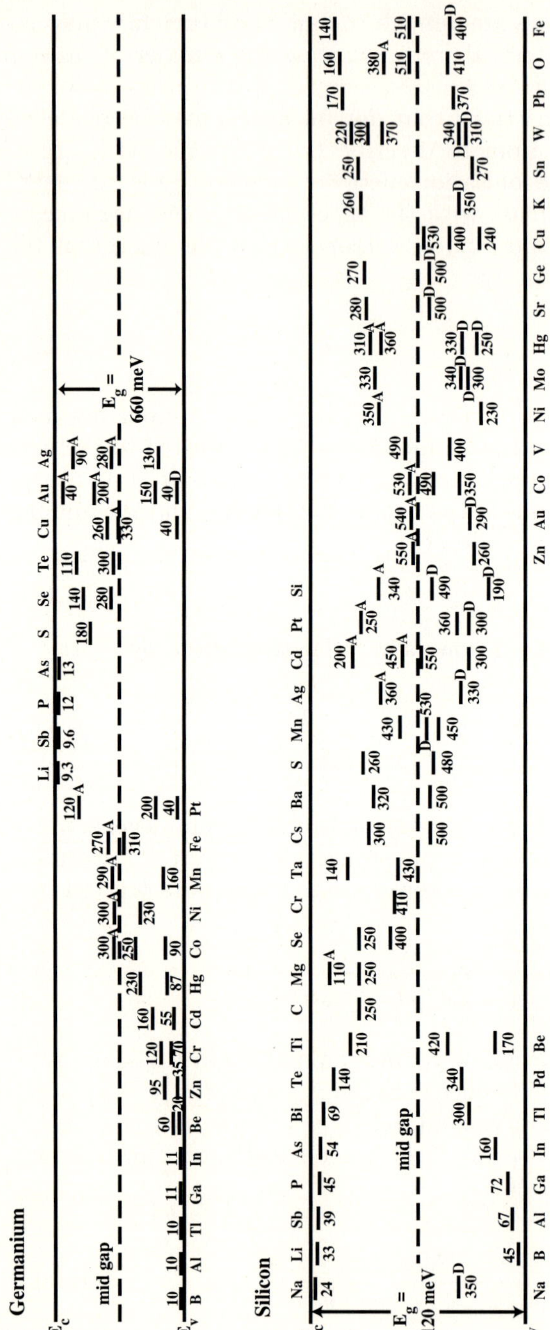

Fig. 3.24. The positions of different impurities in the bandgaps of silicon and germanium

which can be solved to find E_F in the system. If we consider a n-type semiconductor then $N_D \gg N_A$ and the second term in (3.114) can be neglected. If we consider only shallow impurities then at room temperature for moderate concentrations $N_D/N_C \exp\left[\left(E_C - E_D\right)/k_B T\right] \ll 1$ and therefore

$$E_F \cong E_c - k_B T \ln\left[\frac{N_C}{N_D}\right] \tag{3.115}$$

which on substitution can be rewritten in terms of E_i, N_D and n_i as

$$E_F \cong E_i + k_B T \ln\left[\frac{N_D}{n_i}\right] \tag{3.116}$$

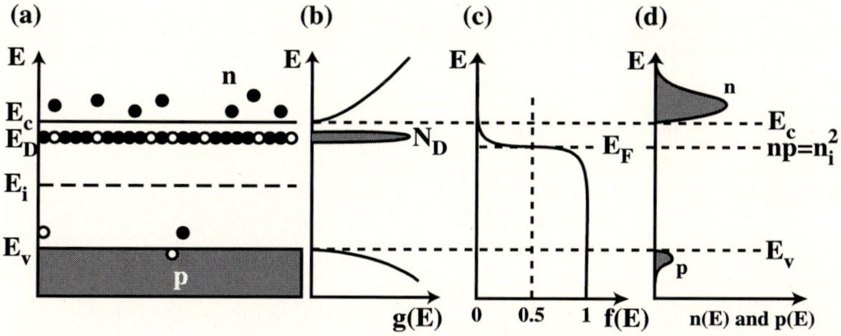

Fig. 3.25. A schematic diagram of (a) the conduction and valence bands of a n-type semiconductor with impurity states from dopants in the band gap (b) the density of states (c) the Fermi-Dirac distribution and the Fermi energy and (d) the number of electrons and holes in the system

All the important parameters discussed above are shown in Fig. 3.25 for a n-type semiconductor. The dopants lie at an energy of E_D in the bandgap and hence $N_D \gg N_A$. The Fermi level lies close to the conduction band and the number of electrons in the conduction band is substantially greater than the number of holes in the valence band.

The electrons in a n-type semicondutor are the majority carriers but there will also be holes in the system which are the minority carrier. To find the number of holes, (3.100) and (3.102) must be multiplied together to give

$$np = n_i^2 = N_C N_V \exp\left[-\frac{E_c - E_v}{k_B T}\right] = N_C N_V \exp\left[-\frac{E_g}{k_B T}\right] \tag{3.117}$$

and the hole density is given by

$$p = \frac{n_i^2}{N_D} \tag{3.118}$$

Similar expressions can be obtained for a p-type material where the holes are the majority carriers and the electrons are the minority carriers.

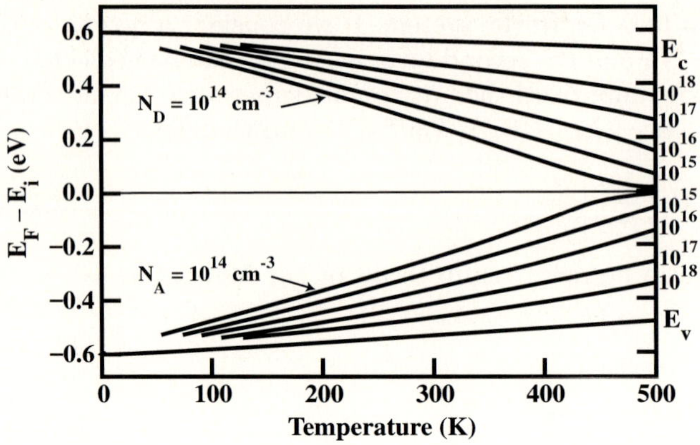

Fig. 3.26. The variation of the Fermi level in Si for different doping densities and different temperatures (after Grove)

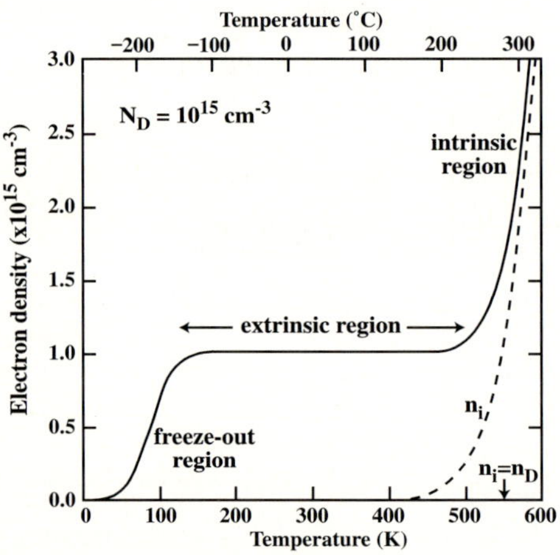

Fig. 3.27. The variation of the carrier density in Si as a function of temperature for a fixed doping concentration (after Sze)

Figure 3.26 plots the temperature variation of the Fermi level for a number of different doping densities as a function of temperature. As may be observed, the Fermi level approaches the intrinsic value near the middle of the bandgap as the temperature rises. When the intrinsic carrier density becomes larger than the doping concentration then the silicon is intrinsic. At the other extreme at very low temperatures, there may not be enough thermal energy in the system to excite the electron or hole to or from the impurity state in the bandgap and therefore the number of carriers in the system is reduced. This regime is termed the freeze-out regime as eventually at 0 K, there will be no carriers in the system since they will all be trapped on the impurity states in the bandgap. The most useful regime is the extrinsic regime which is intermediate to the intrinsic and the freeze-out where all the donors and acceptors are ionised and the carrier density in the system is controlled by the number of dopants. These regimes are shown in Fig. 3.27.

3.3.4 The Two Dimensional Electron Gas (2DEG)

There are two main systems in silicon by which a two dimensional electron gas (2DEG) may be created. The first is an inversion layer as used in the metal oxide semiconductor field effect transistor (MOSFET) where the application of a gate bias forms a thin electron layer in the p-type silicon beside an oxide interface, in effect the electric field creates a potential well at the oxide interface (Fig. 3.28). The second is a modulation doped semiconductor field effect transistor (MODFET) where the dopant atoms are placed a distance from a quantum well created by the different bandgaps in two materials and the electrons (or holes) in the system reside in the quantum well (Fig. 3.28).

Let us consider a quantum well grown in the z-direction so that the x- and y-directions are in the plane of the electron gas. The electrons in either of the quantum well systems are described in the effective mass approximation by the Schrödinger equation

$$-\frac{\hbar^2}{2m^*}\nabla^2\psi + E_c\psi = E\psi \tag{3.119}$$

which has solutions of the form

$$E = \frac{\hbar^2 k_x^2}{2m^*} + \frac{\hbar^2 k_y^2}{2m^*} + E_{z,n} \tag{3.120}$$

where the nth subband energy $E_{z,n}$ is given by the solutions given in Sect. 3.1.3 for the appropriate shape of the quantum well. If only the lowest subband in the system is below the Fermi energy, the device will have a single dynamical 2DEG with subband energy $E_{z,0}$. Having calculated $E_{z,0}$ from the quantum mechanics, it may be regarded as the effective potential energy of the 2DEG and the carrier density n may be calculated from

$$n_{2D} = \int_{E_{z,0}}^{\infty} g_{2D}(E)\, f(E)\mathrm{d}E \tag{3.121}$$

with g_{2D} the 2D density of states.

For both Si and GaAs, the spin degeneracy $g_s = 2$. For GaAs with a single valley $g_v = 1$. While Si has six equivalent valleys, due to the quantisation in the quantum well with the longitudinal and transverse m^* of electrons, only two valleys form the ground state and so the valley degeneracy for a 2DEG is $g_v = 2$. For a compressively strained-$Si_{1-x}Ge_x$ quantum well the ground state subbands are formed from 4 valleys with the transverse m^* giving a degeneracy of $g_v = 4$ while for tensile strained-Si or $Si_{1-x}Ge_x$, the ground

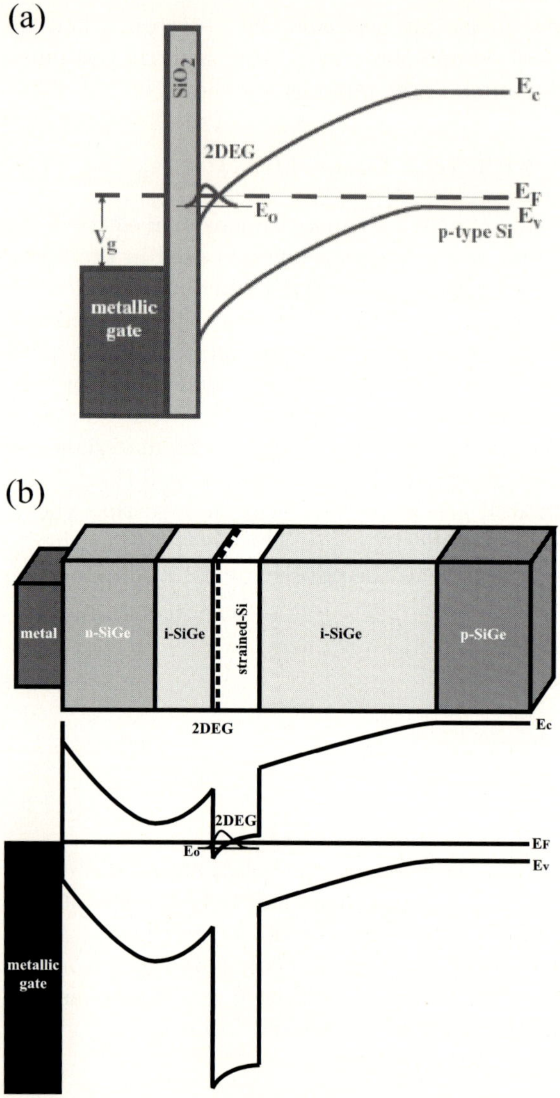

Fig. 3.28. (a) The band structure of a MOSFET with a 2DEG formed by an inversion layer on a p-type silicon substrate. (b) A modulation-doped sample with the 2DEG formed in a quantum well of strained-Si. The donors are spacially separated from the 2DEG and hence scattering is reduced in the modulation-doped scheme

state is composed of two valleys with the longitudinal m^* giving $g_v = 2$. For a 2DEG with an isotropic spectrum, the sheet electron density is given by

$$n_s = \frac{g_s g_v (E_F - E_0) m^*}{2\pi\hbar^2} \tag{3.122}$$

while the Fermi surface is a circle of radius

$$k_F = \sqrt{\frac{4\pi n_s}{g_s g_v}} \tag{3.123}$$

3.4 Electronic Transport in a Semiconductor

3.4.1 The Drift Current

In this section we investigate the effects of electric fields on carriers in a semiconductor. As a small electric field, $\mathbf{F}$ is applied to carriers, the carriers will drift in the field with a drift velocity defined as $\mathbf{v}_d$. It should be noted that generally both the electric field and the drift velocity are vector quantities but commonly the basis states are chosen so that only a 1D problem is required to be solved and only one component is used. To relate the drift velocity to the electric field, the common approach is that due to Drude which states that in the steady-state, the rate at which electron receive momentum from the external electric field is equal to the rate at which the electrons lose momentum due to scattering forces.

$$\left.\frac{dm^*\mathbf{v}_d}{dt}\right|_{scattering} = \left.\frac{dm^*\mathbf{v}_d}{dt}\right|_{field} \tag{3.124}$$

Therefore by defining a momentum relaxation time, τ_m which is the mean time between scattering collisions (which may be due to scattering from the lattice, impurities, other carriers, phonons, etc. (see Fig. 3.29(b)) the force on an electron balances as

$$\frac{m^*\mathbf{v}_d}{\tau_m} = q\mathbf{F} \qquad \Rightarrow \qquad \mathbf{v}_d = \frac{q\tau_m}{m^*}\mathbf{F} \tag{3.125}$$

The mobility (sometimes called the drift mobility) is defined as the ratio of the drift velocity to the applied electric field such that

$$\mu_n = \left|\frac{\mathbf{v_d}}{\mathbf{F}}\right| = \frac{q\tau_m}{m^*} \tag{3.126}$$

Hence the mobility is a crude measure of the scattering processes in a semiconductor which prevent the carriers from accelerating indefinitely (Fig. 3.29(b)). It is clear that carriers have higher drift velocities in high mobility materials and hence there is great interest in high mobility semiconductors as electrons are transported across the device faster and the transit time of the device

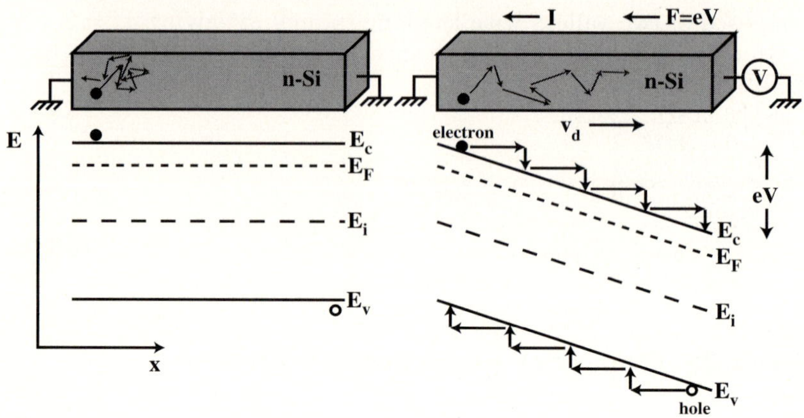

Fig. 3.29. A schematic diagram of a n-type semiconductor in (a) thermal equilibrium and (b) the electron transport under an applied electric field, **F**

is reduced which in a correctly designed transistor producing higher speed switching. Eventually the velocity saturates when the scattering balances the force from the electric field.

If we consider a 2D system such as the 2DEG in a MOSFET or MODFET then the same equations may be considered from the k-space picture of the semiconductor. Since the quantum mechanical definition of momentum is $\mathbf{p} = \hbar \mathbf{k}$, the k-space equivalent of (3.125) is

$$\delta \mathbf{k} = -\frac{q\tau_m}{\hbar}\mathbf{F} \tag{3.127}$$

For the case of a 2DEG, the Fermi surface which divides the filled from the unfilled states is a circle in k-space (Fig. 3.30). The effect of the electric field is to move the Fermi circle by the amount $\delta \mathbf{k}$ in the opposite direction to the applied field, **F** (because the electrons have negative charge). This demonstrates that at sufficiently low temperatures it is only electrons within $k_B T$ of the Fermi circle which contribute to electron transport in the semiconductor.

The current density in the system is defined as

$$\mathbf{J} = nq\mathbf{v}_d = \sigma_n \mathbf{F} \tag{3.128}$$

From these equations the conductivity of electrons is defined as

$$\sigma_n = qn\mu_n = \frac{nq^2\tau_m}{m^*} \tag{3.129}$$

One may also define the resistivity as

$$\rho_n = \frac{1}{\sigma_n} \tag{3.130}$$

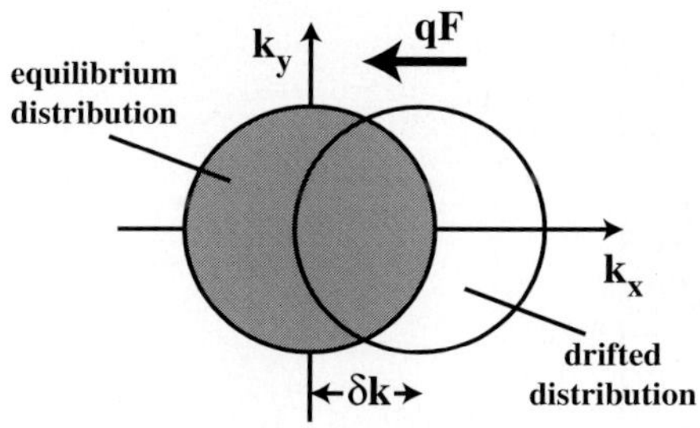

Fig. 3.30. A schematic diagram of a 2DEG with a Fermi circle in an applied electric field, **F**. The electron drift due to the electric field is related to the movement of the Fermi circle by an amount δk

In a measurement it is the conductance or the resistance which is obtained. For a 2D sample of length L and width W, the resistance is related to the resistivity through

$$R = \frac{L}{W}\rho \tag{3.131}$$

and in a similar fashion the conductance, G is defined as

$$G = \frac{W}{L}\sigma_n \tag{3.132}$$

For the performance of devices, the carrier velocity as a function of electric field is important. Figure 3.32 shows the velocity-field curves for a number of significant semiconductors. The curve for Si saturates at $> 10^5\,\mathrm{Vm^{-1}}$. In detailed studies of mobility, the principal energy loss mechanism for electrons in high electric fields is optical phonon emission (see Sect. 3.6 for a complete description of phonons in semiconductors), characterised by an energy $E_{opt} \sim 62$ meV and 37 meV for Ge. Assuming this mechanism is efficient, the maximum drift velocity becomes saturated at

$$v_{sat} = \sqrt{\frac{8E_{opt}}{3\pi m^*}} \tag{3.133}$$

In GaAs, the decrease in velocity above $4000V/cm$ (Fig. 3.32) is explained by a two valley model. At low electric fields, the electrons reside in the lower valley with higher mobility while at higher electric fields, the electrons get heated and reside in the upper valley. The negative differential mobility translates into negative differential resistance and may be used to form microwave oscillators such as Gunn diodes. At very high electric fields, the carriers get sufficient excess energy to generate electron-hole pairs through impact ionisation. The electron therefore acquires enough energy to break a chemical bond and promote an electron from the valence band to the conduction band. For high electric fields dropped over a sufficiently thick piece of semiconductor, this process may multiply and then the process is known as avalanching. It

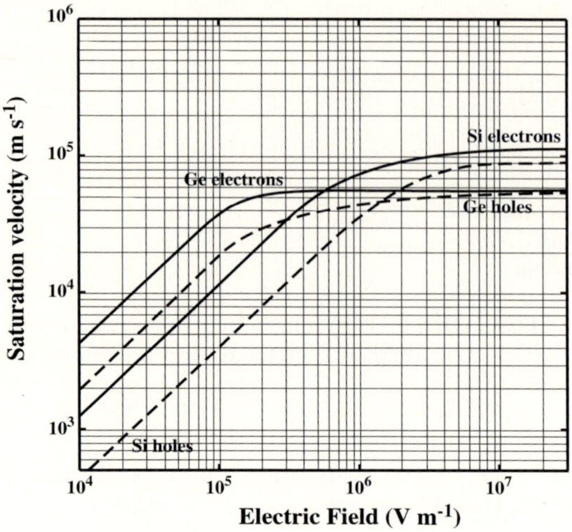

Fig. 3.31. The saturation velocity for both electrons (solid lines) and holes (dashed lines) in silicon and germanium

is equivalent to dielectric breakdown in other situations. In layered semiconductor structures, it is sometimes possible to control this avalanche process to form impact ionisation avalanching and transit time (IMPATT) diodes, the most powerful but noisy source of microwaves. Such processes are also used to increase the sensitivity of photodiode detectors.

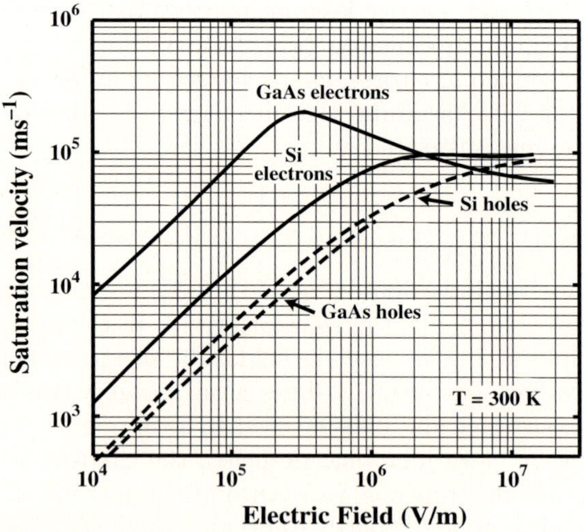

Fig. 3.32. The saturation velocity for both electrons (solid lines) and holes (dashed lines) in GaAs compared to Si

3.4.2 The Diffusion Current and the Einstein Relation

In the previous section the drift current which resulted from an applied electric field was considered. A second way to create a current is from diffusion of charge rather than drift from a field. Diffusion currents are generally not important in metals because the conductivity is extremely high and drift dominates but in semiconductors the carrier density can be non-uniform making diffusion a possible transport mechanism.

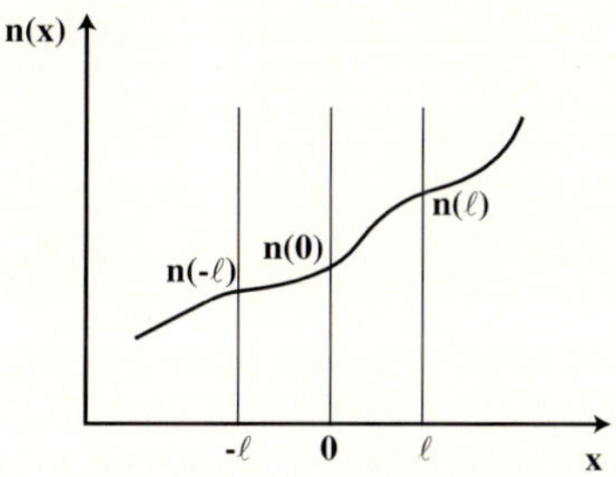

Fig. 3.33. The electron concentration, n as a function of distance, x in a 1D semiconductor

If we have a change in the electron density, $n(x)$ along the length x- of a n-type semiconductor then if we consider the number of electrons crossing the plane at $x = 0$ per unit time, per unit area we can have the random thermal motion of the carriers creating a current without an applied electric field. If we define an electron mean free path, ℓ which is the average distance an electron travels before being scattered then $\ell = v_{th}\tau_m$ where v_{th} is the thermal velocity. The average rate of electrons per unit area crossing $x = 0$ per unit time depends on those that started at $x = -\ell$ and is

$$\frac{1}{2}n(-\ell)\, v_{th} \tag{3.134}$$

where the factor $\frac{1}{2}$ results from the fact that half of the electrons travel from the left and half from the right. The half that travel from the right are

$$\frac{1}{2}n(\ell)\, v_{th} \tag{3.135}$$

so the net flow of electrons from the left to the right is given by

$$\frac{1}{2}v_{th}\left[n(-\ell) - n(\ell)\right] \tag{3.136}$$

Applying a Taylor series expansion to the densities at $x = \pm\ell$ then the flow becomes

$$\frac{1}{2}v_{th}\left(\left[n(0) - \frac{\mathrm{d}n}{\mathrm{d}x}\ell\right] - \left[n(0) + \frac{\mathrm{d}n}{\mathrm{d}x}\ell\right]\right) = -v_{th}\ell\frac{\mathrm{d}n}{\mathrm{d}x} \tag{3.137}$$

Since each electron has a charge $-q$, the particle flow corresponds to a current

$$J_n = -q \, \text{flow} = q\ell v_{th} \frac{\mathrm{d}n}{\mathrm{d}x} \tag{3.138}$$

The diffusion constant for electrons is defined as

$$D_n = \ell v_{th} \tag{3.139}$$

Therefore the diffusion current is normally written in the form

$$J_n = qD_n \frac{\mathrm{d}n}{\mathrm{d}x} \tag{3.140}$$

and in a similar fashion for holes

$$J_p = -qD_p \frac{\mathrm{d}p}{\mathrm{d}x} \tag{3.141}$$

where D_n and D_p are the diffusion constants for electrons and holes respectively. The diffusion current is therefore proportional to the rate of change of the carrier density along a sample which arises due to the random thermal motion of carriers at finite temperature in a concentration gradient.

Using equipartition of energy, the thermal velocity of electrons at a temperature T given by Boltzmann statistics as

$$\frac{1}{2} m^* v_{th} = \frac{3}{2} k_B T \tag{3.142}$$

If this is substituted into (3.138) and also the definition of mobility (3.126) then the current density can be written as

$$J_n = q \left(\frac{k_B T}{q} \mu_n \right) \frac{\mathrm{d}n}{\mathrm{d}x} \tag{3.143}$$

and on comparision with (3.140) the diffusion constant is also

$$D_n = \left(\frac{k_B T}{q} \right) \mu_n \tag{3.144}$$

while for holes

$$D_p = \left(\frac{k_B T}{q} \right) \mu_p \tag{3.145}$$

Equations (3.144) and (3.145) are called the Einstein relations.

3.4.3 The Current-Density Equations

The total current density is the sum of the drift and the diffusion currents. Therefore using (3.128) and (3.140) the total current density is given for electrons by

$$J_n = qn\mu_n F + qD_n \frac{\mathrm{d}n}{\mathrm{d}x} \tag{3.146}$$

and for holes

$$J_p = qp\mu_p F - qD_p \frac{\mathrm{d}p}{\mathrm{d}x} \tag{3.147}$$

3.4.4 The Hall Effect and Mobility Measurements

The Hall effect involves measuring the conductivity of electrons in a weak magnetic field and is important as it allows the carrier density and mobility of a semiconductor material to be accurately measured. Without the magnetic field, the conductivity only gives the mobility and carrier density product. Generally the technique can be used to measure 3D samples but it is most frequently used for 2D systems such as the 2DEG or 2DHG in a MOSFET or MODFET structure. Therefore the geometry that is appropriate and the experimental setup is shown in Fig. 3.34.

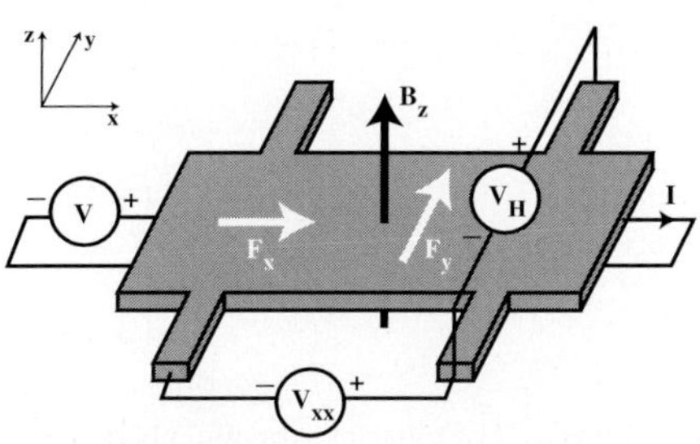

Fig. 3.34. A Hall bar geometry device for measuring the carrier density and mobility showing the directions of the electric fields and the currents

As has been discussed in the electronic transport section, at steady state, the rate at which the electrons receive momentum from the applied electric field is equal to the rate at which this momentum is lost due to scattering. Therefore

$$\left.\frac{\mathrm{d}m^* \mathbf{v}_d}{\mathrm{d}t}\right|_{scattering} = \left.\frac{\mathrm{d}m^* \mathbf{v}_d}{\mathrm{d}t}\right|_{field} \tag{3.148}$$

which rewritten as the Lorentz force with the momentum relaxation time, τ_m is

$$\text{Force} = \frac{m^* \mathbf{v}_d}{\tau_m} = q\left[\mathbf{F} + \mathbf{v}_d \times \mathbf{B}\right] \tag{3.149}$$

Strictly speaking the velocity and the electric field can vary in both the x- and y-directions for the 2D sample in Fig. 3.34 as they are vector properties and so

$$\begin{bmatrix} \frac{m^*}{q\tau_m} & -B \\ B & \frac{m^*}{q\tau_m} \end{bmatrix} \begin{pmatrix} v_x \\ v_y \end{pmatrix} = \begin{pmatrix} F_x \\ F_y \end{pmatrix} \tag{3.150}$$

where the usual definitions have been used for the x- and y-components of the drift velocity, $\mathbf{v}$ and the electric field, $\mathbf{F}$. The current density is related to the sheet electron density (2D density), n_s through the relationship $\mathbf{J} = q\mathbf{v}_d n_s$ and so (3.150) becomes

$$\begin{bmatrix} \frac{m^*}{q\tau_m} & -B \\ B & \frac{m^*}{q\tau_m} \end{bmatrix} \begin{pmatrix} J_x/qn_s \\ J_y/qn_s \end{pmatrix} = \begin{pmatrix} F_x \\ F_y \end{pmatrix} \tag{3.151}$$

which when rearranged defines the conductivity tensor, σ_n through

$$\begin{pmatrix} F_x \\ F_y \end{pmatrix} = \frac{1}{\sigma_n} \begin{bmatrix} 1 & -\mu_n B \\ \mu_n B & 1 \end{bmatrix} \begin{pmatrix} J_x \\ J_y \end{pmatrix} \tag{3.152}$$

with $\mu_n = q\tau_m/m^*$. Since the resistivity tensor, ρ is defined as

$$\begin{pmatrix} F_x \\ F_y \end{pmatrix} = \begin{bmatrix} \rho_{xx} & \rho_{xy} \\ \rho_{yx} & \rho_{yy} \end{bmatrix} \begin{pmatrix} J_x \\ J_y \end{pmatrix} \tag{3.153}$$

(3.152) produces the relationships

$$\rho_{xx} = \frac{1}{\sigma_{xx}} \tag{3.154}$$

$$\rho_{yx} = -\rho_{xy} = \frac{\mu_n B}{\sigma_{xx}} = \frac{B}{qn_s} \tag{3.155}$$

Therefore the simple Drude model of electron transport shows that the longitudinal resistance, ρ_{xx} of the Hall bar sample is a constant while the transverse resistivity, ρ_{yx} varies linearly with the magnetic field and inversely as the carrier density in the sample.

In the experimental setup, a current, I is passed along the sample and two probes on the one side of the Hall bar are used to measure the longitudinal voltage, V_{xx}. Similarly the Hall voltage, V_H is measured using two probes on either side of the Hall bar. If the magnetic field, B is applied in the z-direction then

$$\rho_{yx} = \frac{V_H}{I} \tag{3.156}$$

and

$$\rho_{xx} = \frac{V_{xx}}{I}\frac{W}{L} \tag{3.157}$$

where W is the width of the Hall bar and L is the length. By substituting the appropriate equations and variables, these equations can be rearranged to give the sheet carrier density through

$$\frac{1}{n_s} = q\frac{d\rho_{yx}}{dB} = \frac{q}{I}\frac{dV_H}{dB} \tag{3.158}$$

It should be noted that the direction of the Hall voltage changes if holes rather than electrons are used and therefore the Hall effect can also be used to distinguish whether electrons or holes are the majority carrier in the system. The Hall mobility is also obtained as

$$\mu_H = \frac{1}{qn_s\rho_{xx}} = \frac{I}{q}\frac{L}{n_s V_{xx} W} \tag{3.159}$$

The Hall mobility for electrons is related to the drift mobility, μ_n through the Hall scattering factor, r as $\mu_n = r\mu_H$. In many cases the Hall scattering factor is 1 or close to 1. Care must be taken, however, as many authors ignore the difference and incorrectly quote the Hall mobility as the mobility. Since this technique is the most accurate for measuring the carrier density and mobility, it is typically used for characterising new materials.

3.4.5 Poisson's Equation and Gauss's Law

Clearly all semiconductors must obey Maxwell's equations which explain the electrostatic and electromagnetic properties of materials. One of the most useful and most used in semiconductor devices is Poisson's equation which allows the charge density in any system to be obtained. Conventionally, the electrostatic potential, ψ_i is defined as

$$\psi_i = -\frac{E_i}{q} \tag{3.160}$$

Again the negative sign represents the negative electron charge compared to the definition of positive charge movement for current. The electric field, F is the electrostatic force per unit charge and is equal to the change in the electrostatic potential as

$$F = -\frac{\mathrm{d}\psi_i}{\mathrm{d}x} \tag{3.161}$$

The charge density, $\rho(x)$ is obtained by differentiating (3.161) to obtain Poisson's equation

$$\frac{\mathrm{d}^2\psi_i}{\mathrm{d}x^2} = -\frac{\mathrm{d}F}{\mathrm{d}x} = -\frac{\rho(x)}{\varepsilon_0\varepsilon_r} \tag{3.162}$$

where ε_0 is the permittivity of free space ($\varepsilon_0 = 8.85 \times 10^{-12}\,\mathrm{Fm}^{-1}$) and ε_r is the dielectric constant of the medium. This is a 1D equation but generally a 2D or a 3D equation can be written and solved. The 3D case is given by

$$\nabla^2 \cdot \psi_i = -\nabla \cdot \mathbf{F} = -\frac{\rho(x, y, z)}{\varepsilon_0 \varepsilon_r} \tag{3.163}$$

By integrating (3.162) we get Gauss's law which gives

$$F = \frac{1}{\varepsilon_0 \varepsilon_r} \int \rho(x)\, \mathrm{d}x = \frac{Q}{\varepsilon_0 \varepsilon_r} \tag{3.164}$$

where Q is the total charge in the system.

In any semiconductor, charge can be fixed or mobile. The mobile charge is either electrons or holes with densities n and p respectively. The fixed charges are the donors and the acceptors with densities of N_D and N_A. Therefore for any semiconductor Poisson's equation (3.162) can be rewritten as

$$\frac{\mathrm{d}^2\psi_i}{\mathrm{d}x^2} = -\frac{\mathrm{d}F}{\mathrm{d}x} = -\frac{q}{\varepsilon_0 \varepsilon_r}\left[p(x) - n(x) + N_D(x) - N_A(x)\right] \tag{3.165}$$

If there is no applied electric field ($F = 0$) then the right hand side of (3.165) is set to zero and the potential is a constant throughout the semiconductor.

3.4.6 Carrier Concentrations

It will become useful when calculating the carrier densities in many devices to define the carrier density as a function of the electrostatic potential in the semiconductor. For a n-type semiconductor, (3.116) can be rewritten in terms of the electrostatic potential as

$$|\psi_F - \psi_i| = \frac{k_B T}{q} \ln\left[\frac{N_D}{n_i}\right] \tag{3.166}$$

where $\psi_F = -E_F/q$ is the Fermi potential. A similar expression can be written for the holes. (3.100) and (3.102) can also be rewritten in terms of electrostatic potentials rather than energies as

$$n = n_i \exp\left[\frac{E_F - E_i}{k_B T}\right] = n_i \exp\left[\frac{q(\psi_i - \psi_F)}{k_B T}\right] \tag{3.167}$$

and

$$p = n_i \exp\left[\frac{E_i - E_F}{k_B T}\right] = n_i \exp\left[\frac{q(\psi_F - \psi_i)}{k_B T}\right] \tag{3.168}$$

These equations are often called the Boltzmann relations and are valid for both n- and p-type semiconductors and for moderate doping levels.

3.4.7 The Debye Length

The bands and the intrinsic Fermi level in inhomogeneous semiconductors usually follow the variations of (3.166). If the doping concentration, however, changes abruptly then the bands may not follow this change as quickly as the doping profile. This is because the doping profile may be discontinuous while ψ_i and the first derivatives must be continuous from any thermal diffusion effects. There is therefore a length scale, called the Debye length, L_D which is the distance over which the bands of a semiconductor respond to a change of doping of N_D for a n-type semiconductor. To obtain this length, (3.167) is substituted into the (3.165) form of Poisson's equation to give

$$\frac{\mathrm{d}^2\psi_i}{\mathrm{d}x^2} = -\frac{q}{\varepsilon_0 \varepsilon r}\left[N_D(x) - n_i \exp\left(\frac{q(\psi_i - \psi_F)}{k_B T}\right)\right] \tag{3.169}$$

To find the Debye length, consider an incremental change in the doping concentration, $\Delta N_D(x)$ with respect to an uniform background doping. The change in the intrinsic potential, $\Delta\psi_i$ can be found by expanding the exponential term in (3.169) and removing the constants which have no spacial variation.

$$\frac{\mathrm{d}^2(\Delta\psi_i)}{\mathrm{d}x^2} - \frac{q^2 N_D}{\varepsilon_0 \varepsilon_r k_B T}\Delta\psi_i = -\frac{q}{\varepsilon_0 \varepsilon_r}\Delta N_D(x) \tag{3.170}$$

This is a second order differential equation with the solution for $\Delta\psi_i$ of the form $\exp(-x/L_D)$ with

$$L_D = \sqrt{\frac{\varepsilon_0 \varepsilon_r k_B T}{q^2 N_D}} \tag{3.171}$$

While in many conventional electronic devices, such as transistors, the Debye length is usually much smaller than any lateral device dimensions, when quantum devices are considered, it may become an important parameter.

3.5 Low Dimensional Physics: Quantum Wires and Dots

3.5.1 Important Length Scales

Up to this point in this chapter, the vast majority of physics described has been aimed either at 3D or 2D systems. While these are the easiest to realise experimentally, it is possible to confine electrons to lower dimensions using a number of different techniques. Techniques include electrostatic confinement using gates which realised some of the first 1D transport along with self-organised growth where 0D quantum dots may be formed. Care must be taken in defining the dimensionality of the sample as the length scales for different phenomena may be substantially different. A good example is disordered transport in narrow 2DEGs at low temperatures where the sample may be electrically 2D, the weak localisation due to disorder is typically 2D but the electron-electron interactions in the system may be 3D since the interactions depends on the thermal length which can be very small compared to the sample. If a sample is made with dimensions, L_x, L_y and L_z then the dimensionality of the system with respect to different transport regimes may be inferred by comparing the sample dimensions to the various scattering and characteristic lengths defined below (Fig. 3.35). Three main regimes for a device of length, L_x and width L_y for 2D system exists; the diffusive regime where scattering dominates, the quasi-ballistic regime with very few scattering events and the ballistic regime where there is no scattering inside the device on average. The forth main transport regime is quantum mechanical tunnelling through a barrier.

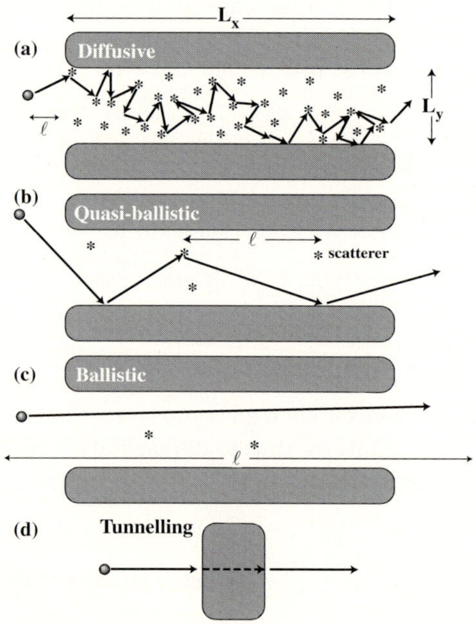

Fig. 3.35. The electron transport regimes for small mesoscopic devices of length, L_x and width, L_y: (a) the diffusive case where $\ell < L_y < L_x$, (b) the quasi-ballistic case where $L_y < \ell < L_x$, (c) the ballistic case where $L_y < L_x < \ell$, (d) the quantum mechanical tunnelling regime

The distance an electron travels at constant kinetic energy is

$$l_{in} \qquad \text{the inelastic scattering length} \tag{3.172}$$

Examples include electron-electron scattering (at low temperature) and electron-phonon scattering (at higher temperature e.g. room temperature). The distance an electron travels at constant wave vector is

$$l_e \qquad \text{the elastic scattering length} \tag{3.173}$$

The effects from Coulombic impurity potentials are an example of elastic scattering. The minimum of the elastic and inelastic scattering lengths is

$$\ell \qquad \text{the mean free path} \tag{3.174}$$

In most semiconductor systems, elastic scattering is almost always the shorter length. In 2D this may be approximated by the expression

$$\ell = \frac{\mu_n \hbar}{q} \sqrt{\frac{4\pi n_{2D}}{g_s g_v}} \tag{3.175}$$

The distance an electron travels before the phase of the wavefunction is lost is

$$l_\phi \qquad \text{the phase coherence length} \tag{3.176}$$

The relation between a scattering length l_x and the equivalent scattering time t_x are defined through the diffusion constant D by

$$l_x = \sqrt{D\tau_x} \tag{3.177}$$

and the Einstein relation relates the electron (or hole) mobility to the diffusion constant at an absolute temperature T as

$$\mu_n = \frac{qD_n}{k_B T} \tag{3.178}$$

In most conduction processes in semiconductors, only electrons close to the Fermi energy need to be considered for which the relevant length scale is the Fermi wavelength

$$\lambda_F = \frac{2\pi}{k_F} = \sqrt{\frac{2\pi}{n_{2D}}} = \frac{h}{\sqrt{2m^* E_F}} \tag{3.179}$$

Length scales associated with the physical quantities of temperature, magnetic and electric field must also be considered. Thermal smearing and the associated phase randomisation of an electron of the Fermi distribution produces an energy uncertainty of order $k_B T$. This defines

$$l_T = \sqrt{\frac{D\hbar}{k_B T}} \tag{3.180}$$

The magnetic field produces the characteristic length scale

$$l_B = \sqrt{\frac{\hbar}{qB}} \tag{3.181}$$

while for an electric field F,

$$l_F = \left(\frac{D\hbar}{qF}\right)^{1/3} \tag{3.182}$$

In the strongly localised regime, the radius of a hydrogen atom

$$a_B = \frac{4\pi\varepsilon_0\hbar^2}{q^2 m_0} \tag{3.183}$$

and the effective Bohr radius which is the radial extent of the wavefunction of a hydrogen atom like donor in a host crystal is

$$a_B^* = \frac{4\pi\varepsilon_0\varepsilon_r\hbar^2}{q^2 m^*} \tag{3.184}$$

3.5.2 1D Wires

According to Ohm's law, the conductance of a large sample is given by $G = \frac{\sigma L_y}{L_x}$. If, however, the length of the sample is much shorter than the mean free path (i.e. $L_x \ll \ell$ - see Fig. 3.35) then it is found that instead of the conductance going towards infinity as L_x is increased, the conductance reaches a limiting value. This appears counter intuitive because we have just described a short 1D wire where there is no scattering and hence one might expect the wire to have zero resistance. Landauer described the conductance of such systems in terms of transmission probabilities of mode T_n, such that

$$G = \frac{g_s g_v q^2}{h} \sum_{n=1}^{N} T_n \tag{3.185}$$

Solving this for a short 1D channel with N 1D subbands such that no scattering on average occurs in the channel (i.e. the channel length $\ll \ell$) gives

$$G = g_s g_v \frac{q^2}{h} N \tag{3.186}$$

This transport has been termed ballistic transport and the ballistic resistance is therefore quantised in units of q^2/h.

3.6 Lattice Vibrations and Phonons

3.6.1 The Vibrations of a 1D Monatomic Lattice

The way that heat interacts with electrons and a crystal lattice are extremely important in determining the electronic transport in a semiconductor but these properties also strongly effects the device performance. The heat properties of crystals was not fully understood until the work of Born and von Karman in 1912. Their model is extremely simple in that it treates the heat in a crystal lattice as the vibrations of the atoms in the lattice which allows the problem to be treated as if the atoms are connected by springs.

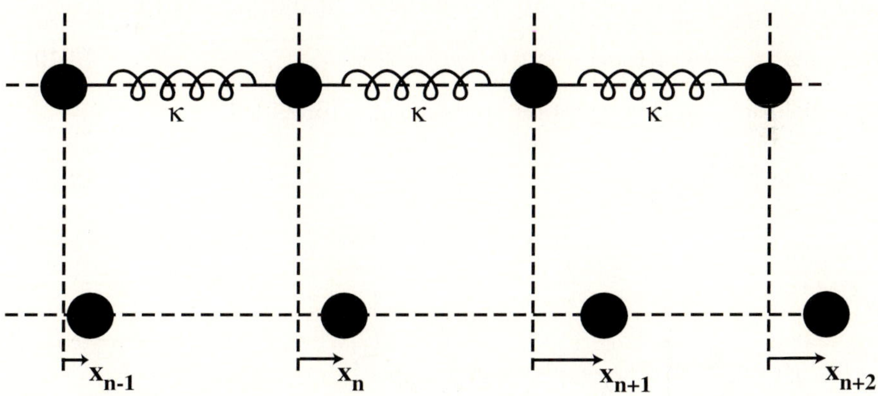

Fig. 3.36. A schematic diagram of a linear chain of atoms coupled together by springs of spring constant κ and with displacement of the nth atom by x_n

The model has the following assumptions:-
1. Each atom is located at an equilibrium position.
2. Each atom can oscillate about an equilibrium position with the amplitude of oscillation small compared to the internuclear distance.
Since the force in Newton's laws depends linearly on the displacement, the atoms can therefore be modelled as if they are connected by springs (Fig. 3.36). Let the spring constant be κ and only the nearest neighbour interactions will be considered. If the displacement of the nth atom is x_n then the potential energy in the system is

$$E = \sum_{n=1}^{N} \frac{\kappa}{2} \left(x_{n+1} - x_n\right)^2 \tag{3.187}$$

where there are N atoms in the lattice chain which are a distance a apart when the atoms are not moving (i.e. at 0 K). The force in the system is

$$\text{force} = -\frac{dE}{dx} \tag{3.188}$$

The equation of motion for the atoms is therefore

$$m\frac{d^2x_n}{dt^2} = \kappa\left[(x_{n+1} - x_n) - (x_n - x_{n-1})\right] \tag{3.189}$$

All N atoms in then lattice have the same equation of motion. Each of the terms on the right-hand side of (3.189) represents the spring interacting with the atoms on each side of the atom being considered. Since springs oscillate, the obvious solution to try is a wave solution of the form

$$x_n = x\exp\left[\mathrm{i}\left(nqa - \omega t\right)\right] \tag{3.190}$$

This corresponds to a wave along the chain of $\exp\left[\mathrm{i}\left(nqa\right)\right]$ with amplitude, x, and time dependence, $\exp\left[\mathrm{i}\omega t\right]$. Substituting (3.190) into (3.189) gives

$$-m\omega^2 x = \kappa\left[\exp\left(\mathrm{i}qa\right) + \exp\left(-\mathrm{i}qa\right) - 2\right]x \tag{3.191}$$

The solutions to this equation holds for all x provided

$$\omega = 2\sqrt{\frac{\kappa}{m}}\left|\sin\frac{qa}{2}\right| \tag{3.192}$$

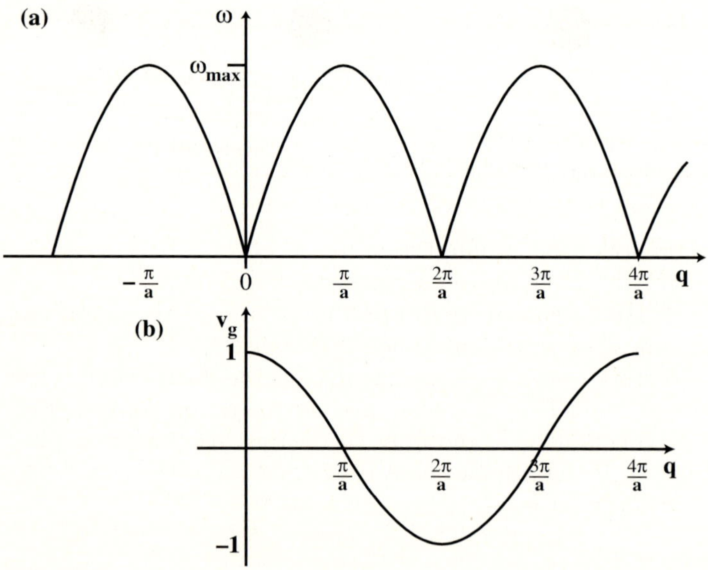

Fig. 3.37. (a) The dispersion curves of ω versus the wavevector q for longitudinal waves along a linear monatomic chain. (b) The group velocity for the linear monatomic chain of (a)

The dispersion relation is plotted in Fig. 3.37. The points to note are:

1. there is symmetry about $q = 0$ i.e. $-q$ is equivalent to $+q$ or in other words, the waves are identical in both directions.
2. the results repeat with period $\frac{2\pi}{a}$.
3. there is a maximum frequency or cut-off frequency above which atoms cannot sustain travelling waves.
4. the index n does not appear in the dispersion relation so only q is required to find ω.

The group velocity of a wavepacket is given by $v_g = \frac{d\omega}{dk}$ and so the value for the lattice waves between $q = 0$ and $\frac{\pi}{a}$ is

$$v_g = \sqrt{\frac{\kappa}{m}} a \cos\left(\frac{qa}{2}\right) \tag{3.193}$$

and $v_g = 0$ for $q = \frac{\pi}{a} + \frac{m2\pi}{a}$ for all integers m.

The above solutions are not changed by much if non-nearest neighbour interactions are also included. Some additional fine structure is obtained but the periodicity and general properties shown above are retained. With regard to the periodicity, all the information required to describe the lattice vibrations of the crystal are contained in an interval of $q = \frac{2\pi}{a}$. The first Brillouin zone of the 1D lattice is therefore used in a similar manner to that for electrons and so only values between $-\frac{\pi}{a} \leq q \leq \frac{\pi}{a}$ are plotted.

In the present 1D crystal, there are N atoms with 1 degree of freedom for longitudinal motion along the 1D chain and therefore N q-values in this zone. If transverse directions are also included (i.e. a 3D crystal) then there are $3N$ degrees of freedom. If the crystal is considered to have rotational symmetry like most semiconductors then there are 2 transverse polarised vibrations which are degenerate. For both of these the ω versus q curves are identical and these have the same shape as the longitudinal curves shown in Fig. 3.37. In a 3D crystal with one atom at each of N lattice points there are $3N$ degrees of freedom and $3N$ q=values which can all be confined to the first Brillouin zone with 1 longitudinal and 2 transverse branches.

3.6.2 The 1D Diatomic Chain

If we have a material such as SiGe or GaAs then there are two different masses on the lattice and so the next system to calculate is a 1D lattice with two different masses. The system will be similar to the monatomic chain except two masses, m and M with $m < M$ will be used (Fig. 3.38). The displacement of the smaller masses will be x_n and that of the large masses is X_n. Now the lattice spacing a is twice the distance between nearest-neighbour atoms and the unit cell has 2 atoms. Again let the system have N unit cells so that the potential energy in the system is

$$E = \sum_{n=1}^{N} \frac{\kappa}{2} (x_n - X_n)^2 + \frac{\kappa}{2} (X_n - x_{n+1})^2 \tag{3.194}$$

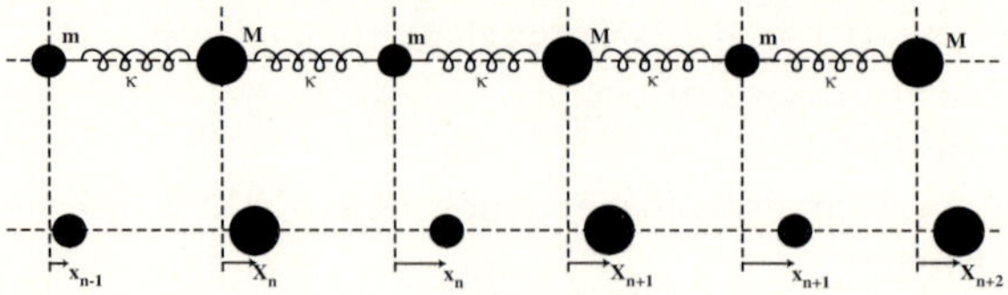

Fig. 3.38. A schematic diagram of a linear chain of atoms coupled together by springs of spring constant κ and with displacement of the nth atom by x_n

and in a similar fashion to the single monatomic chain, the equations of motion are

$$m\frac{d^2x_n}{dt^2} = \kappa\left(X_n + X_{n-1} - 2x_n\right) \tag{3.195}$$

$$M\frac{d^2X_n}{dt^2} = \kappa\left(x_{n+1} + x_n - 2X_n\right) \tag{3.196}$$

Again the solutions to try are travelling waves of the form

$$x_n = x\exp\left[\mathrm{i}\left(2nqa - \omega t\right)\right] \tag{3.197}$$

$$X_n = X\exp\left[\mathrm{i}\left((2n+1)\,qa - \omega t\right)\right] \tag{3.198}$$

where x and X are the displacement amplitudes for the masses m and M respectively. Substituting these equations into (3.195) and (3.196) the solutions are

$$-m\omega^2 x = \kappa\left[X\left(1 + \exp\left[iqa\right]\right) - 2x\right] \tag{3.199}$$

$$-M\omega^2 X = \kappa\left[x\left(\exp\left[-iqa\right] + 1\right) - 2X\right] \tag{3.200}$$

To solve these simultaneous equations, the determinant of the coefficients of x and X are required to be set to zero so that

$$\begin{vmatrix} m\omega^2 - 2\kappa & \kappa\left(1 + \exp\left[-iqa\right]\right) \\ \kappa\left(1 + \exp\left[iqa\right]\right) & M\omega^2 - 2\kappa \end{vmatrix} = 0 \tag{3.201}$$

which when reduced gives

$$mM\omega^4 - 2\kappa\left(m + M\right)\omega^2 + 2\kappa^2\left(1 - \cos qa\right) = 0 \tag{3.202}$$

Solving for ω^2 gives

$$\omega^2 = \kappa\frac{m+M}{mM} \pm \kappa\sqrt{\left(\frac{m+M}{mM}\right)^2 - \frac{2\left(1 - \cos\left[qa\right]\right)}{mM}} \tag{3.203}$$

The dispersion relation is plotted in Fig. 3.39. The solutions fall into two different branches which are normally named as the optic and the accoustic

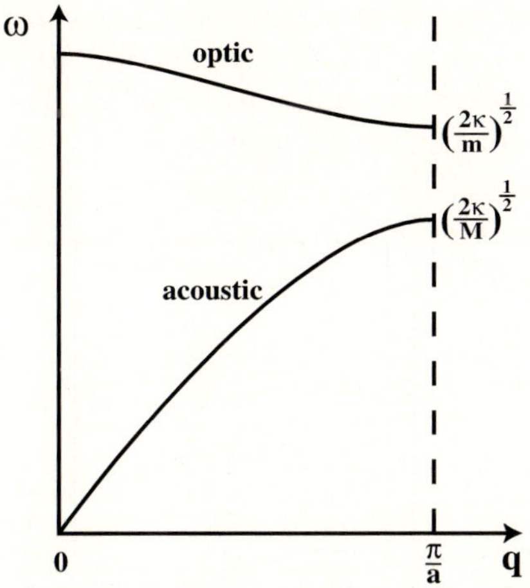

Fig. 3.39. (a) The dispersion curves of ω versus the wavevector q for a linear diatomic chain. The dispersion is split into the optic branch and the acoustic branch

branches with the optic branch having finite values for $q = 0$ and the accoustic branch being the same as that observed in the 1D monatomic chain.

If there are N primitive unit cells each with z atoms per cell then in 3D there are $3zN$ different modes. This is what you might expect and is very similar to the band structure of the electrons in the lattice i.e. the system is very complicated if you look at individual modes but due to the periodicity in the system, the modes of the system are much more simple. The solutions to the simple harmonic oscillator are normally termed normal modes. The normal modes of the present system are the linear combinations of atomic displacements that completely diagonalise the Hamiltonian terms up to at least the second order displacements. Once the individual modes are known, the Hamiltonian of the whole system is the sum of the individual simple harmonic Hamiltonians, one for each $3zN$ normal modes.

Since the wave motion of the lattice vibrations produces a simple harmonic oscillator, the energy of each mode is given by

$$E = \left(n_q + \frac{1}{2}\right)\hbar\omega_q \qquad \text{where } n_q = 0, 1, 2, \ldots \tag{3.204}$$

The total energy in the system is the sum of all the individual modes. It is clear from (3.204) that the lattice vibrations of the system are quantised and equally spaced in energy. These lattice quanta are called phonons (named so as to provide a comparison with photons, the quanta of light).

The phonon dispersion relations for real systems can be calculated and also measured. They are shown for Si, Ge and GaAs in Fig. 3.40. In all cases the phonon spectra are split into longitudinal optic (LO), transverse optic (TO), longitudinal acoustic (LA) and transverse acoustic (TA). The optical phonon energy, E_{opt} at $q = 0$ is important for a number of transport phenomena and electron scattering process. It is this energy that was used in Sect. 3.4.1 in the velocity saturation of electrons in semiconductors. The values of E_{opt} are 62 meV for Si, 37 meV for Ge and 35 meV for GaAs.

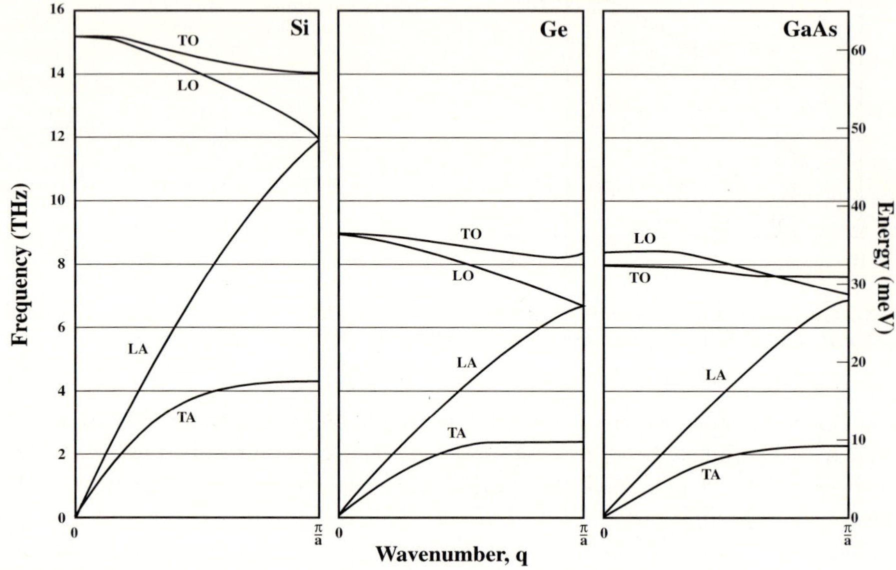

Fig. 3.40. The measured phonon spectra along the (100) direction for Si, Ge and GaAs. The phonon spectra are split into longitudinal optic (LO), transverse optic (TO), longitudinal acoustic (LA) and transverse acoustic (TA)

Phonons are bosons and as such obey Bose-Einstein statistics unlike electrons which are fermions and obey Fermi-Dirac statistics. Unlike fermions which can only singly occupy quantum states, for bosons there is no limit to the number which can occupy a single quantum states. The thermodynamic average occupation number, $\langle n_q \rangle$ of phonons in the qth mode of a simple harmonic oscillator in contact with a heat bath at a temperature T is

$$\langle n_q \rangle = \frac{\sum\limits_{n=0}^{\infty} n_q \exp\left[\frac{-n_q \hbar\omega_q}{k_B T}\right]}{\sum\limits_{n=0}^{\infty} n \exp\left[\frac{-n_q \hbar\omega_q}{k_B T}\right]} \tag{3.205}$$

which reduces to

$$\langle n_q \rangle = \frac{1}{\exp\left[\frac{\hbar\omega_q}{k_B T}\right] - 1} \tag{3.206}$$

This expression is the thermal occupation factor sometimes called the Bose-Einstein factor. It is worthwhile noting that unlike electrons, the number of phonons in a system is not conserved.

The average thermal energy for phonons in a system, $\langle E \rangle = \langle n_q \rangle \hbar\omega_q$ and is

$$\langle E \rangle = \frac{\hbar\omega_q}{\left[\exp\left[\frac{\hbar\omega_q}{k_B T}\right] - 1\right]} \tag{3.207}$$

which is known as Planck's law and was originally obtained for photons. At very low temperatures when $\hbar\omega_q \gg k_B T$ then both $\langle E \rangle$ and $\langle n_q \rangle \equiv \exp\left[-\frac{\hbar\omega_q}{k_B T}\right]$. The probability of phonons to be excited is therefore exponentially low at low temperatures. At high temperatures where $\hbar\omega_q \ll k_B T$

$$\langle n_q \rangle \approx \frac{k_B T}{\hbar\omega_q} \tag{3.208}$$

$$\langle E \rangle = \langle n_q \rangle \hbar\omega_q \approx k_B T \tag{3.209}$$

3.7 Optical Properties of Semiconductors

3.7.1 Blackbody Radiation

The simplest way to generate light from any material is to heat the material to a temperature, T where incandescence light or black body radiation will be emitted. To find the states available to a thermal photon in a cubic box of length, L, again periodic boundary conditions are imposed this time on the radiation field so that the wavevector, $\mathbf{k}$ is indexed by the integers, n_x, n_y and n_z such that

$$\mathbf{k} = \frac{2\pi}{L}\left(n_x\mathbf{a} + n_y\mathbf{b} + n_z\mathbf{c}\right) \tag{3.210}$$

and

$$\mathbf{k}^2 = \left(\frac{2\pi}{L}\right)^2 \mathbf{r}^2 \tag{3.211}$$

Provided the box is large enough then the variable, $\mathbf{r}$ can be considered to be continuous. Therefore in terms of the frequency

$$\frac{2\pi}{c}\nu = |\mathbf{k}| = \frac{2\pi}{L}|\mathbf{r}| \tag{3.212}$$

where c is the speed of light and therefore

$$|\mathbf{r}| = \frac{L}{c}\nu \tag{3.213}$$

For the next section the direction of the vectors is not important and only the magnitude of the vectors will be used. The total number of states in the volume $V = L^3$ between r and $r + dr$ is given by the density of states per volume, N by a spherical shell

$$VN(r)\,dr = 2 \cdot 4\pi r^2 dr \tag{3.214}$$
$$= 8\pi r^2 dr \tag{3.215}$$
$$= 8\pi \left(\frac{L}{c}\right)^2 \nu^2 \frac{L}{c} d\nu \tag{3.216}$$
$$= 8\pi \frac{V}{c^3} \nu^2 d\nu \tag{3.217}$$
$$\tag{3.218}$$

$$N(\nu) = \frac{8\pi}{c^3} \nu^2 d\nu \tag{3.219}$$

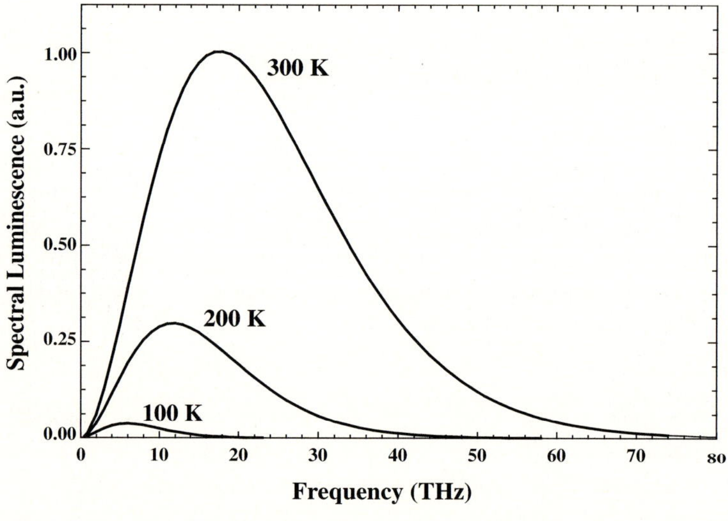

Fig. 3.41. The distribution of the power radiated from a black body emitter for three different temperatures

The first factor of 2 corresponds to the two possible transverse photon polarisations that are possible in any isotropic system. The thermal photon energy per volume, U in the box is then found by multiplying the density of states by the photon energy $h\nu$ and the Bose-Einstein distribution for the photon occupancy of the available states

$$u = h\nu \frac{8\pi}{c^3}\nu^2 \frac{1}{\exp\left[\frac{h\nu}{k_B T}\right] - 1} \tag{3.220}$$

$$= \frac{8\pi h\nu^3}{c^3\left(\exp\left[\frac{h\nu}{k_B T}\right] - 1\right)} \tag{3.221}$$

The spectrum of this function is shown for three different temperatures in Fig. 3.41

3.7.2 Generation and Recombination Processes

As has been suggested already, provided enough thermal energy is applied to an electron in the valence band of a semiconductor, the electron may be excited to the conduction band leaving a hole behind in the valence band. The complete description of such phenomena requires time-dependent perturbation theory and in particular Fermi's Golden Rule. An electron can also be excited from the valence band to the conduction band by absorption of a photon in direct band gap materials. In indirect band gap materials, however, phonons are also required for the excitation process. The alternative way of describing such processes is in terms of the balance of electrons and holes in a system in equilibrium such that $n_i^2 = np$. The generation or recombination process can then be thought of as a non-equilibrium process where $n_i^2 \neq np$.

Fermi's Golden Rule describes the transitions rate, W for a transitions from an initial state ψ_i with energy, E_i to a final state ψ_f with energy, E_f when a Hamiltonian, $\hat{H}$ describes the transition. If photons are also included with energy, $\hbar\omega$ then

$$\begin{aligned} W_{i\to f} &= \frac{2\pi}{\hbar}\left|\langle\psi_f|\,\hat{H}\,|\psi_i\rangle\right|^2 \delta\left(E_f - E_i - \hbar\omega\right) \\ &+ \frac{2\pi}{\hbar}\left|\langle\psi_f|\,\hat{H}\,|\psi_i\rangle\right|^2 \delta\left(E_f - E_i + \hbar\omega\right) \end{aligned} \tag{3.222}$$

Generally the generation and recombination processes can be divided into radiative and non-radiative. The non-radiative transitions do not involve photons: they may involve the interaction of the electron with phonons or the exchange of energy or momentum with another electron or hole. Both energy and momentum must be conserved in any process to be discussed in this section with Fermi's Golden Rule describing the transition. The transitions may also be divided into band-to-bound state transitions and band-to-band transitions.

3.7.3 Intrinsic Band-to-Band Generation-Recombination Processes

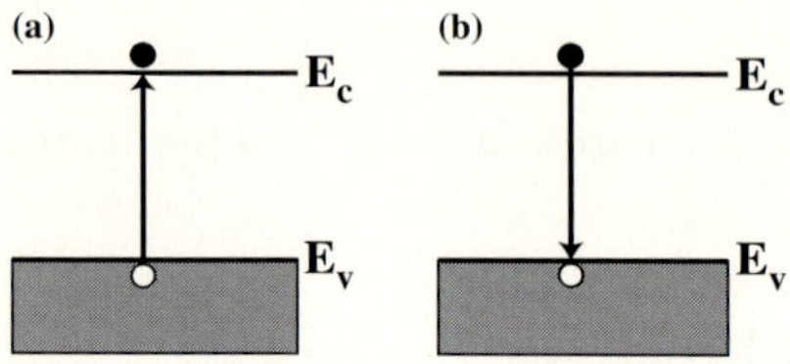

Fig. 3.42. The energy band diagram for (a) generation and (b) recombination of an electron-hole pair across the bandgap of a semiconductor

For band-to-band transitions as shown in a schmatic diagram of Fig. 3.42, the rate of recombination for electrons, R_n and holes, R_p is proportional to the product of the electron, n and hole, p concentrations such that

$$R_n = R_p = \gamma np \tag{3.223}$$

(see Fig. 3.42 (b)) where γ is the capture coefficient and the generation rate for electrons, G_n or holes, G_p may be written in terms of the emission rate, ϵ as

$$G_n = G_p = \epsilon \tag{3.224}$$

(see Fig. 3.42 (a)) The net recombination rate is

$$R = R_n - G_n = R_p - G_p = \gamma np - \epsilon \tag{3.225}$$

If no electric or optical excitation of carriers is creating an external perturbation to the system then in thermal equilibrium the net recombination rate should be zero giving

$$\gamma n_0 p_0 - \epsilon = 0 \tag{3.226}$$

where n_0 and p_0 are the electron and hole concentrations at thermal equilibrium i.e. $n_0 p_0 = n_i^2$. The net recombination rate can therefore be rewritten as

$$R = \gamma \left(np - n_0 p_0 \right) \tag{3.227}$$

If the carrier concentrations for electrons and holes deviate from the equilibrium values by δn and δp respectively,

$$n = n_0 + \delta n \tag{3.228}$$

$$p = p_0 + \delta p \tag{3.229}$$

and therefore for low levels of injections (i.e. where $\delta p \ll (n_0 + p_0)$) the recombination rate is

$$R = \gamma\left(n_0 + p_0\right)\delta n = \frac{\delta n}{\tau_R} \tag{3.230}$$

where $\delta n = \delta p$ since the electrons and holes are created in pairs for interband transitions and the recombination lifetime has been defined as

$$\tau_R = \frac{1}{\gamma\left(n_0 + p_0\right)} \tag{3.231}$$

3.7.4 Extrinsic Shockley-Read-Hall Generation-Recombination Processes

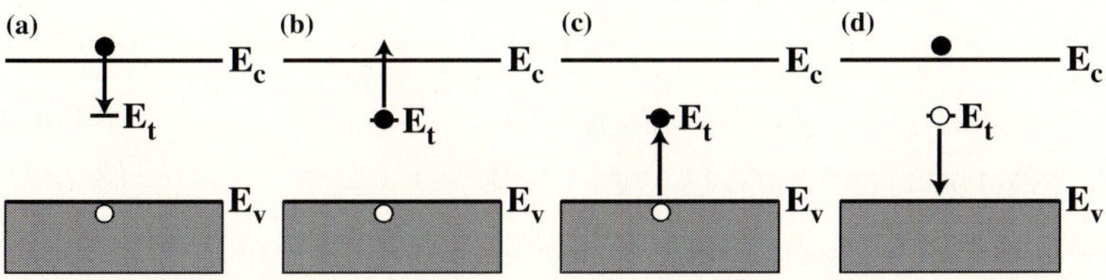

Fig. 3.43. The energy band diagram for Shockley-Read-Hall generation and recombination process (a) electron capture, (b) electron emission, (c) hole capture and (d) hole emission

The Shockley-Read-Hall processes are related to the emission or absorption of a phonon in the system and the four basic processes are shown in Fig. 3.43

- 1. The first process is electron capture (Fig. 3.43 (a)) where the recombination rate for the electron is proportional to the density of the electrons and the concentration of traps in the band gap, N_t multiplied by the probability that the trap is empty $(1 - f_t)$ where f_t is the occupation function of the trap.

$$R_n = \gamma_n n N_t \left(1 - f_t\right) \tag{3.232}$$

  where γ_n is the capture coefficient for the electrons.
- 2. The second process is electron emission (Fig. 3.43 (b)) where the generation rate of the electron is

$$G_n = \epsilon_n N_t f_t \tag{3.233}$$

  where ϵ_n is the emission coefficient and $N_t f_t$ is the density of traps that are occupied by electrons.

- 3. The third process is hole capture (Fig. 3.43 (c)) where the recombination rate for holes is given by the capture of holes by occupied traps. This number is $N_t f_t$ and

$$R_p = \gamma_p p N_t f_t \tag{3.234}$$

  where γ_p is the capture coefficient for the holes.
- 4. The final process is hole emission (Fig. 3.43 (d)) where the generation rate is proportional to the density of the traps that are empty (i.e. occupied by holes)

$$G_p = \epsilon_p N_t (1 - f_t) \tag{3.235}$$

  where the emission coefficient for holes is ϵ_p.

Since at thermal equilibrium the net recombination and generation rates are zero, by using the principle of detailed balancing,

$$R_n - G_n = \gamma_n n_0 N_t (1 - f_{t0}) - \epsilon_n N_t f_{t0} = 0 \tag{3.236}$$
$$R_p - G_p = \gamma_p p_0 N_t f_{t0} - \epsilon_p N_t (1 - f_{t0}) = 0 \tag{3.237}$$

Here the values with a $_0$ subscript refer to the equilibrium values. Therefore

$$\epsilon_n = \gamma_n n_0 \frac{1 - f_{t0}}{f_{t0}} \tag{3.238}$$
$$\epsilon_p = \gamma_p p_0 \frac{f_{t0}}{1 - f_{t0}} \tag{3.239}$$

For degenerate semiconductors, the intrinsic concentrations may be replaced by an effective concentration, n_{ie}.

The net recombination rate is

$$R_n - G_n = R_p - G_p = \frac{(\gamma_n \gamma_p np - \varepsilon_n \varepsilon_p) N_t}{\gamma_n n + \varepsilon_n + \gamma_p p + \varepsilon_p} \tag{3.240}$$
$$= \frac{\gamma_n \gamma_p \left(np - n_i^2\right) N_t}{\gamma_n \left(n + n_0 \frac{1 - f_{t0}}{f_{t0}}\right) + \gamma_p \left(p + p_0 \frac{f_{t0}}{1 - f_{t0}}\right)} \tag{3.241}$$

for e.g. an n-type semiconductor with $n_0 \gg p$, $\Delta p \gg p$, the net recombination rate, $R_n - G_n$ is roughly given by $\Delta p / \tau_p$, where

$$\tau_p = \frac{1}{\gamma_p N_t} \tag{3.242}$$
$$\tau_n = \frac{1}{\gamma_n N_t} \tag{3.243}$$

which are the lifetimes associated with the holes and electrons respectively.

3.7.5 Auger Generation-Recombination Processes

The Auger recombination process involves the transfer of momentum and energy released by the recombination of an electron-hole pair to another particle (either an electron or hole). Usually the Auger process is important when the carrier concentration is high or the injection level of carriers is high.

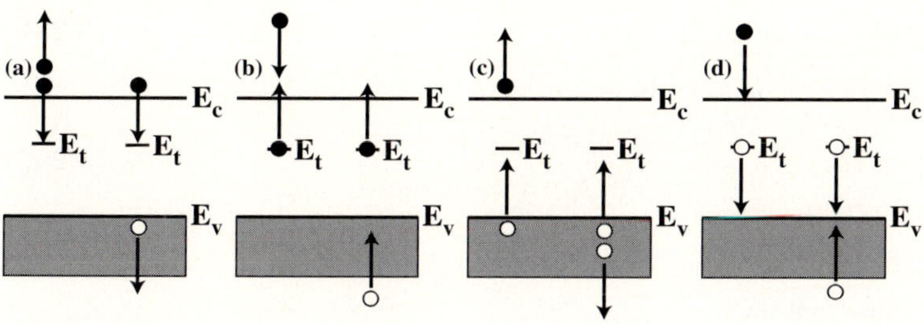

Fig. 3.44. The energy band diagram for band-to-bound state Auger generation and recombination process (a) electron capture, (b) electron emission, (c) hole capture and (d) hole emission

The Auger generation and recombination processes need to be divided into band-to-bound state transitions and band-to-band state transitions. The following band-to-bound state transitions are possible:

- 1. Electron capture with the energy released being absorbed by an electron or a hole (Fig. 3.44(a)). The recombination rate requires two capture coefficients to account for either an electron or hole gaining energy from the released energy

$$R_n = (\gamma_n^n n + \gamma_n^p p)\, n N_t\, (1 - f_t) \tag{3.244}$$

- 2. Electron emission with an energetic electron or hole supplying the energy (Fig. 3.44(b))

$$G_n = (\varepsilon_n^n n + \varepsilon_n^p p)\, N_t f_t \tag{3.245}$$

- 3. Hole capture with an electron or hole absorbing the released energy (Fig. 3.44(c))

$$R_p = \left(\gamma_p^n n + \gamma_p^p p\right) p N_t f_t \tag{3.246}$$

- 4. Hole emission with an energetic electron or hole supplying the energy (Fig. 3.44(d))

$$G_p = \left(\varepsilon_p^n n + \varepsilon_p^p p\right) N_t\, (1 - f_t) \tag{3.247}$$

If all processes for impurity assisted transitions due to thermal, optical and Auger-impact ionisation mechanisms exists then

$$\gamma_n = \gamma_n^t + \gamma_n^0 + \gamma_n^n n + \gamma_n^p p \tag{3.248}$$

$$\varepsilon_n = \varepsilon_n^t + \varepsilon_n^0 + \varepsilon_n^n n + \varepsilon_n^p p \tag{3.249}$$

$$\gamma_p = \gamma_p^t + \gamma_p^0 + \gamma_p^n n + \gamma_p^p p \tag{3.250}$$

$$\varepsilon_p = \varepsilon_p^t + \varepsilon_p^0 + \varepsilon_p^n n + \varepsilon_p^p p \tag{3.251}$$

The net recombination rate is given by

$$R_n - G_n = R_p - G_p = \frac{\gamma_n \gamma_p np - \varepsilon_n \varepsilon_p}{\gamma_n n + \gamma_p p + \varepsilon_n + \varepsilon_p} N_t \tag{3.252}$$

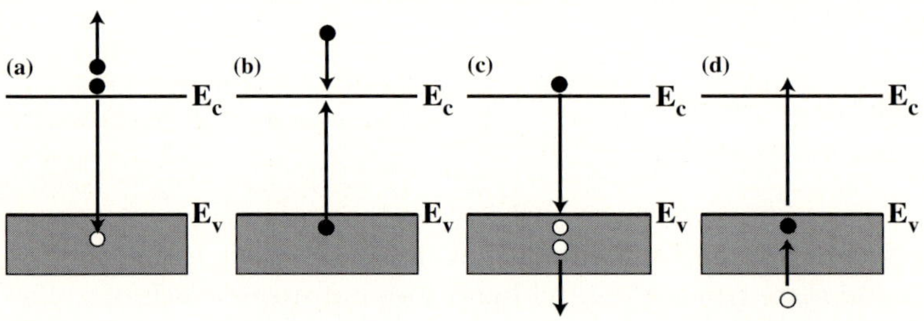

Fig. 3.45. The energy band diagram for band-to-band state Auger generation and recombination process (a) electron capture, (b) electron emission, (c) hole capture and (d) hole emission

For band-to-band Auger-impact ionisation processes, there are again four different processes.

- 1. Electron capture with an electron in the conduction band recombining with a hole in the valence band and a nearby electron absorbing the energy (Fig. 3.45(a)). An electron-hole pair is destroyed by this process.

$$R_n = \gamma_n^{Au} n^2 p \tag{3.253}$$

- 2. Electron emission with an electron in the valence band being excited to the conduction band because of impact ionisation of an incident electron in the conduction band which breaks a bond and creates an electron-hole pair (Fig. 3.45(b)).

$$G_n = \varepsilon_n^{Au} n \tag{3.254}$$

- 3. Hole capture where an electron in the conduction band recombines with a hole in the valence band with the released energy absorbed by a nearby hole destroying an electron-hole pair (Fig. 3.45(c)).

$$R_p = \gamma_p^{Au} np^2 \tag{3.255}$$

- 4. Hole emission where an electron in the valence band is excited to the conduction band due to the impact of an energetic hole in the valence band (Fig. 3.45(d)). The breaking of a bond to create an electron-hole pair may also provide enough energy.

$$G_p = \varepsilon_p^{Au} p \tag{3.256}$$

Since each process creates or destroys one electron-hole pair, the total net Auger recombination rate is the sum of the net rates for electrons and holes.

$$\begin{aligned} R &= R_n - G_n + R_p - G_p \\ &= \left(\gamma_n^{Au} n + \gamma_p^{Au} p\right)\left(np - n_i^2\right) \end{aligned} \tag{3.257}$$

Typical values for the Auger coefficient, $\gamma_n^{Au} n$ in silicon are of the order of 1 to $2{\times}10^{-31}\,\mathrm{cm^6 s^{-1}}$.

3.7.6 Impact Ionisation Generation-Recombination Processes

The final generation and recombination mechanisms to be considered are those by impact ionisation which typically occur when high electric fields are involved. The process is similar to the reverse Auger processes discussed in the previous section except the rates depend on the current densities rather than the concentrations of carriers. The generation rates for electrons and holes are

$$G_n = \alpha_n \frac{|\mathbf{J}_n|}{q} \tag{3.258}$$

$$G_p = \alpha_p \frac{|\mathbf{J}_p|}{q} \tag{3.259}$$

where α_n and α_p are the ionisation coefficients for electrons and holes respectively. α_n is the number of electron-hole pairs generated per unit distance due to an incident electron.

The total net recombination rate is

$$R = -G_n - G_p \tag{3.260}$$

$$= -\alpha_n \frac{|\mathbf{J}_n|}{q} - \alpha_p \frac{|\mathbf{J}_p|}{q} \tag{3.261}$$

3.8 The Continuity Equations Including Recombination and Generation

The current densities equations from Sect. 3.4.3 must now be added to the continuity equations for a semiconductor to include the generation and recombination rates so that

$$\frac{\partial n}{\partial t} = G_n - R_n + \frac{1}{q}\nabla \cdot \mathbf{J}_n \tag{3.262}$$

$$\frac{\partial p}{\partial t} = G_p - R_p - \frac{1}{q}\nabla \cdot \mathbf{J}_p \tag{3.263}$$

describes the electrons and holes in a semiconductor where

$$\mathbf{J}_n = q\mu_n n\mathbf{F} + qD_n \nabla n \tag{3.264}$$

$$\mathbf{J}_p = q\mu_p p\mathbf{F} - qD_p \nabla p \tag{3.265}$$

$$\mathbf{J} = \mathbf{J}_n + \mathbf{J}_p \tag{3.266}$$

J is the total current in the system. Once the generation and recombination rates are obtained for the dominant processes (Sect. 3.7), the current density equations can be applied to any semicoductor structure or device as will be shown in later chapters.

3.9 Further Reading

1. S.L. Chuang, Physics of Optoelectronic Devices, John Wiley and Sons, New York (1995)
2. S. Datta, Electronic Transport in Mesoscopic Systems, CUP, Cambridge (1995)
3. J.H. Davies, The Physics of Low Dimensional Semiconductors, CUP, Cambridge (1998)
4. M.J. Kelly, Low Dimensional Semiconductors: Materials, Physics, Technology, Devices, OUP, Oxford (1995)
5. O. Madelung, Introduction to Solid State Theory, Springer, Berlin (1981)
6. K. Seeger, Semiconductor Physics: An Introduction, 6th Edition, Springer, New York (1997)
7. S.M. Sze, The Physics of Semiconductor Devices, 2nd Edition, John Wiley and Sons, New York (1981)
8. S.M. Sze, Semiconductor Devices: Physics and Technology, 2nd Edition, John Wiley and Sons, New York (2002)

4. Realisation of Potential Barriers

The function of a device is essentially based on internal potential barriers which may be modified by external voltages. The traditional way to create potential barriers is performed by p/n-junctions or metal/ insulator/ semiconductor-junctions. The internal barriers created are usually high –in the order of the band gap–, but distributed across several tenths of nanometers to micrometers which is too broad for utilisation of quantum effects. For heterostructures the internal barriers are smaller –several hundreds of millivolts–, but confined to –ideally– atomic distances. Quantum device designs, therefore, may combine internal barriers created by p/n- and heterojunctions.

4.1 Depletion layer and built in voltage

The barrier created by a p/n junction will now be derived but a fuller derivation of all the transport properties of a p/n is given in Sect. 5.1. Consider a one sided abrupt junction (Fig. 4.1) between a heavily n-doped side (marked n^+) and a p-doped side (marked p). The diffusion across the junction and recombination of carriers will result in the formation a depletion layer with width W_D. The depletion region extends mainly in the lower doped part because the sum over all charges has to be zero (neutrality).

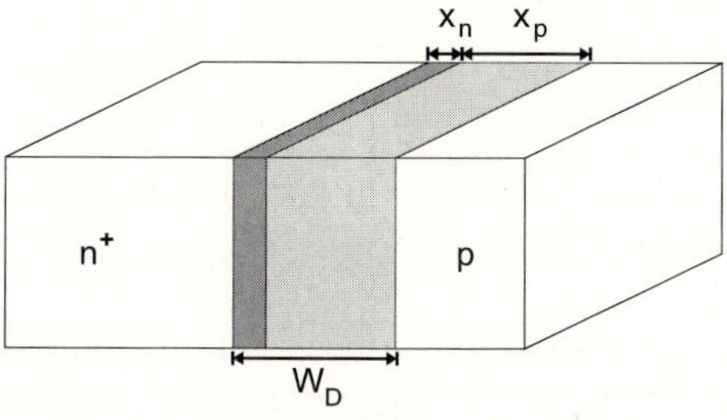

Fig. 4.1. The depletion layer of width W_D at the junction between a high doped n^+–region and a lower doped p–region

$$N_D \times x_n = N_A \times x_p \tag{4.1}$$

$$W_D = x_n + x_p \cong x_p \tag{4.2}$$

In our case $N_D \gg N_A$, and $x_p \gg x_n$.

Without any external voltage these sheets of positive (n - side, N_D^+ - ions) and negative (p - side, N_A^- - ions) charges create an electrical field and a built in potential ψ_{bi} the height of which may best be determined by considering a uniform Fermi energy level at equilibrium (Fig. 4.2). The built in potential ψ_{bi} is given by

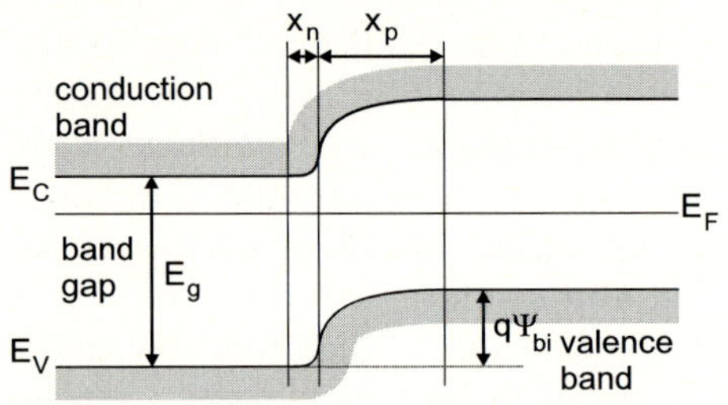

Fig. 4.2. The band diagram of the p/n-junction of Fig. 4.1. An energy barrier $q \times \psi_{bi}$ is created in the absence of an external voltage

$$\psi_{bi} = V_T \ln \left(\frac{N_A \times N_D}{n_i^2} \right) \tag{4.3}$$

with N_A, N_D doping levels (acceptor, donor), n_i intrinsic carrier density ($10^{10}\,\mathrm{cm}^{-3}$ for Si, $T = 300\,\mathrm{K}$), $V_T = k_B T/q$ thermal voltage (25 mV at room temperature). The typical built in voltage is somewhat lower than the band gap, e.g. around 0.8V – 1 V in Si. Application of an external voltage either decreases the potential barrier ($\psi_{bi}-$V) in the forward direction (positive voltage with voltage reference at the n - side) and increases it in the reverse direction. The corresponding depletion width W_D is given by

$$W_D = \sqrt{\frac{2\varepsilon_0 \varepsilon_r (\psi_{bi} - V)}{q N_A}} \tag{4.4}$$

(where ε_0 is the permittivity in a vacuum and ε_r is the relative dielectric constant of the material) resulting in the high nonlinear forward characteristics and the low reverse current of a biased p/n-junction.

Similarly a metal/semiconductor creates a depletion layer and a built in potential which is determined by the Schottky barrier energy ϕ_B and the doping level N_D (n-type semiconductor).

$$\psi_{bi} = \frac{\phi_B}{q} - V_T \ln \left(\frac{N_C}{N_D} \right) \tag{4.5}$$

(N_C conduction band effective density of states).

In a metal/insulator/semiconductor structure three regimes can be observed. A depletion layer extends from the insulator/semiconductor interface in the bulk for gate voltages between the flat band voltage, V_{FB}, and the threshold voltage, V_T. Beyond the threshold voltage the depletion layer extension stops and a minority carrier inversion channel adds to the interface. Beyond the flat band voltage an accumulation layer of majority carriers replaces the depletion region. Typical depletion widths of abrupt junction (using (4.4)) are shown in Table 4.1.

Table 4.1. Depletion layer width for a one sided abrupt junction $\psi_{bi} - V = 1$ V

Doping N_A (cm^{-3})	10^{16}	10^{17}	10^{18}	10^{19}
Layer width W_D (nm)	360	110	36	11

4.2 δ-Doping and n-i-p-i Structures

As shown in Table 4.1 the depletion width shrinks at very high dopings to values below 10 nm. Uniform doping is limited to roughly one percent (5×10^{20} cm^{-3}) of all the lattice sites occupied by the dopants. Much higher local occupation maybe obtained with delta doping (δ). Theoretically, in a monolayer δ-doping, all the atomic sites of the sheet of atoms are occupied by the dopants. In (100) Si the atomic density of a monolayer (ML) sheet is given by 6.783×10^{14} atoms cm^{-2}. In reality the doping profile is broader (Fig. 4.3), because of diffusion and segregation of atoms during growth or because of the dopant supply during growth. Usually growth is interrupted for the supply of the δ-dopant with a sheet concentration, N_S, and then continued with a lower temperature to reduce dopant segregation. The potential perpendicular to the delta sheet is defined by the electric field between the dopant ions and the mobile carriers smeared around the δ-sheet. Carriers are smeared out, because of carrier diffusion (Debye length) and the distribution of the wave function. A coupled Poisson- Schrödinger equation solution is necessary to describe the potential well and the energy levels (Fig. 4.4). In the given example (Fig. 4.4) the potential well of a roughly 1/10 monolayer (ML) sheet of boron is about 600 meV deep. Within the potential well the confined states of heavy holes (hh), light holes (lh) and of the split off band holes (so) are found with the heavy holes at the lowest energy levels **(the ground state)**. The carriers are concentrated within a few nanometers from the δ-doping.

A periodic sequence of n- and p- δ-doping spikes separated by intrinsic regions of length a_i are called a nipi-structure. Let us first consider the symmetric situation with the same donor- and acceptor sheet concentration

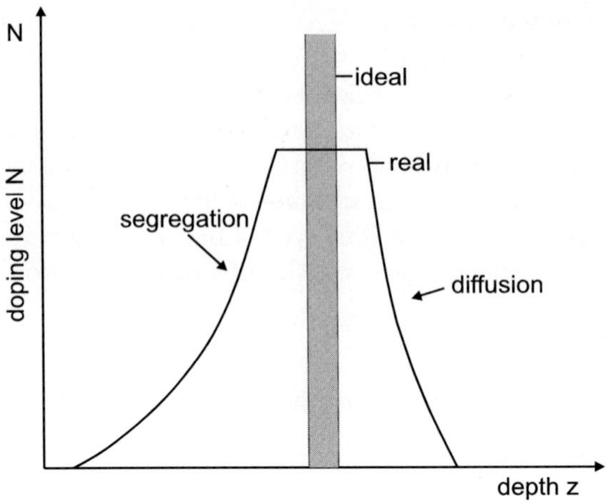

Fig. 4.3. Delta (δ) doping. The ideal case compared with the real doping distribution. Ideally, if the doping sheet concentration N_S is confined within a monolayer (0.138 nm for Si (100)). In reality smearing by segregation and diffusion broadens the profile from 0.1 nm to several nanometers

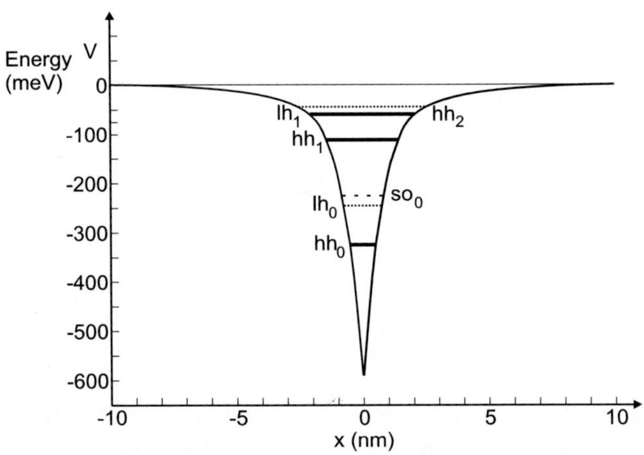

Fig. 4.4. The potential energy V(x) perpendicular to a 10^{14} cm^{-2} boron δ-doping. Included are also the first six eigenstates of the confined hole wave functions (hh heavy hole states, lh light hole states, so split off hole states)

N_S and complete ionization of the impurities (Fig. 4.5). The electric field strength F is then given by:

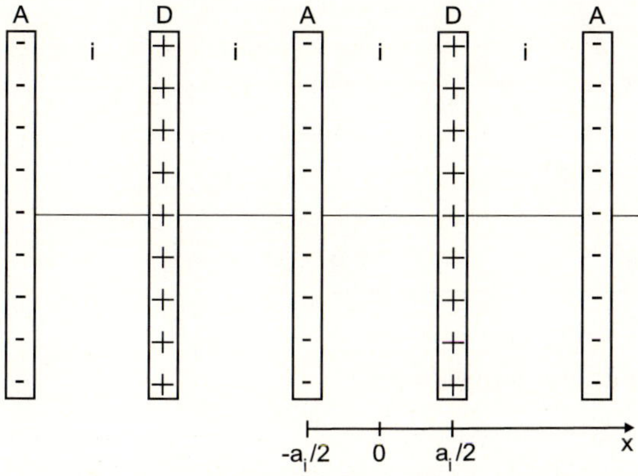

Fig. 4.5. A schematic diagram of a nipi-structure with completely ionised impurities (A, D). The length coordinate origin is chosen to be between the impurities

$$F = \pm \frac{qN_S}{2\varepsilon_0\varepsilon_r} \tag{4.6}$$

(negative sign when the positive charge $-N_D^+-$ is right sided).

The zig-zag potential ϕ of a nipi-structure is given by (Fig. 4.6)

$$\phi(x) = \frac{qN_S}{2\varepsilon_0\varepsilon_r} x \qquad \frac{-a_1}{2} \le x \le \frac{a_1}{2} \tag{4.7a}$$

$$\phi(x) = -\frac{qN_S}{2\varepsilon_0\varepsilon_r}(x - a_i) \qquad \frac{a_i}{2} \le x \le \frac{3a_i}{2} \tag{4.7b}$$

A continuous sequence of nipi-structures is called a doping super-lattice. Height and spatial potential distribution are between classical doping and heterostructures as pointed out first by Gottfried Döhler. The V-shaped quantum well has a depth, V_0, given by the potential

$$V_0 = \pm \frac{qN_S a_i}{2\varepsilon_0\varepsilon_r} \tag{4.8}$$

The energy qV_0 cannot increase to more than approximately the band gap energy E_g otherwise all the impurities cannot contribute to the depletion layer. This gives an upper bound to a meaningful value for the $N_S a_i$ product of a depleted symmetrical nipi-structure.

$$N_S a_i < \frac{2\varepsilon_0\varepsilon_r}{q} \frac{E_g}{q} \tag{4.9}$$

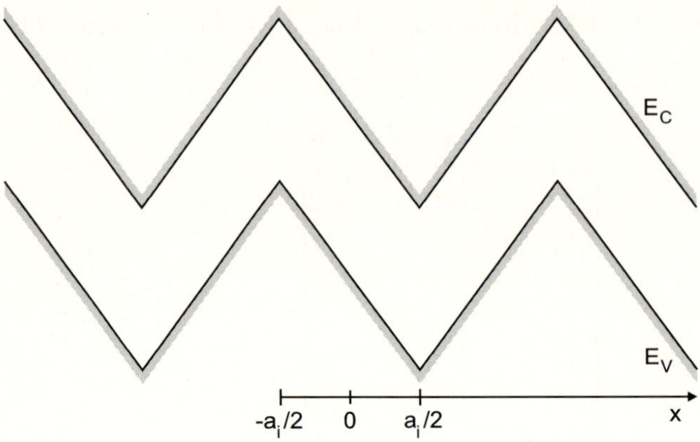

Fig. 4.6. The Zig-Zag band diagram of a nipi-structure

A typical set of data would be a 1/100 ML sheet density $N_S = 7 \times 10^{12}\,\mathrm{cm}^{-2}$ and a sheet distance a_i of 15 nm resulting in 0.84 V potential depth.

The potential depth, V_0, stays constant when the sheet distance, a_i, is decreased and the doping spike N is increased by the same factor ($N_S a_i$ constant).

The set of isolated quantum wells is gradually converted to a doping superlattice, because the energy levels in the narrower wells are increased and the carriers between neighboring wells interact. The exact solution of the Schrödinger equation is based on Airy functions, but for the ground level a good approximation is found by a variational method. The ground wave function $\psi_0(x)$ is approximated by

$$\psi_0(x) = A_0(1 - \alpha_0 x)\exp(\alpha_0 x) \qquad x \geq 0 \tag{4.10a}$$

$$= A_0(1 + \alpha_0 x)\exp(-\alpha_0 x) \qquad x < 0 \tag{4.10b}$$

with $\alpha_0^3 = \left(\frac{9}{4\epsilon} q^2 N_S \frac{m^*}{\hbar^2}\right)$ and $A_0^2 = \frac{2\alpha_0}{5}$.

The ground state energy E_0 (measured from the bottom of the V shaped well) is then expressed by

$$E_0 = \frac{3}{10}\left[\frac{81}{16}\frac{q^4 N_S^2}{\varepsilon_0^2 \varepsilon_r^2}\frac{\hbar^2}{m^*}\right]^{1/3} \tag{4.11}$$

Typical values for a 1/10 ML sheet are 240 meV for heavy electrons (longitudinal mass m_l), 410 meV for light electrons (transversal mass m_t), 350 meV for heavy holes, and 390 meV for light holes. The spatial extent of the ground state wave function is obtained as inverse proportional to $N_S^{1/3}$. The superlattice type overlap of wave-functions requires very small sheet distances (around 2 nm), because of the relatively high effective masses of electrons and holes in Si.

As soon as the sheet density of donors and acceptors are not equal, the system is non-compensated and the mobile carriers screen and modify the perfect V-shaped potential considered until now. The simplest approach to this problem is to apply the Thomas-Fermi approximation which described the screening of the potential by the free carriers. The density of carriers is taken to be that of a quasi-free carrier gas in thermal equilibrium with the local value of the potential thereby ignoring the quantisation of electron states. More refined schemes use the self consistent computation of electronic states. The Thomas-Fermi approximation works rather well at room temperature when several subbands are occupied.

4.3 Heterointerfaces (type I, type II), Abruptness and Height of Barriers

Let us consider an atomically abrupt interface between two materials A and B. The basic properties can be explained by the Anderson model whereas quantitative descriptions need more refined treatments. In a piece of material, the vacuum electron energy level is several electron volts above the Fermi energy, E_F. This energy difference is called the work function ϕ, which for metals is a material constant. For semiconductors the work function is additionally doping dependent. We assume as a constant of each material the electron affinity χ. The work function ϕ_W is then given by

$$\phi_W = \chi + E_{CF} \tag{4.12}$$

where $E_{CF} = E_C - E_F$ is the conduction band edge to Fermi energy difference, which for a non degenerate semiconductor maybe expressed as

$$E_{CF} = E_g - E_i - k_B T \ln\left(\frac{n}{n_i}\right) \tag{4.13}$$

The intrinsic energy level E_i is roughly equal to the half band gap value (exactly $E_i = E_g/2 - (kT/2)\ln(N_C/N_V)$). At room temperature the electron concentration, n, is roughly equal to the net doping ($N_D - N_A$) for n-Si and $n = n_i^2/(N_A - N_D)$ for p-Si.

In thermodynamic equilibrium the Fermi energy in both materials is equal creating an energy barrier (a Schottky barrier), ϕ_B, in a metal-semiconductor couple (Fig. 4.7). At a semiconductor interface the band line up requires offsets ΔE_V and ΔE_C at the valence band and conduction band edges, respectively. In the most simple case as assumed in the Anderson model the vacuum energy varies smoothly across the interface which means no dipole moment at the interface.

Then the band offsets are defined by

$$\Delta E_V = \chi_A + E_{gA} - (\chi_B + E_{gB}) \tag{4.14}$$

$$\Delta E_C = \chi_B - \chi_A \tag{4.15}$$

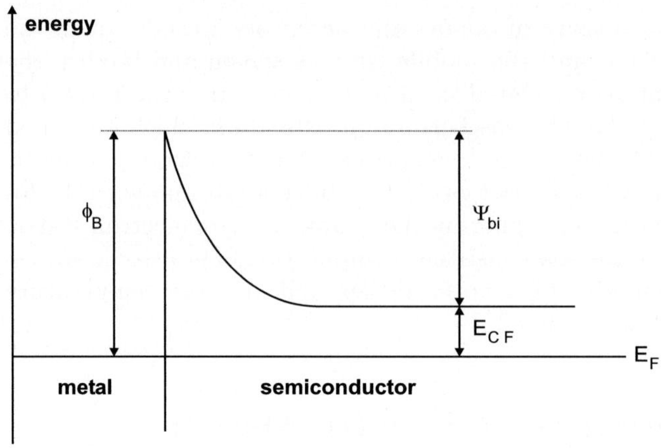

Fig. 4.7. The Schottky barrier (height ϕ_B) between a metal and a n-type semiconductor

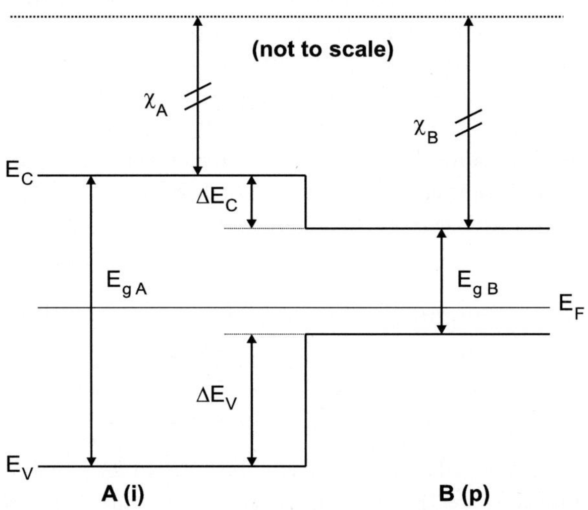

Fig. 4.8. The Anderson model. The band line up of a type I semiconductor interface. For clarity of presentation the doping (A intrinsic, B p-type) was chosen to avoid built in fields

In this notation we have chosen positive values for ΔE_V and ΔE_C for the most frequent situation (interface type I) where the band gap of the smaller band gap material B fits into the band gap of the material A (Fig. 4.8). A staggered band alignment (Fig. 4.9) is called a type II interface. In the notation of (4.14) and (4.15) this means a positive ΔE_V and a negative ΔE_C. There are more generally other possible configurations of the small band gap outside the other band gap but this will not be considered here as it cannot be realised in silicon based systems. In the given notation the sum

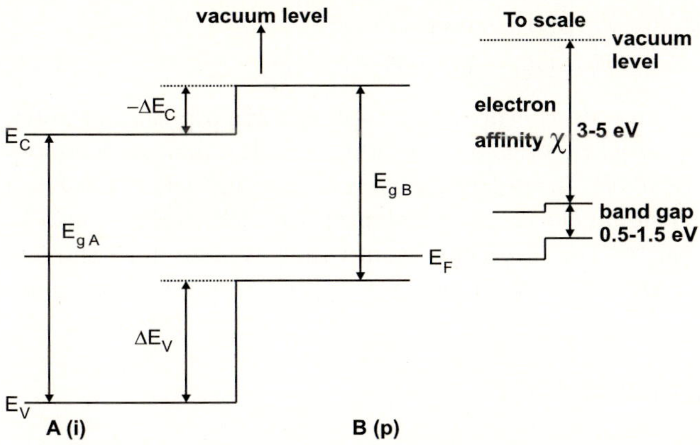

Fig. 4.9. A type II semiconductor interface with a staggered band line up. In our notation the conduction band offset, ΔE_C, is chosen negative. Within the framework of the Anderson model this interface type is obtained when the electron affinity χ_B of the small band gap semiconductor is smaller than χ_A

of the band offsets equals the difference in the band gaps

$$\Delta E_g = \Delta E_V + \Delta E_C \tag{4.16}$$

The numerical values of the band offsets will often differ from that predicted by the Anderson model. The main reason for the discrepancy is related to dipole moments caused by interfaces states. For metal semiconductor interfaces, already Bardeen postulated Fermi level pinning by dipole charges as a possible mechanism to explain rather weak doping levels and metal work function dependence of Schottky barrier heights. Later, Heine proved, that interface states in the forbidden band gap are created by wave functions penetrating into the opposite material. For metal semiconductor interfaces these are called MIGS – metal induced gap states. A rather high density of charges is needed however, to produce a measurable potential offset with a plate capacitor with atomic distances, e.g. 0.1 ML ($7 \times 10^{13}\,\mathrm{cm}^{-2}$) charges at atomic distance (0.14 nm) result in 160 meV dipole potential which has to be added to the Anderson model.

Flat band conditions were chosen in Fig. 4.8 and Fig. 4.9 to demonstrate the band offsets clearly. But generally a doping structure is superposed which in a conventional way is modeled by a built in electric field. For discussion, let us consider a highly n-doped wide gap side (n^+-A) and a p-doped side (p-B). Exchange of charge and recombination of carriers leads to a depletion layer mainly on the lower doped p-side. Exchange is stopped when the connected built in voltage V_{bi} leads to a flat Fermi energy level E_F (see Figs. 4.10 and 4.11).

The built in voltage has to be as high as given by (4.17) to meet the flat Fermi level conditions.

$$q\psi_{bi} = E_{gA} - \Delta E_V - E_{CF}(A) - E_{FV}(B)$$
$$= E_{gB} + \Delta E_C - E_{CF}(A) - E_{FV}(B) \quad (4.17)$$

Certain combinations of heterostructures and dopings form a potential notch at the interface, e.g. the n^+-wide band gap combined with a p-small bandgap at a type I interface (Fig. 4.10). In the given example the conduction potential notch collects two-dimensional electrons at the interface. For applications like the hetero-bipolar-transistor the electrons should rapidly emit into the low band side. Therefore, a potential as given by the type II interface is more favourable.

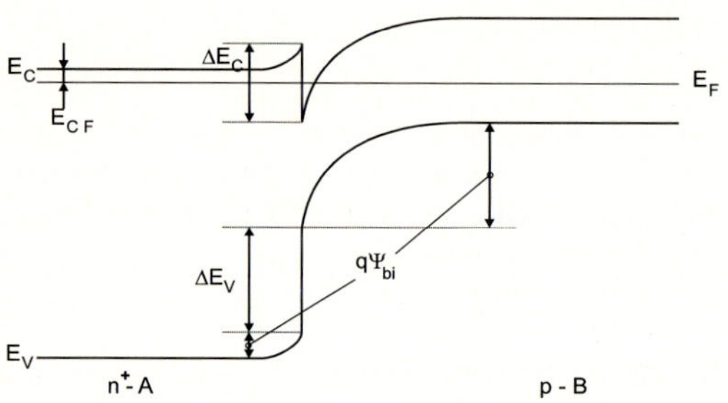

Fig. 4.10. A doped heterojunction with n^+-A and p-B semiconductor (type I interface). A flat Fermi energy is obtained with a depletion layer with ψ_{bi} voltage drop

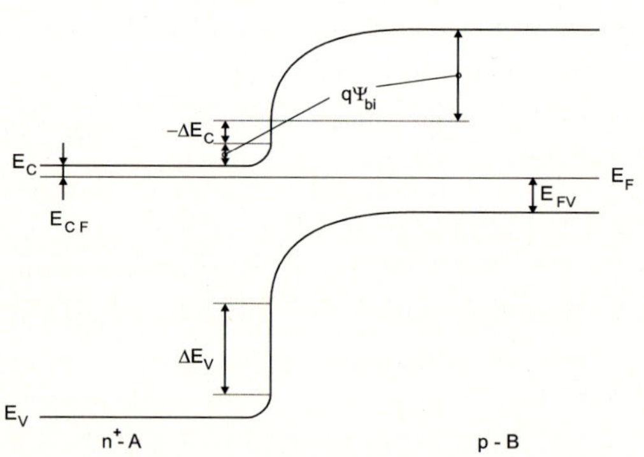

Fig. 4.11. A $n^+ - A/p - B$ heterostructure (with the same doping as in Fig. 4.10) with type II interface

4.3.1 Modulation Doping

In order to increase the mobility of two-dimensional carriers the doping was further refined. The realization of a high mobility two-dimensional electron gas has been made possible thanks to the modulation-doping technique, which was first applied by Dingle and Störmer to GaAs/GaAlAs heterostructures. Section 7.4.1 will review modulation doping in the Si/SiGe materials system. The idea of selective doping of heterostructures has proved to be very powerful. The idea is best explained by δ-doping spike distant from the interface by a spacer distance d_i (Fig. 4.12).

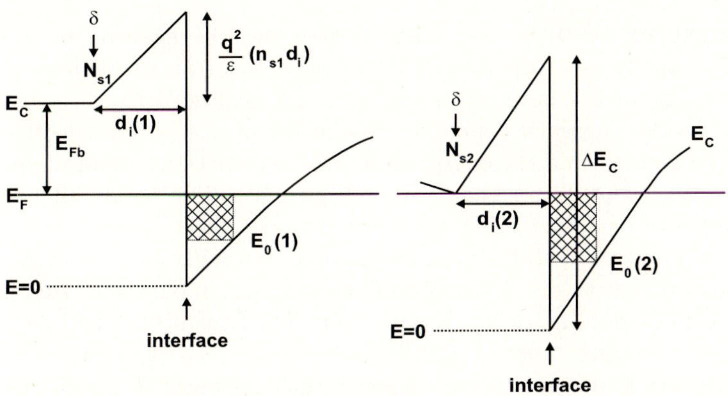

Fig. 4.12. Modulation doping. Demonstration of the effect of higher modulation doping ($N_{S2} > N_{S1}$) on the conduction band edge E_c. Cross hatched area (between E_0 and E_f) is proportional to the electron states in a two-dimensional carrier system

Let us first assume for simplicity that all the donors (of sheet density N_s) are ionised and the electrons (sheet density n_s; $n_s = N_s$ for flat band conditions) are collected in the potential notch at the other side of the interface. For complete ionisation we assume the Fermi level to be held below the conduction band neglecting high doping effects such as band gap shrinkage. We discuss qualitatively the influence of sheet doping, N_s, of spacer thickness, d_i, and of an applied gate voltage. In Fig. 4.12 the conduction band edge of an n-modulation doped heterostructure is shown. The electric field ($F = qn_s/\varepsilon_0\varepsilon_r$) between the positive donor ions (δ-doping with sheet density N_{s1}) and the electrons in the well build up an energy barrier with height $q^2 n_s \times d_i/\varepsilon_0\varepsilon_r$. The band offset, ΔE_C, at the interface forms a well whose width is roughly proportional to the inverse of n_s. This is because the field strength (gradient of the potential) at the interface follows n_s leading to an increasing energy position E_0 of the first confined electron state above the well bottom. The Fermi energy level lies above the electron state E_0 at an energetic position also proportional to the density n_s, because the density of states is constant in a two-dimensional carrier system.

$$E_F - E_0 = n_s/g_{2D} \tag{4.18}$$

The ground energy E_0 may be approximated with reasonable accuracy by the first confined state of a one sided triangular potential well (see Chap. 3). The 2D density of states g_{2D} is also given in Chap. 3. The simple formulation of (4.18) is strictly valid only for Fermi statistics at 0 K but it is also a reasonably good approximation at room temperature. More important is to check the energetic position of the next higher energy levels. The partly occupation of the next energy level E_1 has to be considered if the Fermi energy is higher than E_1. Equation (4.18) reads then

$$(E_F - E_0) + (E_F - E_1) = n_s/g_{2D} \quad \text{with} \quad (E_F > E_1) \tag{4.19}$$

Equation (4.18) (or (4.19) if $E_F > E_1$) determines the position of the Fermi energy, E_F, solely as a function of the channel electron sheet density n_s. On the other side of the heterojunction (left side in Fig. 4.12) also two parts contribute to the energetic difference to the Fermi energy, the electric field barrier as given above and the energy difference E_{Fb} between conduction band in the bulk and the Fermi energy (see Fig. 4.12). The height of the electric field barrier $q^2/\varepsilon_0\varepsilon_r(n_s d_i)$ is proportional to the sheet concentration n_s. Both effects lead to an upshift of the Fermi level E_F with increasing N_S and they mark together with the downshift of the conduction band left side of the barrier the highest doping available with complete ionization (right side of Fig. 4.12). Higher doping suffers from the incomplete transfer of electrons and parallel conduction in the δ-layer. The conduction band is above the Fermi level by the amount E_{Fb} at the position of the δ-doping.

$$E_{Fb} = \Delta E_C - E_0 - \frac{n_s}{g_{2D}} - \frac{q^2}{\varepsilon_0\varepsilon_r}(n_s d_i) \tag{4.20}$$

where g_{2D} is the constant density of states of a 2-dimensional (2D) carrier system. This equation (4.20) is valid for complete charge transfer. For incomplete transfer instead of N_S the transferred (channel) density has to be taken. In our simplifying approximation the maximum channel density is obtained when $\Delta E_{Fb} = 0$. The structure of (4.20) is most easily seen for the case of varying spacer thickness where only the last term is changed. Figure 4.13 demonstrates the band structure when the spacer thickness is reduced by a factor of 2. The left side shows the diagram for the maximum spacer width d_i (2) at which complete transfer is possible. By reducing the width nothing changes on the channel side, only at the spacer side does the potential barrier shrink by a factor 2.

For a given heterostructure (given ΔE_C, m^*, ϵ_s) the set of equations (4.18) -(4.20) describes completely the Fermi energy, E_F, and the bulk conduction band, E_{Fb}, as function of n_s and d_i. With indirect semiconductors the proper choice of the effective mass is often the practical problem, its solution is shown in the next section when the influence of strain on the bandstructure is treated (Sect. 4.4).

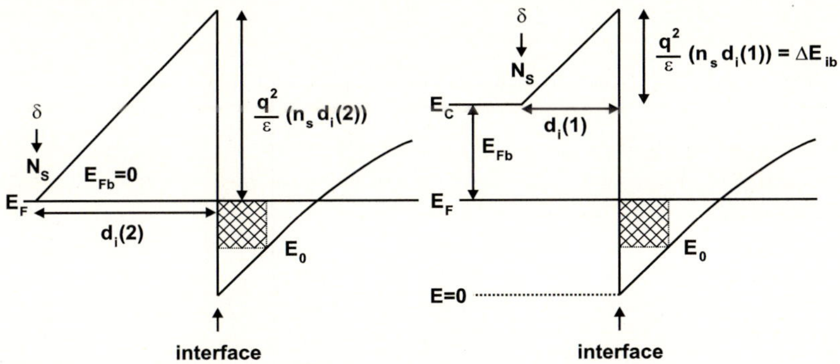

Fig. 4.13. Modulation doping. Demonstration of the effect of the spacer width d_i. Left: large spacer. Right: thin spacer

Let us consider an important special case (a strained-Si channel) that electrons with the longitudinal mass m_l^* are confined by the potential well (by this mass therefore the energy E_0 is determined) and perpendicular wave vectors with transversal mass m_t^* contribute to the 2D density of states (g_{2D} is determined by m_t^*). In the next figure (Fig. 4.14) the maximum sheet carrier density, $n_{s,max}$, is calculated using the above given choice of material constants and using the simplifying assumption $E_{Fb} = 0$.

Up to now we have assumed an ideal $\delta-$doping where the electric field, F, stops immediately at the sheet doping. What is changing when a thin (thickness d_d) layer with high doping density, N_D, provides the carrier? The only change is on the left side of the heterostructure where the electric field barrier, ΔE_{ib}, (fourth term in (4.20)) increases because the field now penetrates from the spacer into the doping by a width n_s/N_D.

The electric field barrier, ΔE_{ib}, reads now

$$\Delta E_{bi}/q = \frac{q}{\varepsilon_0 \varepsilon_r} n_s \left(d_i + \left(\frac{n_s}{2N_D} \right) \right) \tag{4.21}$$

For $N_D \to \infty$ the term converges to $\frac{q}{\varepsilon_0 \varepsilon_r}(n_s d_i)$ as used in (4.20).

4.3.2 Gated Channel

Up to now we have assumed flat band conditions at the surface with no electric field between the δ-doping and the surface. This condition is changed by applying a proper gate voltage in a gated structure. Let us consider a gated Schottky barrier which adjusts the electron surface potential ψ_s (Fig. 4.15).

Now, only a fraction α_{2D} of the electrons are transferred to the channel, the other fraction (1-α_{2D}) is transferred to the gate electrode. On the heterostructure side again (4.18) may be used but with $n_s = \alpha_{2D} N_S$ as the electron density in the channel. On the gate side the relation holds

$$q\psi_s + E_{Fb} = \phi_B + (-qV) \tag{4.22}$$

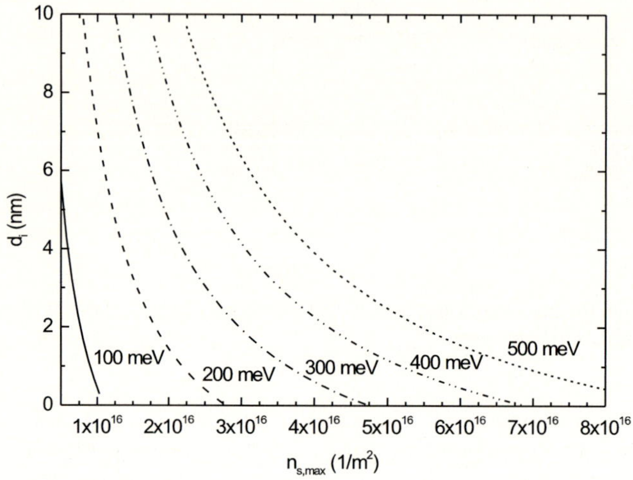

Fig. 4.14. The spacer width, d_i, and maximum sheet concentration, $n_{s,max}$, for various band offsets, ΔE_C. For calculation of E_0 the approximation (4.26) was used ($m_l^* = 0.92m_0$). For calculation of g_{2D} (3.91) was used ($g_s = 2$, $g_v = 2$, $m_t^* = 0.19m_0$)

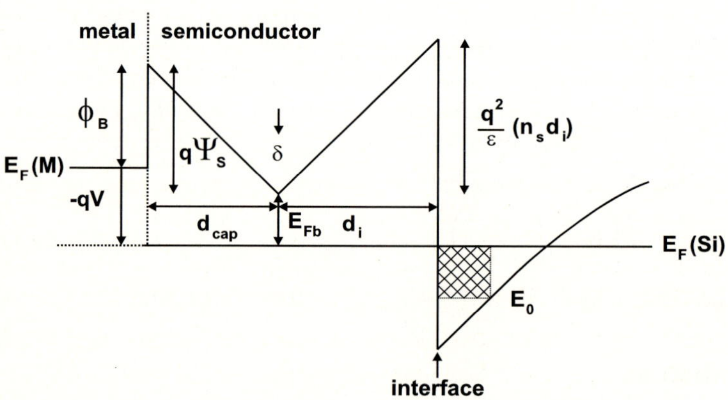

Fig. 4.15. A gated modulation doped heterojunction. An applied voltage ($-V$) to a metal gate (Schottky barrier ϕ_B) shifts the surface energy by an amount ψ_S compared to the flat band conditions ($\psi_S = 0$)

The surface potential ψ_s is given by the product of the electric field $\frac{q}{\varepsilon_0 \varepsilon_r} N_S(1-\alpha_{2D})$ and the cap thickness d_{cap}.

$$\psi_s = \frac{q}{\varepsilon_0 \varepsilon_r} N_S(1-\alpha_{2D}) d_{cap} \tag{4.23}$$

For flat band conditions the surface potential $\psi_s = 0$. The voltage is the flat band voltage V_{FB}

$$V_{FB} = \frac{\phi_B}{q} - \frac{\Delta E_C}{q} + \frac{q}{\varepsilon_0 \varepsilon_r} N_S d_i + \frac{N_S}{q g_{2D}} + Const. N_S^{2/3} \tag{4.24}$$

The surface potential ψ_s is again somewhat larger (same reason as explained in (4.21)) if the $\delta-$doping is replaced by a thin (thickness d_d), highly doped (doping density N_D) layer.

$$\psi_s = \frac{q}{\varepsilon_0 \varepsilon_r} N_S(1-\alpha_{2D}) \left(d_{cap} + \frac{1}{2} \frac{N_S(1-\alpha_{2D})}{N_D} \right) \tag{4.25}$$

Exact solutions of equations 4.18- 4.25 need a self consistent solution of the energy level E_0 as function of $N_S \alpha_{2D}$. For a rough assessment a trial function in a variational solution may be used leading to the following carrier dependence for the triangular potential.

$$\frac{E_0}{q} \cong 2 \left(\frac{\hbar^2 q}{m^* \varepsilon_0^2 \varepsilon_r^2} \right)^{1/3} (N_S \alpha_{2D})^{2/3} \tag{4.26}$$

$$Const. = 2 \left(\frac{\hbar^2 q}{m^* \varepsilon_0^2 \varepsilon_r^2} \right)^{1/3}$$

The following implicit equation for the fraction α_{2D} is obtained.

$$\frac{\Delta E_C}{q} - \frac{\phi_B}{q} + V + N_S \frac{q}{\varepsilon_0 \varepsilon_r} d_{cap} = \tag{4.27}$$

$$(N_S \alpha_{2D}) \left[\frac{q}{\varepsilon_0 \varepsilon_r} (d_{cap} + d_i) + \frac{1}{q g_{2D}} \right] + Const. \, (N_S \alpha_{2D})^{2/3}$$

It should be remembered this equation is valid as long as E_{Fb} is larger than zero, which can be tested with (4.18), and the channel density n_s is given by $N_S \alpha_{2D}$.

With increasing reverse voltage ($V_r < 0$V) the transferred fraction α_{2D} decreases and reaches 0 at the threshold voltage V_T.

$$V_T = \frac{\phi_B}{q} - \frac{\Delta E_C}{q} - \frac{q}{\varepsilon_0 \varepsilon_r} N_S d_{cap} \tag{4.28}$$

That means that the threshold voltage, V_T, is determined by the material system properties ϕ_B, ΔE_C and the technological properties N_S, d_{cap}. A higher sheet doping N_S or a larger cap layer, d_{cap}, shifts the threshold voltage to more negative voltages (for a n-channel device).

For an overview of all the effects let us change the voltage V between the gate and the channel from negative values to positive values. The semiconductor side is considered as a reference. The δ-doping density, N_S, is chosen somewhat lower than the maximum sheet density, $n_{s,max}$.

For strong negative values of the voltage V the doping layer and the channel are depleted. At the threshold voltage V_T the first electrons jump into the channel (Fig. 4.16, above) when the down moving first quantum well level E_0 meets the Fermi energy. In the given example of Fig. 4.16 the Schottky barrier is 0.45 eV and the threshold voltage V_T is -0.15 V. With higher Schottky barrier heights (eg. 0.7 eV) the threshold voltage would shift to positive values with the same geometrical factors (d_i, d_{cap}, N_S). With more positive values of voltage V ($V - V_T > 0\,\mathrm{V}$) the channel will be filled with electrons (density n_s) when the energy level E_0 shifts below the Fermi energy. Do not forget that the energy E_0 by itself is a function of n_s, because of the triangular bottom of the quantum well. In Fig. 4.16 (middle) half of the electrons in the δ-doping jump into the channel ($n_s = N_S/2$) when the voltage, V, reaches 0.125 V. With increasing voltage V_{FB} (flat band voltage) all the electrons from the δ-doping jump into the channel with a vanishing electric field in the cap layer. In the example (Fig. 4.16, below) the flat band voltage V_{FB} amounts to roughly 0.42 V. Further increase of the voltage beyond the Schottky barrier value, Φ_B/q, creates a surface channel like a MOS varactor which creates high leakage currents to the non insulating gate.

The simple picture which has been presented is in reality complicated by several effects, the most important of which we will discuss qualitatively. One has to consider the finite temperature, T, the real doping is spread out by a distance d_d with doping level N_d ($N_d \times d_d = N_S$), and the bulk volume is doped oppositely, that results for the treated case of an n-channel, the bulk must be p-doped to insulate the channel through p/n-junction isolation. The most prominent effect of the temperature is the finite occupation of energy levels even if the Fermi energy is below that energy level. This leads to incomplete vanishing charge in the channel at strong negative voltages (sub threshold) and to incomplete vanishing electrons at the δ-doping even if the density is below the maximum. An exact calculation of the occupation needs consideration of the asymmetric triangular well at the δ-doping (see Fig. 4.16, not to be confused with the one sided triangular well at the channel).

The finite thickness of the doping layer is easily calculated using Poisson's equation. For a uniform doping N_d, with thickness d_d in between the undoped spacer d_i and the undoped cap d_{cap}, the voltage barriers to the spacer and the cap read $\left(\frac{q}{\varepsilon_0\varepsilon_r}\right) n_s \left(d_i + \frac{n_s}{2N_D}\right)$ and $\left(\frac{q}{\varepsilon_0\varepsilon_r}\right)(N_S - n_s)\left(d_{cap} + \frac{N_S - n_s}{2N_D}\right)$, respectively. For increasing doping values N_D we approximate the δ-doping case.

The opposite bulk doping isolates the channel from the substrate, but that means, that a part of the electrons from the δ-doping are used to create the depletion layer to the p-bulk. The number ($N_S \times \alpha'$) used is given by

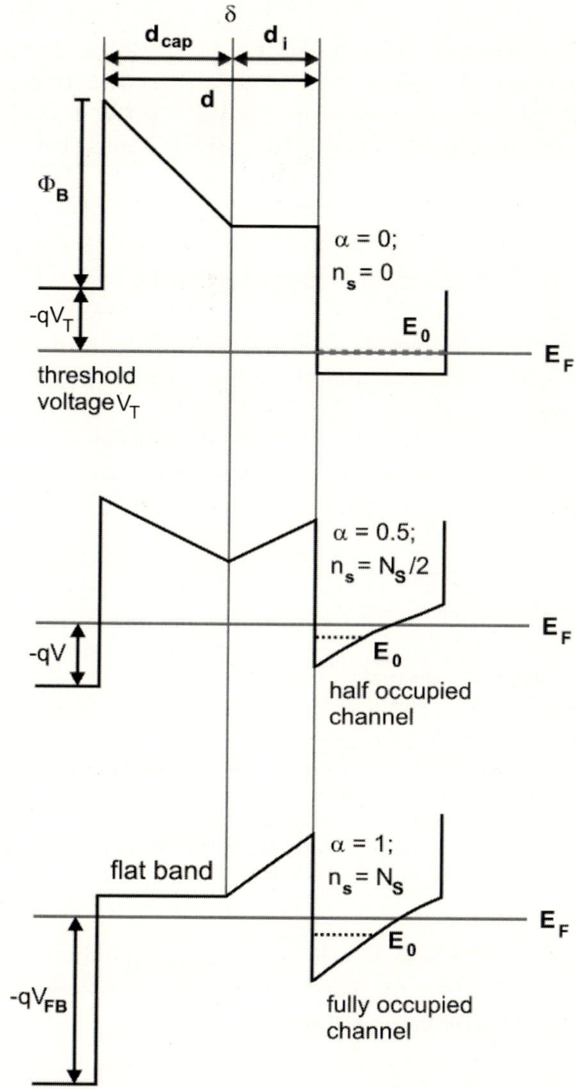

Fig. 4.16. A schematic diagram of the potential energy in a modulation doped quantum well. Shown is the variation with increasing gate voltage, V, from the empty channel (threshold voltage V_T) to the full channel (flat band voltage V_{FB}). The energy E_0 is not constant but a function of the channel charge, because of triangular well bottom

$$N_S\alpha' = N_A \times d_b \left[\left(1 + 2\frac{\varepsilon_0 \varepsilon_r}{q} \frac{V_{bi}}{N_A d_b^2} \right)^{1/2} - 1 \right] \tag{4.29}$$

whereas N_A and d_b are bulk doping density and distance of bulk doping from δ-doping, respectively. A typical value of $(N_S\alpha')$ is $4 \times 10^{11}\,\mathrm{cm}^{-2}$ for $N_A = 10^{17}\,\mathrm{cm}^{-3}$, $d_b = 100\,\mathrm{nm}$ and built in voltage $V_{bi} = 0.8\,\mathrm{V}$. A conceptually good solution to this problem is delivered by a second δ-doping below the quantum well.

4.4 Influence of Strain on Bandstructure

Application of strain is frequently used for the investigation of basic material properties, from such studies we understand the effects of hydrostatic pressure and uniaxial stress.

Strain has two main effects on the band structure: hydrostatic strain shifts the energetic position of a band and uniaxial strain splits degenerate bands (Fig. 4.17). The strain state of the semiconductor can be expressed by

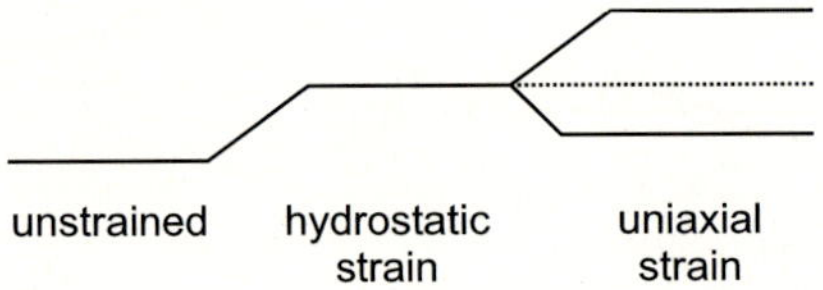

Fig. 4.17. A schematic diagram representation of the strain on a triply degenerate band. Hydrostatic strain shifts the absolute energy position of the band. Uniaxial strain splits the degeneracy; illustrated is a case where the threefold degenerate band is split into a set of twofold degenerate bands at lower energy plus a singly degenerate band at higher energy. Note that the average over the three bands (dotted line) is unaffected by uniaxial strain

the strain tensor $\overline{\boldsymbol{\epsilon}}$. The hydrostatic strain, corresponding to the fractional volume change $\frac{\Delta V}{V}$, is given by the trace of the strain tensor, which amounts for biaxial strained epitaxial layers to

$$\frac{\Delta V}{V} = Tr(\epsilon) = 2\epsilon + \epsilon_z, \tag{4.30}$$

where the normal strain ϵ_z in an isotropic notation given by (2.24b).

For simplicity we assume the z-direction in [001], a threefold degenerate valence band at $k = 0$ (light holes 1, heavy holes 2, spin-orbit split-off holes 3) and an indirect conduction band with sixfold degenerate Δ valleys (in the (100) directions).

The effect of hydrostatic and biaxial strains on the bandstructures is expressed via deformation potentials. These deformation potentials have been determined experimentally and theoretically for Si and Ge; no direct determinations have been performed for SiGe alloys. Linear interpolation tends, however, to be a good approximation. The deformation potentials for hydrostatic strains will be denoted by the symbol a, and the deformation potential for uniaxial strain in tetragonally distorted cubic lattice cells are denoted by b.

4.4.1 Hydrostatic Strain

Hydrostatic strain shifts the average position of the valence and the conduction band by $\Delta E_{v,av}$ and $\Delta E_{c,av}$, respectively.

$$\Delta E_{v,av} = a_v(2\epsilon + \epsilon_z) = 2\epsilon\frac{1-2\nu}{1-\nu}a_v \tag{4.31a}$$

$$\Delta E_{c,av} = a_c(2\epsilon + \epsilon_z) = 2\epsilon\frac{1-2\nu}{1-\nu}a_c \tag{4.31b}$$

where ν is Poisson's ratio. Numerical values of the hydrostatic deformation potentials a_v and a_c are given in Table 4.2. The selected values follow values recommended by C. Van de Walle, who performed the basic studies on band offsets in the strained SiGe system.

Table 4.2. The deformation potentials a_v and a_c for hydrostatic strain in Si and Ge. The deformation potentials b_v and b_c are already given in a notation appropriate for tetragonal strain (E_1, E_2 in Kane's notation, and $\Xi_d + \frac{1}{3}\Xi_u$, Ξ_u in Van de Walle notation).

Material	Deformation potential (eV)			
	a_v	a_c	b_v	b_c
Si	2.46	4.18	-2.35	9.16
Ge	1.24	2.55	-2.55	9.42

4.4.2 Uniaxial Strain

The biaxial strain in the epitaxial plane can be considered as hydrostatic deformation superposed by an uniaxial strain $(\epsilon_z - \epsilon)$ perpendicular to the plane. This uniaxial strain component is written in isotropic notation as

$$\epsilon_z - \epsilon = -\epsilon\frac{1+\nu}{1-\nu}, \tag{4.32}$$

In the valence band the light hole (lh) band is shifted by $3b_v(\epsilon_z - \epsilon)$ compared to the heavy hole and spin-orbit split-off band. Compared to the average the shifts are given by

$$\Delta E_{v1}(lh) = 2b_v(\epsilon_z - \epsilon) = -2b_v\epsilon\frac{1+\nu}{1-\nu} \tag{4.33a}$$

$$\Delta E_{v2} = \Delta E_{v3} = -b_v(\epsilon_z - \epsilon) = b_v\epsilon\frac{1+\nu}{1-\nu} \tag{4.33b}$$

In the absence of strain the threefold degeneracy of the valence band is already lifted by the spin orbit interaction and splits the bands by an amount of Δ_0. With respect to the average band position $E_{v,av}$ two bands are shifted up by an amount of $\frac{\Delta_0}{3}$ (the light and heavy hole bands, which are denoted E_{v1}, E_{v2}), while one band is shifted down by an amount $2\frac{\Delta_0}{3}$ (the spin-orbit split-off band, which we denote E_{v3}).

For completeness the relations are given for lifting of the valence band degeneracy by strain and band orbit interaction. The superposition of both effects is nonlinear for the light hole and spin orbit split-off band.

$$\begin{aligned}\Delta E_{v1} = &-\frac{1}{6}\Delta_0 + \frac{b}{2}(\epsilon_z - \epsilon) + \\ &\frac{1}{2}\sqrt{\Delta_0^2 + 2\Delta_0 b(\epsilon_z - \epsilon) + 9b^2(\epsilon_z - \epsilon)^2}\end{aligned} \tag{4.34a}$$

$$\Delta E_{v2} = \frac{1}{3}\Delta_0 - b(\epsilon_z - \epsilon) \tag{4.34b}$$

$$\begin{aligned}\Delta E_{v3} = &-\frac{1}{6}\Delta_0 + \frac{b}{2}(\epsilon_z - \epsilon) - \\ &\frac{1}{2}\sqrt{\Delta_0^2 + 2\Delta_0 b(\epsilon_z - \epsilon) + 9b^2(\epsilon_z - \epsilon)^2}\end{aligned} \tag{4.34c}$$

The strain lifts the sixfold (6g) degeneracy of the conduction band into the fourfold (4g) degenerate bands with energy minima lying in-plane and the twofold degenerate band with energy minima lying perpendicular to the (001) plane. For (001) growth and Δ type energy minima the energy shifts ΔE_c with respect to the average are easily written as

$$\Delta E_c(2g) = \frac{2}{3}b_c(\epsilon_z - \epsilon) = -\frac{2}{3}b_c\epsilon\frac{1+\nu}{1-\nu} \tag{4.35a}$$

$$\Delta E_c(4g) = -\frac{1}{3}b_c(\epsilon_z - \epsilon) = \frac{1}{3}b_c\epsilon\frac{1+\nu}{1-\nu} \tag{4.35b}$$

Let us now consider a model material with simple properties $a_v = 2\,\text{eV}$, $a_c = 3\,\text{eV}$, $b_v =$ -2.5 eV, $b_c = 9\,\text{eV}$, $\Delta_0 = 0$, $\nu = \frac{1}{3}$ under either tensile or compressive biaxial strain ϵ of magnitude 0.01. First we investigate the valence band under the influence of strain (Fig. 4.18). The band average $E_{v,av}$ is shifted up by tensile strain (+20 meV) and shifted down by compressive strain. Tensile strain further shifts up the light hole (lh) energy levels by 100 meV while heavy holes (hh) are lowered by 50 meV. In our model material the spin-orbit split-off holes (soh) are degenerate ($\Delta_0 = 0$) with heavy holes at $k = 0$. As a result of tensile strain the light holes energy (E_{v1}) are lifted by 120 meV, while the heavy holes ($E_{v2} = E_{v3}$) are lowered by 30 meV.

Under compressive biaxial strain (growth on a substrate with a smaller lattice constant) the heavy hole band will be at the band edge; under tensile

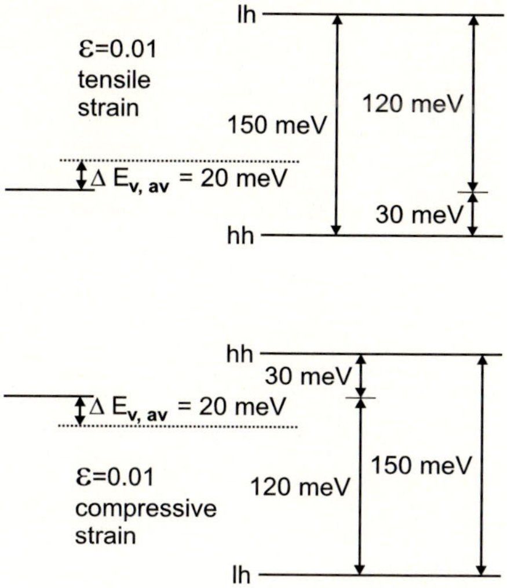

Fig. 4.18. The valence band splitting under the influence of strain for $|\epsilon| = 0.01$. For the properties of the model material see the text

strain (growth on a substrate with a larger lattice constant) the light hole band will be preferentially occupied. The different effective masses of the light and heavy hole bands lead to different behaviour in terms of the mobility of carriers or confinement shifts in the quantum wells. One can therefore use strain to tailor the band structure according to the needs of a particular application.

Now, we investigate the Si-like conduction band (indirect, Δ-minimum) under the influence of tensile strain ($\epsilon = 0.01$). The average energy,$E_{c,av}$, is shifted up by 30 meV. The twofold (2g) electrons perpendicular to the plane are shifted down by $\Delta E_c(\text{2g}) = 120\,\text{meV}$, the fourfold (4g) in-plane electrons are shifted up by $\Delta E_c(\text{4g}) = 60\,\text{meV}$ (Fig. 4.19) resulting in a final energy position of $E_c(\text{2g}) = -90\,\text{meV}$, $E_c(\text{4g}) = +90\,\text{meV}$ with 180 meV splitting. The dominant effect of band splitting compared to the shift of the average energy results in the band gap being reduced by the strain. With tensile strain the smallest gap is between the light hole valence band and the twofold (2g) electrons with a band gap shrinkage of 210 meV (120 meV and 90 meV contribution from valence band and conduction band, respectively). With compressive strain the lowest gap is between hh states in the valence band and 4g electron states in the conduction band. The band gap shrinkage amounts to 120 meV (30 meV and 90 meV from the valence band and conduction band, respectively).

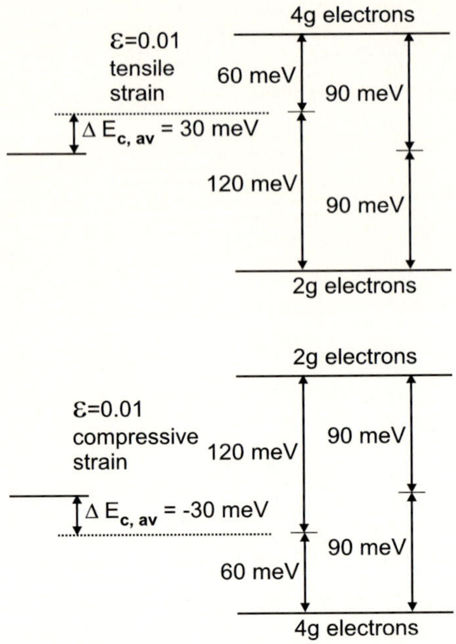

Fig. 4.19. The conduction band splitting under the influence of strain for $|\epsilon| = 0.01$. The same model material has been used as in Fig. 4.18

4.5 Band Alignment of Strained SiGe

When two semiconductor are joined at a heterojunction, discontinuities occur in the valence band and in the conduction band. In the absence of strain, i.e. for a lattice matched interface, the alignment simply requires one to determine how the band structures of the two materials line up at the interface; the line up then produces values for the valence band discontinuity, ΔE_v, and the conduction band discontinuity, ΔE_c. When the materials are strained, the strains will produce additional shifts (due to hydrostatic strain) and splittings (due to biaxial strain) as described in the forgoing section.

We will follow here the treatment of Van de Walle who first considered the average energy, E_v^0, of the unstrained valence band and then added the influence of strain to obtain the individual energy levels.

4.5.1 Average Valence Band Energy E_v^0

The average valence band energy of Ge is roughly 0.58 eV above that value of Si (Note: Both, the experimental and theoretical values on that property are uncertain by about 0.1 eV). That would result in a valence band offset between unstrained Si and unstrained Ge of 0.662 eV, because the highest valence bands (lh, hh) are shifted up by $\Delta_0/3$. The spin orbit splitting Δ_0

increases strongly from Si ($\Delta_0 = 44\,\text{meV}$) to Ge ($\Delta_0 = 290\,\text{meV}$). For both quantities E_v^0 and Δ_0 we will assume a linear relationship with the chemical composition of an $Si_{1-x}Ge_x$ alloy (expressed by atomic Ge content x).

$$E_{v,av}^0(SiGe) = 580\,\text{meV} \times x - \frac{44}{3}\,\text{meV} \tag{4.36}$$

$$\Delta_0(SiGe) = 44\,\text{meV} + x \times 246\,\text{meV} \tag{4.37}$$

The valence band edge (lh, hh) of unstrained Si is taken as reference energy. The valence band edge $E_{v,1} = E_{v,2}$ of unstrained SiGe is then given by

$$SiGe : E_{v,av}^0 + \Delta_0/3 = 662\,\text{meV} \times x \tag{4.38}$$

(Remember; the notation v1, v2, v3 is used for the light hole, heavy hole and spin-orbit split-off subbands of the valence band). The energetic position of the conduction band is more complicated, because of band bowing with Ge content x and because of Ge-like (eightfold, 8g) band minima for high Ge contents ($x > 0.85$). We consider here only the mainly dominating Si-like (6g, unstrained SiGe) states. The reader is referred to the section on Further Reading, e.g. D. Paul, for the small regime around Ge with Ge-like states.

The band gap E_g^0 of unstrained SiGe is roughly given by a parabolic law

$$E_g^0 = (1.17 - 0.44x + 0.206x^2)\,\text{eV} \tag{4.39}$$

which is obtained by adding the binding energy of the free exciton (14.7 meV for Si, 4.15 meV for Ge) to the excitonic bandgap obtained from low temperature photoluminescence measurements (Weber, Alonso).

4.5.2 Compressive Strain

A typical situation for compressive strain is given by a pseudomorphic SiGe film on a Si substrat.

We consider only the energy offsets for the most occupied subbands (highest valence band, lowest conduction band). For compressive strain these are the heavy hole valence band (E_{v2}) and the (4g) in plane conduction band minima (E_{c4}). The typical contributions for the valence band offset are shown in Fig. 4.20 on the example of strained Ge on an unstrained Si substrate. The Si substrate is on the left side with the average valence band energy $E_{v,av}$ an amount $\Delta_0/3 = 44/3\,\text{meV}$ ($\cong 15\,\text{meV}$) below the valence band edge E_v ($\equiv E_{v1} = E_{v2}$). On the Ge side the hypothetical unstrained average $E_{v,av}^0$ is 0.58 eV above the average Si level (chemical composition shift). The hydrostatic strain (compressive strain has negative sign) shifts down the average level to $E_{v,av}$. The uniaxial component splits the valence bands with the E_{v2} energy (heavy hole states) as the upper level for compressive strain. The conduction band offset may be constructed in the same way by adding the band gap to the average valence band energy and then applying the strain. With compressive strain the fourfold degenerate (4g) in plane conduction

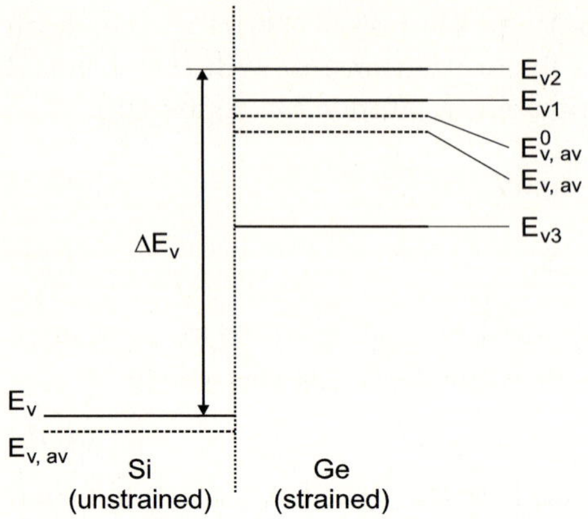

Fig. 4.20. The theoretical valence-band lineups at an interface between unstrained Si and strained Ge (not to scale)

band minima E_{c4} are lowest in energy. Figure 4.21 shows the valence band offsets $\Delta E_v(v2)$ and the conduction band offset $\Delta E_c(4g)$ as function of the Ge content for strained SiGe on Si. For comparison the offsets for the other subbands which do not define the band edge are also shown. The valence band discontinuity ΔE_v increases rather linearly to the strained Ge value of 0.78 eV, while the conduction band discontinuity is clearly nonlinear, because of the band gap bowing and it is quite small (Note: the band offset ΔE_c in Fig. 4.21 is shifted by 1.17 eV to separate it from ΔE_v).

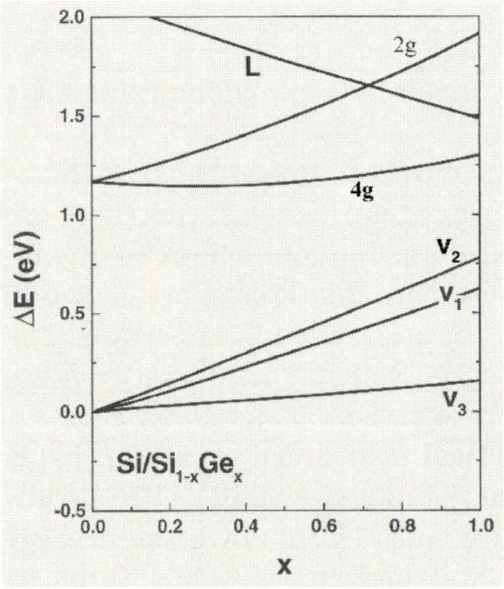

Fig. 4.21. The band offsets ΔE_v, ΔE_c for a pseudomorphic Si/$Si_{1-x}Ge_x$ heterojunction as a function of alloy composition x in the overlayer. ΔE_v is defined by the heavy hole (v2) level, ΔE_c by the in plane (4g) level. For better readability the conduction band offset ΔE_c is shifted up by the bandgap difference of Si (1.17 eV)

4.5.3 Tensile Strain

A typical situation for tensile strain would be a strained-Si channel on a relaxed SiGe buffer (methods to obtain strain adjustment will be discussed in Chap. 2).

Here the light hole (v1) states mark the valence band edge while the perpendicular (2g) states mark the conduction band edge. The treatment is similar to the compressive strain treatment, but one has to consider, that the energy splitting of the light hole (v1) states is more complicated than that of the heavy holes (v2), because of interference with spin orbit interaction (see forgoing section).

The results of the calculation for a strained-Si quantum well on a relaxed $Si_{1-y}Ge_y$ substrate heterojunction are given in Fig. 4.22. The larger band gap overlayer of Si is lower in the valence band edge (v1) which is as expected but is also lower in the conduction band edge (2g) energy which is a characteristics of a type II heterostructure.

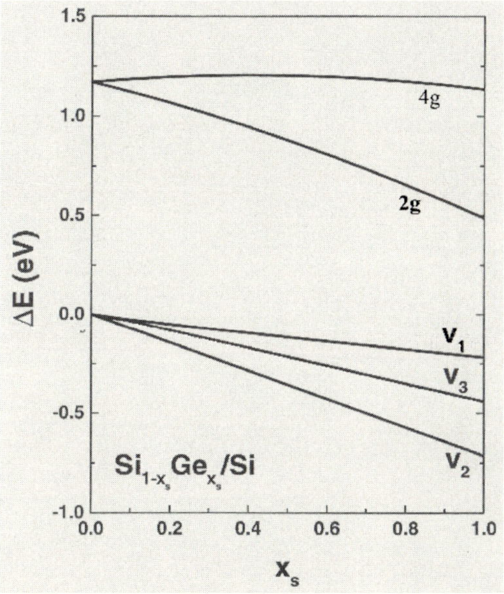

Fig. 4.22. The band offsets ΔE_v, ΔE_c for a pseudomorphic relaxed-$Si_{1-y}Ge_y$/strained-Si heterojunction as a function of alloy composition y of the substrate. The valence band offset ΔE_v is defined by the v1 (light hole) state, the conduction band offset ΔE_c is defined by the 2g (perpendicular to the interface plane) states. ΔE_c is shifted up by 1.17 eV as in Fig. 4.21

4.6 Further Reading

1. D.J. Paul, Semiconductor Science and Technology **19**, R59 (2004)
2. M.J. Kelly, Low Dimensional Semiconductors: Materials, Physics, Technology, Devices" OUP, Oxford (1995)
3. J.H. Davies, The Physics of Low Dimensional Semiconductors, CUP, Cambridge (1998)
4. C.K. Maiti, N.B. Chakrabarti, S.K. Ray, Strained Silicon Heterostructures Materials and Devices, The Institution of Electrical Engineers, London (2001)

5. Electronic Device Principles

This chapter will review the major types of devices which have been demonstrated or suggested as possible devices in the Si/SiGe heterostructure system. At present MOSFETs dominate the semiconductor markets mainly in the circuit form of CMOS but the first transistor action was produced in a germanium bipolar device in 1947. The first SiGe transistor to be on the market place is the heterojunction bipolar transistor (HBT) which will be reviewed in Chap. 6. This chapter will therefore start on bipolar devices before moving to field effect transistors (FETs) such as MOSFETs.

5.1 The p-n Junction

The basic building block and foundation of many electronic devices is the p-n junction or diode. The physics of such junctions was originally developed by Shockley in 1950. The physics is still central in the design of present day ULSI MOSFETs and bipolar transistors.

Let us consider the case of an abrupt junction where the doping in the semiconductor changes from N_A to N_D over a few atomic layers. Figure 5.1 shows the typical structure along with the space charge distribution and the resulting band structure.

The first case to consider is for zero applied potential and no current flow. The current density is given by

$$J_n = q\mu_n nF + qD_n \frac{\partial n}{\partial x} = q\mu_n \left(nF + \frac{k_B T}{q} \frac{\partial n}{\partial x} \right) = 0 \qquad (5.1)$$

for electrons where the electric field is F, the electron mobility is μ_n, q is the electronic charge of an electron, D_n is the electron diffusion constant and n is the electron concentration. Similarly for holes

$$J_p = q\mu_p pF - qD_p \frac{\partial p}{\partial x} = q\mu_p \left(pF - \frac{k_B T}{q} \frac{\partial p}{\partial x} \right) = 0 \qquad (5.2)$$

To solve these equations, (3.167) from Chap. 3 needs to be substituted into (5.1) and (5.2) which gives the result

$$\frac{\partial E_F}{\partial x} = 0 \tag{5.3}$$

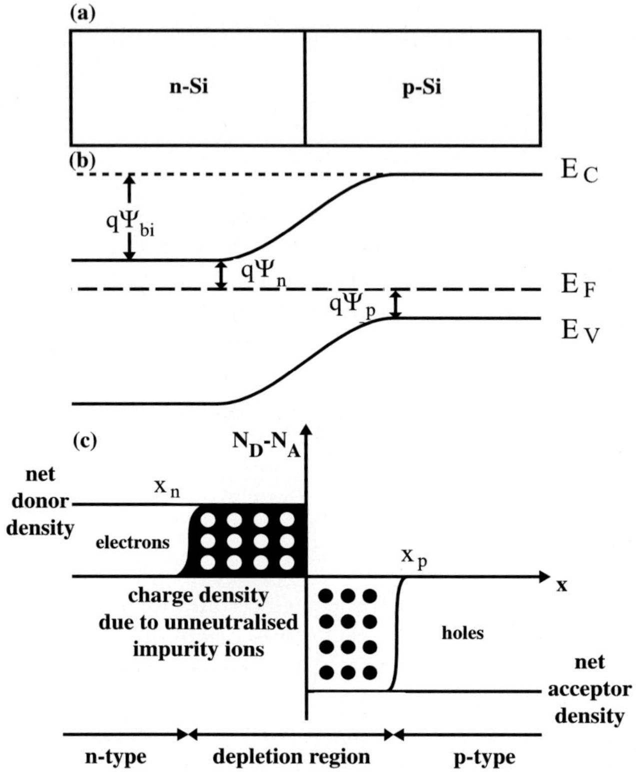

Fig. 5.1. (a) A schematic diagram of an abrupt p-n junction. (b) The band diagram of the p-n junction. (c) The space-charge distribution in the junction at thermal equilibrium

Therefore the requirement of zero current flow in the systems forces the Fermi energy to be constant across the junction. Electrons and holes will therefore redistribute close to the interface forming depletion regions as shown in Fig. 5.1 (c). The built-in potential created by this redistribution is given by

$$q\psi_{bi} = E_g - (q\psi_n + q\psi_p) \tag{5.4}$$

$$= k_B T \ln\left(\frac{N_c N_v}{n_i^2}\right) - \left[k_B T \ln\left(\frac{N_c}{n_n}\right) + k_B T \ln\left(\frac{N_v}{p_p}\right)\right]$$

$$\cong k_B T \ln\left(\frac{N_D N_A}{n_i^2}\right) \tag{5.5}$$

where n_p is the electron density in the p-semiconductor and p_p is the hole concentration in the p-semiconductor. At equilibrium

$$n_n p_n = n_p p_p = n_i^2 \tag{5.6}$$

therefore

$$\psi_{bi} = \frac{k_B T}{q} \ln\left(\frac{p_p}{p_n}\right) = \frac{k_B T}{q} \ln\left(\frac{n_n}{n_p}\right) \tag{5.7}$$

One can turn (5.7) around to find the relationship between the electron and hole densities on either side of the junction which are

$$p_n = p_p \exp\left(-\frac{q\psi_{bi}}{k_B T}\right) \tag{5.8}$$

$$n_p = n_n \exp\left(-\frac{q\psi_{bi}}{k_B T}\right) \tag{5.9}$$

Since no current flows, the electric field in the neutral regions well away from the junction must be zero. Therefore the negative charge per unit area in the p-side must equal the positive charge per unit area in the n-side and so

$$N_A x_p = N_D x_n \tag{5.10}$$

Next we need to apply Poisson's equation to the abrupt junction relating the electrostatic potential, ψ to the charge density, ρ through

$$\frac{\partial^2\psi}{\partial x^2} = -\frac{\partial F}{\partial x} = -\frac{\rho(x)}{\varepsilon_o\varepsilon_r} = -\frac{\rho}{\varepsilon_o\varepsilon_r}\left[p(x) - n(x) + N_D^+(x) - N_A^-(x)\right] \tag{5.11}$$

and so with approximations in the regime $0 < x \leq x_n$ gives

$$\frac{\partial^2\psi}{\partial x^2} \cong -\frac{q}{\varepsilon_o\varepsilon_r} N_D \tag{5.12}$$

while in the regime $-x_p \leq x < 0$ the result becomes

$$\frac{\partial^2\psi}{\partial x^2} \cong \frac{q}{\varepsilon_o\varepsilon_r} N_A \tag{5.13}$$

By integrating (5.12) and (5.13), the appropriate electrics fields can be found as

$$F(x) = -\frac{qN_A(x + x_p)}{\varepsilon_0\varepsilon_r} \tag{5.14}$$

for the region $-x_p \leq x < 0$ and

$$F(x) = \frac{qN_D(x - x_n)}{\varepsilon_0\varepsilon_r} \tag{5.15}$$

for the region $0 < x \leq x_n$.

The maximum electric field occurs at $x = 0$ and is given by

$$|F_m| = \frac{qN_D x_n}{\varepsilon_0\varepsilon_r} = \frac{qN_A x_p}{\varepsilon_0\varepsilon_r} \tag{5.16}$$

Integrating either of (5.14) or (5.15) produces the potential distribution

$$\psi(x) = F_m\left(x - \frac{x^2}{2W_D}\right) \tag{5.17}$$

where W_D is the depletion width which may be defined as

$$\psi_{bi} = \frac{1}{2}F_m W_D = \frac{1}{2}F_m(x_n + x_p) \tag{5.18}$$

W_D may be given explicitly by eliminating F_m from (5.16) and (5.18) which gives for an abrupt two sided junction

$$W_D = \sqrt{\frac{2\varepsilon_0\varepsilon_r}{q}\left(\frac{N_A + N_D}{N_A N_D}\right)\psi_{bi}} \tag{5.19}$$

If the junction is much more heavily doped on one side, i.e. $N_A \gg N_D$ or vice versa then

$$W_D = \sqrt{\frac{2\varepsilon_0\varepsilon_r\psi_{bi}}{qN_D}} \simeq \sqrt{\frac{2\varepsilon_0\varepsilon_r k_B T}{q^2 N_D}\ln\left[\frac{N_D N_A}{n_i^2}\right]} \tag{5.20}$$

The depletion width decreases as the doping density in the semiconductor increases (Fig. 5.2). This becomes extremely important when designing advanced bipolar or FET transistors where the typical dimensions will be comparable to depletion widths. The doping concentration must therefore be increased as sizes are scaled down to reduce the depletion widths and therefore the active size of the transistor.

5.1.1 The Current Voltage Characteristics of a p-n Junction

In calculating the current voltage characteristics of a p-n junction, as voltages are applied across the junction, the chemical potentials and Fermi energies are different in different parts of the device. It is easier to define quasi-Fermi potentials and use these in place of the intrinsic potential. Therefore, rewriting the current density from (5.1) and (5.2)

$$J_n = -qn\mu_n\frac{\mathrm{d}\phi_n}{\mathrm{d}x} \tag{5.21}$$

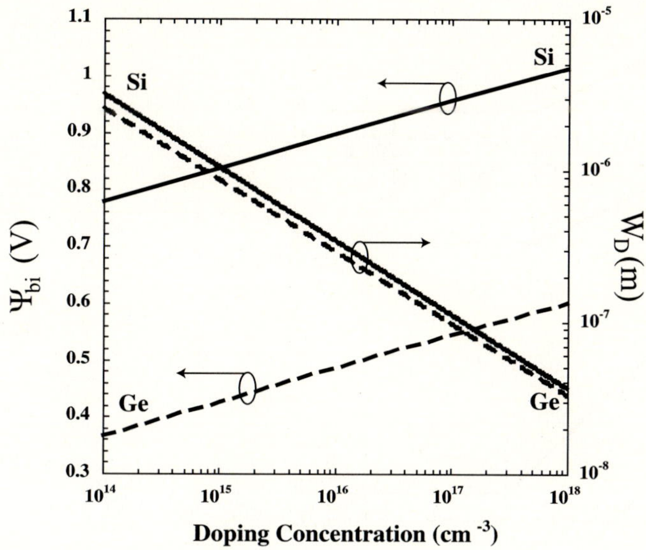

Fig. 5.2. The built in potential ψ_{bi} and the depletion width W_D for abrupt p-n junctions in both Si (line) and Ge (dotted line) where one side is heavily doped. The depletion width corresponds to the lightly doped side

$$J_p = -qp\mu_p \frac{\mathrm{d}\phi_p}{\mathrm{d}x} \tag{5.22}$$

where the quasi-Fermi energies for electrons and holes are defined as

$$E_{F_n} = E_i + k_B T \ln\left(\frac{n}{n_i}\right) \tag{5.23}$$

$$E_{F_p} = E_i - k_B T \ln\left(\frac{p}{n_i}\right) \tag{5.24}$$

where ψ_i is the intrinsic potential defined in terms of the intrinsic Fermi level, E_i as

$$\psi_i = -\frac{E_i}{q} \tag{5.25}$$

The pn product may be derived in terms of the quasi-Fermi energies as

$$pn = n_i^2 \exp\left[\frac{E_{F_n} - E_{F_p}}{k_B T}\right] \tag{5.26}$$

The spacial variation of the quasi-Fermi energies in a p-n junction are illustrated in Fig. 5.3. Far from the junction, $n = n_0$ and $p = p_0$ and so $E_{F_n} = E_{F_p} = E_F$. As is shown in Fig. 5.3, the applied voltage across the diode, V is

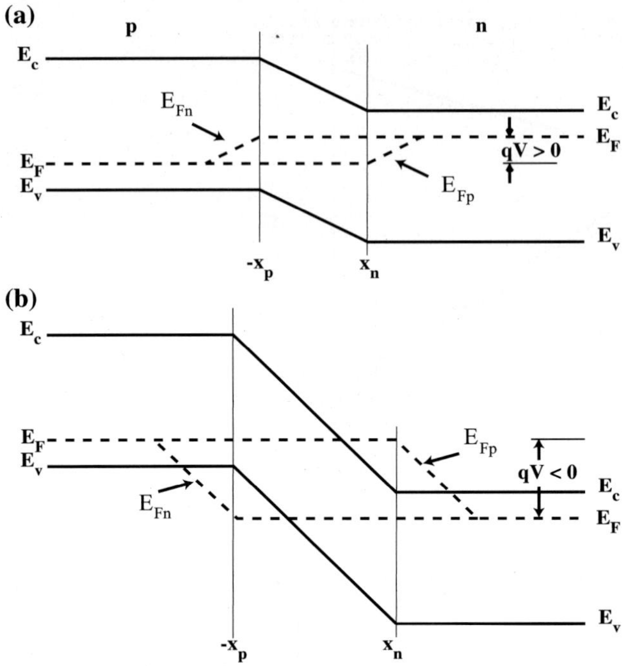

Fig. 5.3. The quasi-Fermi energies E_{F_n} and E_{F_p} in a p-n diode for (a) forward bias ($V > 0$) and for reverse bias ($V < 0$)

$$qV = E_{F_n} - E_{F_p} \qquad \text{for } -x_p \leq x \leq x_n \tag{5.27}$$

The diode is described as being forward biased when a positive voltage is applied to the n-region if the p-region is earthed and the diode is reverse biased if a negative voltage is applied. Figure 5.3 shows how the bands change with the two applied biases. Using (5.26) and (5.27), the electron density at the depletion layer edge on the p-side (at $x = -x_p$) is given by

$$n_p(-x_p) = \frac{n_i^2}{p_p(-x_p)} \exp\left[\frac{qV}{k_B T}\right] \tag{5.28}$$

$$= \frac{n_{p0}(-x_p)\, p_{p0}(-x_p)}{p_p(-x_p)} \exp\left[\frac{qV}{k_B T}\right] \tag{5.29}$$

$$\approx n_{p0}(-x_p) \exp\left[\frac{qV}{k_B T}\right] \tag{5.30}$$

The approximation in the last line is that the majority carrier density is large compared to the minority carrier density and therefore the majority carrier density is equal to its equilibrium value. Using similar arguements, the hole concentration at $x = x_n$ can be shown to be

$$p_n(x_n) \approx p_{p0}(x_n) \exp\left[\frac{qV}{k_B T}\right] \tag{5.31}$$

The continuity equation for electrons is

$$\frac{\partial n}{\partial t} = \frac{1}{q}\frac{\partial J_n}{\partial x} - R_n + G_n \tag{5.32}$$

$$= \frac{1}{q}\frac{\partial J_n}{\partial x} - \frac{n - n_0}{\tau_n} \tag{5.33}$$

where R_n is the electron recombination rate, G_n is the electron generation rate (see Sect. 3.7.2) and τ_n is the electron lifetime defined as

$$\tau_n = \frac{n - n_0}{R_n - G_n} \tag{5.34}$$

and n_0 is the electron concentration at thermal equilibrium.

Let us now consider the electrons in a uniform p-region of the p-n diode where the electric field is zero so that $F = 0$ and ${}^{\partial F}/_{dx} = 0$. Substituting (5.1) into (5.33) and solving for the diode in the steady state gives

$$\frac{\partial n_p}{\partial t} = D_n \frac{\partial^2 n_p}{\partial x^2} - \frac{n_p - n_{p0}}{\tau_n} = 0 \tag{5.35}$$

This equation is typically rewritten using the electron diffusion length defined as

$$L_n = \sqrt{\tau_n D_n} \tag{5.36}$$

and then solved using the boundary conditions of (5.30) and $n_p(x = \infty) = n_{p0}$ to give

$$n_p - n_{p0} = n_{p0}\left(\exp\left[\frac{qV}{k_B T}\right] - 1\right) \exp\left[\frac{-(x_p + x)}{L_n}\right] \tag{5.37}$$

and the electron current density is

$$J_n = \frac{qD_n n_{p0}}{L_n}\left(\exp\left[\frac{qV}{k_B T}\right] - 1\right) \tag{5.38}$$

Similarly, the n-type side of the device has

$$J_p = \frac{qD_p p_{n0}}{L_p}\left(\exp\left[\frac{qV}{k_B T}\right] - 1\right) \tag{5.39}$$

The total current is given by the sum of (5.38) and (5.39) which is

$$J = \left(\frac{qD_n n_{p0}}{L_n} + \frac{qD_p p_{n0}}{L_p}\right)\left(\exp\left[\frac{qV}{k_B T}\right] - 1\right) \tag{5.40}$$

This is the famous Shockley equation for an ideal p-n junction and is plotted in Fig. 5.4

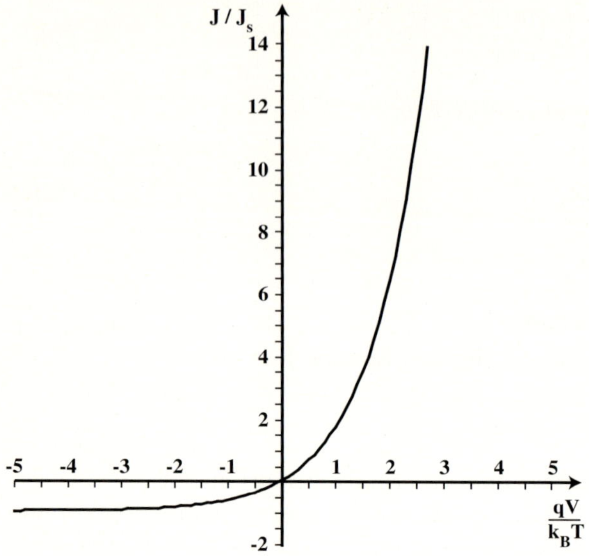

Fig. 5.4. The ideal current voltage characteristics of a p-n junction. J_s is the prefactor in (5.40)

5.2 The Silicon Bipolar Transistor

The silicon bipolar device is still heavily used in many applications such as analogue and high speed digital electronics. While it typically runs much faster than equivalent design-rule MOSFETs, it consumes more power than MOSFETs and therefore its integration density is smaller. The basic physics relies heavily on the p-n junction (Sect. 5.1) as the transistor consists of two back-to-back p-n junctions. While both p-n-p and n-p-n varieties can be designed and fabricated, we will concentrate on the n-p-n device as this is the easiest to modify into a heterojunction bipolar process which will be discussed in Sect. (6.4). Most semiconductor text books provide an overview of the operation of the bipolar transistor but the reference by Taur and Ning in Sect. 5.4 is recommended as it covers in most detail the latest advanced bipolar designs used in present day ULSI.

In the following section for equations, the subscripts, E,C and B will be used for the emitter, collector and base regions of the transistor respectively. The subscripts n and p will correspond to the semiconductor type so that n_n will correspond to the majority electron concentration in an n-doped region while n_p will correspond to the minority electron concentration in a p-doped region.

If Kirchoff's circuit laws are applied to a bipolar transistor then there are only two independent voltages and two independent currents (Fig. 5.5). There are a number of different ways a bipolar transistor can be configured

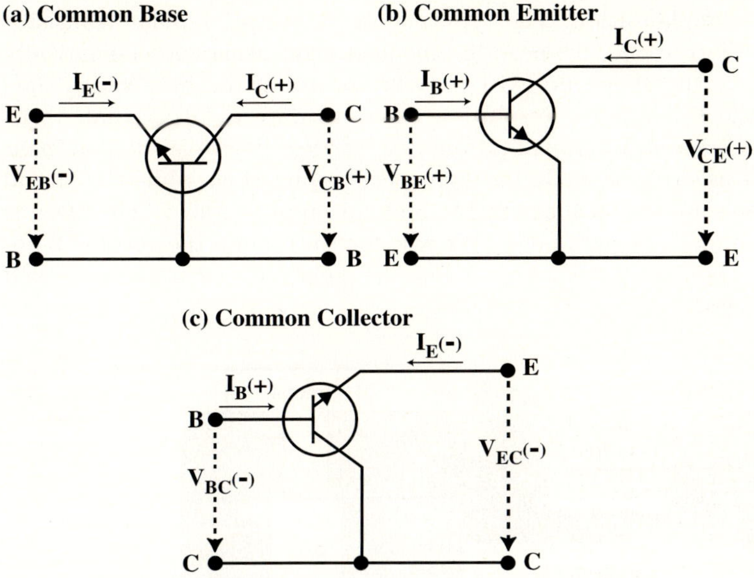

Fig. 5.5. The schematic diagram of different bipolar transistor configurations

in a circuit as an amplifier. These are the common base (Fig. 5.5(a)), common emitter (Fig. 5.5(b)) and the common collector (Fig. 5.5(c)). The most frequently used configuration is the common emitter that can be used for current, voltage or power amplification.

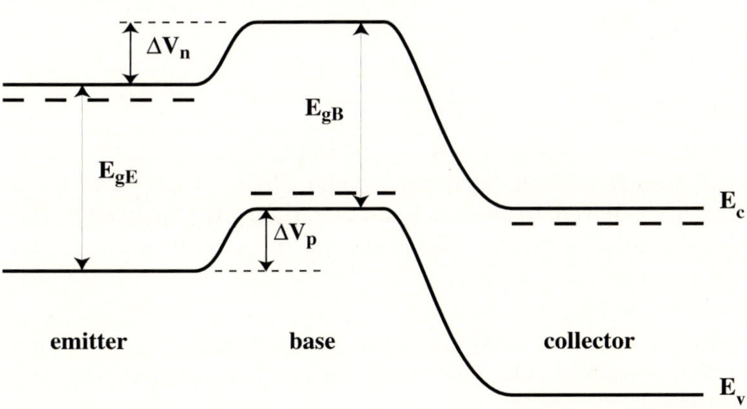

Fig. 5.6. The schematic energy band diagram of a bipolar transistor

Figure 5.6 shows a schematic diagram of the typical n-p-n structure along with the band structure and important parameters for operation. Bipolar transistors are normally vertical devices with the base and collector being formed by ion implantation. The typical poly-Si emitter bipolar transistor structure used in present advanced Si bipolar designs is shown schematically in Fig. 5.7. Lateral version do exist but the control of the base width then depends on the lithographic resolution which to date is larger than vertical dimensions created by implantation. The vertical device is the dominant version and is used particularly for high speed applications while the lateral device is mainly used in specific BiCMOS circuit applications where speed is not the most important parameter. We will concentrate on the vertical bipolar devices in the rest of this chapter but the equations can also be applied to lateral devices.

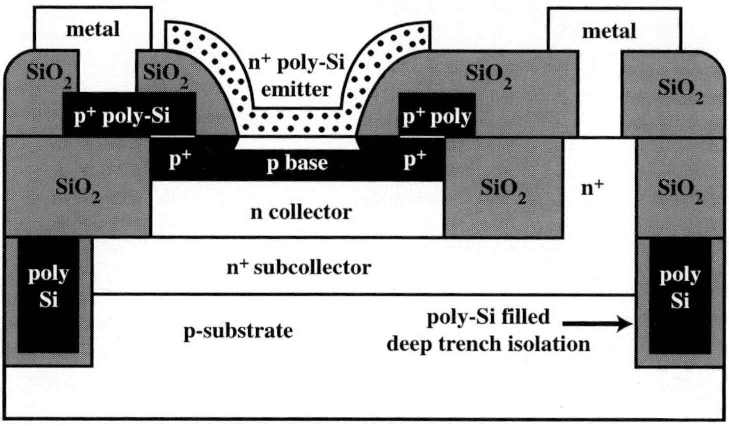

Fig. 5.7. The schematic diagram of a poly-Si emitter bipolar transistor which is typically used in production

The bipolar has a number of natural advantages:

1. Being a vertical transistor, the control of the important dimensions which determine the transit time of electrons through the transistor and hence the speed can be tailored to be much smaller than present lithographically defined lateral dimensions. Most bipolars use implantation for the formation of the base but epitaxy may also be used to form narrower base widths.
2. The entire area of the device conducts current unlike a FET where only a small 2D region conducts. Therefore large output currents per unit area can be maintained.
3. The output current is directly controlled by the input voltage through the relation $I_c = \exp\left(\frac{qV}{\gamma k_B T}\right)$ with $\gamma \approx 1$. The transconductance is therefore $g_m = qI_c/k_BT$ which is the highest obtainable in any three terminal device where the carrier density is modulated by the voltage (almost all conventional devices). The high transconductance produces low input

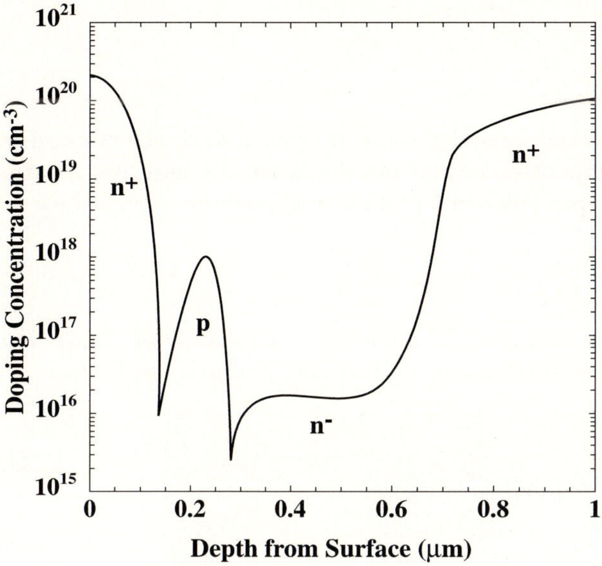

Fig. 5.8. The schematic diagram of the typical doping used in a silicon bipolar transistor

voltage swings which is essential for low power-delay products in logic applications.

4. The turn-on voltage of bipolar transistors, V_{BE}, is relatively independent of device size and also process variations since it is directly related to the built-in potential of a p-n junction, ψ_{bi}. This reduces variations in V_{BE} across a wafer as required for manufacture especially of logic circuits with small voltage swings.
5. The input capacitance of the bipolar transistor scales with the operating current so that the input capacitance adjusts to appropriate values at high or low current operation. Therefore device sizes do not need to be varied to match the driven load in many circuit designs.

Figure 5.8 shows a schematical diagram of the doping concentration normally found in a modern silicon bipolar transistor. The device is quite complicated and is designed with a self-aligned poly-silicon emitter to reduce parasitic capacitances in the system, thereby improving high speed performance.

It will become useful later if general expressions for the bipolar transistor operation are derived here. To start with, the current density equations which also include the drift effects of an electric field, F in addition to diffusion components as also used in the p-n junction ((5.1) and (5.2) are the starting point).

$$J_n(z) = qn\mu_n F + qD_n \frac{\mathrm{d}n}{\mathrm{d}z} \tag{5.41}$$

and

$$J_p(z) = qp\mu_p F - qD_p \frac{\mathrm{d}p}{\mathrm{d}z} \tag{5.42}$$

The electric field must be derived for the p-type base of the transistor. The same technique as that used for the p-n diode will be used so that the quasi-Fermi level in a p-type bulk semiconductor is given by

$$E_{Fp} = E_i - k_B T \ln\left[\frac{p_p}{n_i}\right] \tag{5.43}$$

using the definition of ψ_i in terms of the intrinsic Fermi energy of (5.25). The electric field is the negative gradient of the intrinsic potential and hence

$$qF = \frac{\mathrm{d}E_i}{\mathrm{d}z} \tag{5.44}$$

$$F = \frac{k_B T}{qp_p}\frac{\mathrm{d}p_p}{\mathrm{d}z} + \frac{J_p}{qp_p\mu_p} \tag{5.45}$$

Quasi-neutrality is required in the p-type base when the junction is forward biased since an electron current is flowing in the system and hence

$$p_p(z) = N_A(x) + n_p(z) \tag{5.46}$$

Taking the first derivative

$$\frac{\mathrm{d}p_p}{\mathrm{d}x} = \frac{\mathrm{d}N_A}{\mathrm{d}x} + \frac{\mathrm{d}n_p}{\mathrm{d}z} \tag{5.47}$$

the built-in electrical field in the base can be found by setting $n_p = 0$ and $p_p = N_A$ in (5.45) and realising that in a p-region that J_p is small so the second term of (5.45) is much smaller than the first term. The field is therefore given for vanishing injection as

$$F_0 = \frac{k_B T}{qN_A}\frac{\mathrm{d}N_A}{\mathrm{d}z} \tag{5.48}$$

The electron current in a nonuniformly doped p-type base region can be found by using (5.45) with low J_p, (5.41), (5.47) and (5.48) to give

$$J_n(z) = qn\mu_n F_0 \frac{N_A}{n_p + N_A} + qD_n\left(\frac{2n_p + N_A}{n_p + N_A}\right)\frac{\mathrm{d}n_p}{\mathrm{d}z} \tag{5.49}$$

The current density in the base as injected from the emitter can be found for low values of current where $n_p \ll N_A$ from (5.49) gives

$$J_n(z) \approx qn\mu_n F_0 + qD_n \frac{\mathrm{d}n_p}{\mathrm{d}z} \tag{5.50}$$

It is clear that this represents a drift component from the built-in electric field due to non-uniform doping in the base and a diffusion component which is due to the gradient in the electron concentration. If a large number of electrons is being injected from the emitter into the base then $n_p \gg N_A$ must be inserted into (5.49) which gives

$$J_n(z) \approx 2qD_n \frac{\mathrm{d}n_p}{\mathrm{d}z} \tag{5.51}$$

Therefore the built-in electric field is negligible compared to the gradient of the electron concentration which dominates in this regime.

The electron current density for those injected from the emitter into the base can also be calculated in terms of the carrier concentration in the base. We again take the current density as a function of the quasi-Fermi potentials as defined in (5.21)

$$J_n = n_p \mu_n \frac{\mathrm{d}E_{F_n}}{\mathrm{d}z} \tag{5.52}$$

The effective ionisation energy for impurities in a heavily doped semiconductor decreases as the doping increases which results in the effective bandgap decrease. Therefore as the band gap changes and therefore the density of states changes as the doping increases, the intrinsic carrier density in the base will also change. The standard method of incorporating this is to define an effective intrinsic carrier density, n_{ie} and define it such that all the effects of heavy doping are incorporated into the effective bandgap narrowing, ΔE_g such that

$$n_{ie}^2 = n_i^2 \exp\left[\frac{\Delta E_g}{k_B T}\right] \tag{5.53}$$

While there are many reported experimental results of bandgap narrowing, the scatter in the data is large and empirical expressions are typically used. For $N_D \simeq 7 \times 10^{23}\,\mathrm{m}^{-3}$, del Alamo has suggested the following empirical relation

$$\Delta E_g(N_D) = 18.7 \ln\left(\frac{N_D}{7 \times 10^{23}}\right) \qquad \text{meV} \tag{5.54}$$

where N_D is in units of m^{-3} and zero for lower doping concentrations.

We have already shown for the p-n junction that the quasi-Fermi energy for holes is approximately constant for a p-type region and therefore the gradient is zero in the base i.e.

$$\frac{\mathrm{d}E_{Fp}}{\mathrm{d}z} \approx 0 \tag{5.55}$$

Combining (5.52) and (5.55) before substituting for $E_{Fn} - E_{Fp}$ from (5.23) and (5.24) gives the electron current density in the base as a function of electron and hole concentrations as

$$J_n(z) = qD_n \frac{n_{ie}^2}{p_p} \frac{\mathrm{d}}{\mathrm{d}z} \left(\frac{n_p p_p}{n_{ie}^2} \right) \tag{5.56}$$

In a similar fashion, the hole current density in the emitter can be derived as

$$J_p(z) = qD_p \frac{n_{ie}^2}{n_n} \frac{\mathrm{d}}{\mathrm{d}z} \left(\frac{n_n p_n}{n_{ie}^2} \right) \tag{5.57}$$

Since there is very little recombination in the thin base layer of a transistor, the collector current in the system is mainly due to the electrons being injected from the emitter into the base region of the transistor where they are swept into the collector. Referring to Fig. 5.6, the start of the quasi-neutral base is denoted by $z = 0$ and the end by W_B (i.e. the base width). If there is negligible recombination in the base then J_n is independent of z and (5.56) can be integrated.

$$\frac{J_n}{q} \int_0^{W_B} \frac{p_p}{D_n n_{ie}^2} \mathrm{d}z = \left[\frac{n_p p_p}{n_{ie}^2} \Big|_{z=W_B} - \frac{n_p p_p}{n_{ie}^2} \Big|_{z=0} \right] \tag{5.58}$$

At $z = 0$ the electron concentration is given by (5.30)

$$n_p(0) = n_{p0} \exp \left[\frac{qV_{BE}}{k_B T} \right] \tag{5.59}$$

with V_{BE} the emitter-base, forward-biased voltage. At $z = W_B$, $n_p(W_B)$ is negligible compared to $n_p(0)$ and so the term evaluated at W_B in (5.58) can be ignored producing

$$\frac{J_n}{q} \int_0^{W_B} \frac{p_p}{D_n n_{ie}^2} \mathrm{d}z = \left[\frac{n_{p0}(0)\, p_p(0)}{n_{ie}^2(0)} \exp \left(\frac{qV_{BE}}{k_B T} \right) \right] \tag{5.60}$$

In modern bipolar designs, the base doping concentration peaks at the emitter-base junction. Providing this peak concentration is much larger than the minority carrier concentration, the hole concentration is approximately the thermal equilibrium value of $p_p(0) \approx p_{p0}(0)$ and so using $n_{p0}(0)\, p_{p0}(0) = n_{ie}^2(0)$. We therefore have the collector current density which is equal to J_n as

$$J_C = -\frac{q \exp \left(\frac{qV_{BE}}{k_B T} \right)}{\int_0^{W_B} \frac{p_p}{D_n n_{ie}^2} \mathrm{d}z} \tag{5.61}$$

To calculate the base current density, we need to integrate (5.57). If we assume that there is negligible recombination in the emitter except at the emitter contact $z = W_E$ as is normal for modern bipolar devices with a shallow or transparent emitter then the minority current density is independent of z. We therefore have

$$\frac{J_p}{q} \int_{-W_E}^{0} \frac{n_{n0}}{D_p n_{ie}^2} \mathrm{d}z = -\left[\frac{n_{n0} p_n}{n_{ie}^2} \Big|_{z=0} - \frac{n_{n0} p_n}{n_{ie}^2} \Big|_{z=-W_E} \right] \tag{5.62}$$

It is standard practice to define the surface recombination velocity of holes, S_p and use this in the hole current density as

$$J_p\left(-W_E\right) = -qS_p\left(p_n - p_{n0}\right)|_{z=-W_E} \tag{5.63}$$

The hole concentration at the emitter-base junction is given by (5.62)

$$p_n\left(0\right) = p_{n0}\left(x_n\right)\exp\left[\frac{qV_{BE}}{k_BT}\right] \tag{5.64}$$

with V_{BE} the base-emitter voltage. Substituting (5.63) and (5.64) into (5.62) and reducing produces

$$J_p \approx -\frac{q\exp\left(\frac{qV_{BE}}{k_BT}\right)}{\int_{-W_E}^{0}\frac{n_{n0}}{D_p n_{ie}^2}\mathrm{d}z + \frac{n_{n0}(-W_E)}{n_{ie}^2(-W_E)S_p}} \tag{5.65}$$

Realising that the $n_{n0} \approx N_E$, the emitter doping level, the base current density is written as

$$J_B = -\frac{q\exp\left(\frac{qV_{BE}}{k_BT}\right)}{\int_{-W_E}^{0}\frac{N_E}{D_{pE} n_{ie}^2}\mathrm{d}z + \frac{N_E(-W_E)}{n_{ieE}^2(-W_E)S_p}} \tag{5.66}$$

5.2.1 Operating Parameters and Important Figures of Merit

In this section we collect all the important parameters in the operation of a bipolar transistor and define a number of performance factors important for the operation of the transistor in a circuit.

The collector current density is obtained from (5.61). The collector current is a function of only the base region parameters and the emitter area, A_E.

$$I_C = A_E J_{C0}\exp\left(\frac{qV_{BE}}{k_BT}\right) = A_E\frac{qn_i^2}{G_B}q\exp\left(\frac{qV_{BE}}{k_BT}\right) \tag{5.67}$$

where J_{C0} is the saturated collector current density defined by

$$J_{C0} = \frac{Q}{\int\limits_0^{W_B}\frac{N_B}{D_{nB} n_{ieB}^2}\mathrm{d}z} \tag{5.68}$$

The base Gummel number, G_B is another parameter which is frequently used and defined as

$$G_B = \int\limits_0^{W_B}\frac{n_i^2}{n_{ieB}^2}\frac{p_p}{D_{nB}}\mathrm{d}z \tag{5.69}$$

For a uniformly doped base with width, W_B, the base Gummel number reduces to $N_BW_B \;/\; D_{nB}$ for low injection currents and a uniform bandgap.

The base current is obtained from (5.66) as

$$I_B = A_E J_{B0} \exp\left(\frac{qV_{BE}}{k_B T}\right) = A_E \frac{qn_i^2}{G_E} q \exp\left(\frac{qV_{BE}}{k_B T}\right) \quad (5.70)$$

with the saturated base current density

$$J_{B0} = \frac{q}{\int\limits_{-W_E}^{0} \frac{N_E}{D_{pE} n_{ieE}^2} \mathrm{d}z + \frac{N_E(-W_E)}{n_{ieE}^2(-W_E) S_p}} \quad (5.71)$$

and the emitter Gummel number

$$G_E = \int\limits_{-W_E}^{0} \frac{n_i^2}{n_{ieE}^2} \frac{N_E}{D_{pE}} \mathrm{d}z + \frac{n_i^2 N_E(-W_E)}{n_{ieE}^2(-W_E) S_p} \quad (5.72)$$

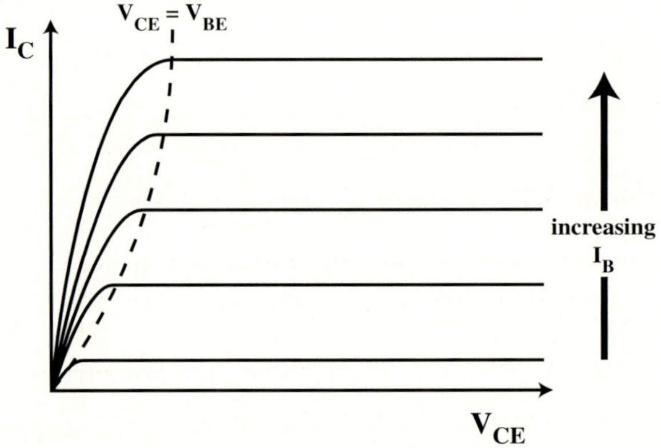

Fig. 5.9. A schematic diagram of the ideal collector current versus collector-emitter bias for different base currents. The dash line is where $V_{CE} = V_{BE}$ in the transistor

The ideal current voltage operation of a n-p-n bipolar transistor is plotted in Fig. 5.9. As the base current is increased, corresponding to increasing V_{BE}, the collector current in the diode increases for a given V_{CE} value. In the region of $V_{CE} < V_{BE}$, the collector-base diode is forward biased and the transistor is saturated. The collector current is predominantly determined by the difference between the electron current injector from the emitter into the base and the electron current from the collector into the base. For $V_{CE} > V_{BE}$, the collector-base diode is reversed biased and the transistor is said to be in the non-saturated region or normal forward-active mode. All the electrons injected into the base from the emitter reach the collector and the current is constant.

There are many factors which prevent the transistor characteristics from being ideal. Only part of the base and collector are involved in the transistor operation, the remaining parts are used to contact the rest of the circuit. These parts for contacts have parasitic resistances and contact resistances resulting from connecting the transistor to metal interconnects. The derivation of the ideal bipolar characteristics has also made a substantial number of approximations and many of these will result in additional terms and effects.

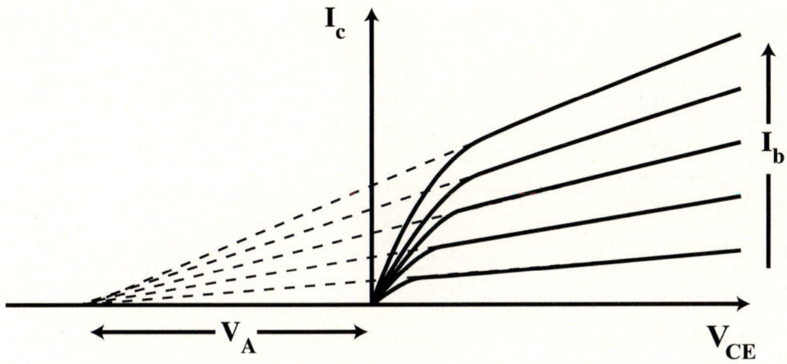

Fig. 5.10. A schematic diagram of the extraction of the Early voltage from the current voltage characteristics of a bipolar transistor

The Early voltage is another parameter used to rate the performance of bipolar transistors. In circuit models, a linear relationship is normally assumed between the collector current in the nonsaturation region and the collector-emitter voltage. The voltage at which the extrapolation of the collector current becomes zero is defined as the Early voltage, $-E_A$ as shown in Fig. 5.10. The complete definition is

$$V_A + V_{CE} = I_C \left[\frac{\partial I_C}{\partial V_{CE}}\right]^{-1} \tag{5.73}$$

For most transistors, V_{CE} is negligible compared to V_A and so

$$V_A \simeq I_C \left[\frac{\partial I_C}{\partial V_{CE}}\right]^{-1} \simeq \frac{qW_B N_B A_{BC}}{C_{BC}} = \frac{qN_B W_B W_C}{\varepsilon_0 \varepsilon_{Si}} \tag{5.74}$$

for a uniformly doped base and fully depleted collector of width, W_C.

The amount of amplification or gain is another important area of transistor performance and will now be defined. The static common-emitter current gain β is defined as

$$\beta = \frac{\partial I_C}{\partial I_B} = \frac{I_C}{I_B} = \frac{J_{C0}}{J_{B0}} = \frac{G_E}{G_B} \tag{5.75}$$

while the static common-base current gain is

$$\alpha = \frac{\partial I_C}{\partial (-I_E)} = \frac{I_C}{-I_E} \tag{5.76}$$

Since from charge conservation $I_C + I_B + I_E = 0$, it can be shown that the gains are related by

$$\alpha = \frac{\beta}{1+\beta} \tag{5.77}$$

$$\beta = \frac{\alpha}{1-\alpha} \tag{5.78}$$

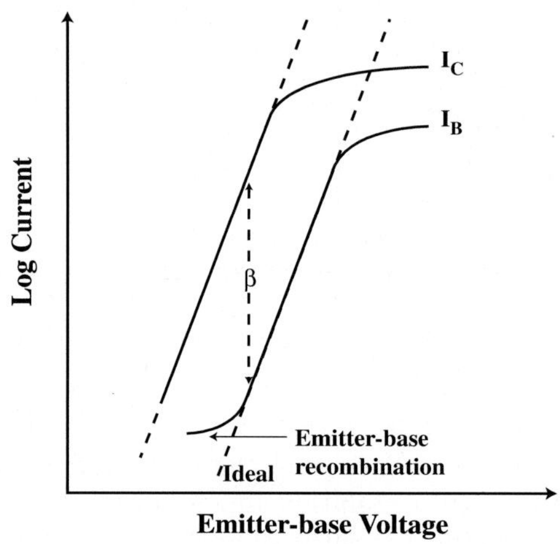

Fig. 5.11. A drawn Gummel plot showing typical I_C and I_B behaviour from the solid curves. The dotted lines show the ideal behaviour. At low V_{BE} the base current deviates from ideality due to recombination at the emitter-base interface. The common emitter current gain, β is shown as the difference between I_C and I_B

As an example for typical values in real transistors, if we assume that the emitter width is much larger than the diffusion length of holes in the emitter and the hole density in the base is much larger than the base electron concentration and equal to the base doping then an approximate value of the gain in a uniformly doped transistor is

$$\beta = \frac{n_{ieB}^2 D_{nB} N_E L_{pE}}{n_{ieE}^2 D_{pE} N_B W_B} \tag{5.79}$$

In modern bipolar transistors, N_E is typically $10^{26}\,\mathrm{m}^{-3}$, N_B is $10^{24}\,\mathrm{m}^{-3}$ and W_B is about $0.1\,\mu$m so that β normally results in values between about 100 and 1000.

The gain, β of the transistor can easily be obtained from a Gummel plot where I_C and I_B are plotted logarithmically against the emitter-base voltage (Fig. 5.11). At low V_{CE} the base current can saturate due to recombination at the emitter-base interface.

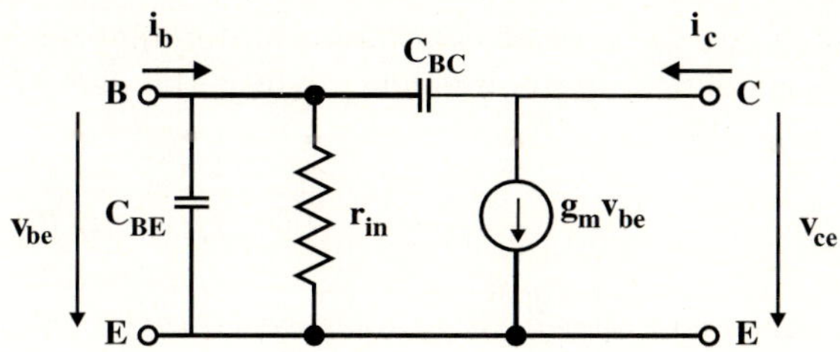

Fig. 5.12. The small signal equivalent circuit for determining the cutoff frequency of a bipolar transistor excluding parasitic resistances

We now want to consider the small signal equivalent circuit of the transistor to calculate the remaining important parameters. Fig. 5.12 shows the equivalent model where v_{be} is the small signal (ac) input voltage applied in series with the dc V_{BE}. i_c and i_b are the small changes in the collector and base current, respectively, and r_{in} is the input resistance. In this model, we ignore any parasitic resistance which in reality may limit the performance of the transistor, especially at high frequencies. The transconductance relates i_c to v_{be} and is defined as

$$g_m = \frac{\partial I_C}{\partial V_{BE}} = \frac{qI_C}{k_B T} \tag{5.80}$$

Bipolar transistors have the highest g_m of any present three terminal device where the carrier density is modulated by the voltage.

The small signal collector current and base currents are obtained from Fig. 5.12 as ($V_{ce} = 0$, short circuit)

$$i_c = g_m v_{be} - \mathrm{i}\omega C_{BC} v_{be} \tag{5.81}$$

and

$$i_b = \left(\frac{1}{r_{in}} + \mathrm{i}\omega C_{BE} + \mathrm{i}\omega C_{BC} \right) v_{be} \tag{5.82}$$

where C_{BC} is the base collector junction depletion layer capacitance, C_{BE} is the sum of the base-emitter junction depletion layer capacitance and the emitter diffusion capacitance and $\mathrm{i} = \sqrt{-1}$. For a frequency ω, the small signal common-emitter current gain is given by

$$\beta(\omega) = \frac{i_c}{i_b} = \frac{g_m - \mathrm{i}\omega C_{BC}}{1/_{r_{in}} + \mathrm{i}\omega \left(C_{BC} + C_{BE} \right)} \tag{5.83}$$

For most frequencies $\omega C_{BC} \ll g_m$ and so

$$\beta(\omega) \approx \frac{g_m r_{in}}{1 + \mathrm{i}\omega \left(C_{BC} + C_{BE} \right) r_{in}} \tag{5.84}$$

At high frequencies, the imaginary part of the denomimator dominates so the cut-off frequency, f_T which is defined as the frequency where the gain of the transistor equals 1 is given by

$$f_T \approx \frac{1}{2\pi} \frac{g_m}{C_{BC} + C_{BE}} \tag{5.85}$$

The analysis based on the equivalent circuit (Fig. 5.12) explains the current dependence of the transistor speed as load effect of the capacitances C_{BE} and C_{BC} but fails to explain the inner speed of the transistor as long as g_m and r_{in} are assumed frequency independent.
The cut-off frequency can also be defined in terms of the transit time of electrons across the device from the emitter to the collector as

$$\frac{1}{2\pi f_T} = \tau_{EC} = \tau_E + \tau_B + \tau_{hC} + \tau_C \tag{5.86}$$

where τ_{EC} is the emitter to collector transit time. This has constituent components of τ_E, the emitter transit time, τ_B, the base transit time, τ_C, the collector transit time and finally τ_{hC}, the delay due to the hole charge required to neutralise the charge of electrons traversing the base-collector depletion region.

The incremental gain at high frequencies can be estimated using a charge model as

$$\beta = \left|\frac{\mathrm{d}I_C}{\mathrm{d}I_B}\right| = \left|\frac{\mathrm{d}I_C}{\mathrm{d}Q_p}\frac{1}{\mathrm{i}\omega}\right| = \frac{f_T}{f} \tag{5.87}$$

and so it follows that

$$\tau_{EC} = \frac{\mathrm{d}Q_p}{\mathrm{d}I_C} \tag{5.88}$$

where $\mathrm{d}Q_p$ is the hole charge associated with an incremental input voltage and $\mathrm{d}I_C$ is the corresponding increment in the output current. It now follows from this result that the delay components in (5.86) correspond to charge stored in each of the different regions of the transistor.

The first component of emitter transit time is associated with the charges $\mathrm{d}Q_{EB}$ stored at the emitter-base depletion region edges which correspond to the electrostatic fields resulting at the p-n junction. If C_{BE} is the base-emitter capacitance then

$$\tau_{E1} = \frac{\mathrm{d}Q_{EB}}{\mathrm{d}I_C} = \frac{C_{BE}\mathrm{d}V_{BE}}{\mathrm{d}I_C} = \frac{C_{BE}}{g_m} \tag{5.89}$$

The second contribution is due to holes that are stored in the quasi neutral emitter

$$Q_E = \int p(x)\mathrm{d}x \tag{5.90}$$

Q_E increases as the emitter efficiency decreases, the surface-recombination velocity decreases, the emitter width increases or the bandgap is decreased. If the emitter width is quite large then it represents the minority charge stored in a p-n junction which is $Q_E = I_B \tau_{pE}$ where τ_{pE} is the lifetime of holes in the emitter. Since the gain of a transistor is the collector current divided by the base current, this therefore can be rearranged to give

$$\tau_{E2} = \frac{\tau_{pE}}{\beta} \tag{5.91}$$

The base delay, τ_B is predominatly associated with the hole charge stored in the quasineutral base region which neutralises the electrons traversing from the emitter to the collector. Again if we assume that only a diffusion element exists in this delay time and no built in electric field is present, τ_B is given by

$$\tau_B = \frac{Q_B}{J_C} = \frac{q \int_0^{W_B} n_p(z) \mathrm{d}z}{J_C} \tag{5.92}$$

which may be approximated by

$$\tau_B = \frac{W_B^2}{2D_n} + \frac{W_B}{v_e} \tag{5.93}$$

where v_e is the exit velocity for the electrons traversing the base. v_e may be approximated by the thermionic emission given by

$$v_e = \sqrt{\frac{k_B T}{2\pi m^*}} \tag{5.94}$$

The delay τ_{hC} resulting from the hole charge which neutralises the traversing electrons in the base-collector depletion region is given by

$$\tau_{hC} = \frac{1}{2} \int \frac{\mathrm{d}x}{v(x)} \tag{5.95}$$

where the integration must be carried out over the base-collector depletion region, W_{BC}. If the electrons traversing the base are travelling at saturation velocity, v_{sat} then

$$\tau_{hC} = \frac{W_{BC}}{2v_{sat}} \tag{5.96}$$

τ_C is the charging time of the base-collector capacitance is therefore an RC time constant. For the incremental base voltage, $\mathrm{d}V_{BE}$, the base-collector junction voltage for short circuited output terminals is given by $\mathrm{d}I_C(R_E + R_C + k_B T/qI_C)$ where the resistances correspond to the appropriate emiiter, base, collector region. It follows that

$$\tau_C = C_{BC}\left(R_E + R_C + \frac{k_B T}{q I_C}\right) \tag{5.97}$$

The various terms of τ_{EC} may be grouped in different ways, e.g. to give a current dependent part from (5.89) and (5.97) (see also (5.85)).

The cut-off frequency provides a good idea of the transit time of electrons through a bipolar transistor at low currents but it does not include the effects from the base resistance which are important in determining the transient response of the transitor. The maximum oscillation frequency, f_{max} is the frequency at which the unilateral power gain of the transistor becomes unity and is defined as

$$f_{\max} = \sqrt{\frac{f_T}{8\pi r_B C_{BC}}} \tag{5.98}$$

where r_B is the base resistance.

Both f_T and f_{max} should only be considered as qualitative indicators of the performance of bipolar transistors. The precise requirements of the transistor in different circuit can be substantially diverse and depend on the other circuit elements.

5.3 Metal Oxide Semiconductor Field Effect Transistors MOSFETs

While the bipolar transistor was the first solid state transistor amplifier to be fabricated and measured, today it is the MOSFET which is the dominant semiconductor device in manufacture. The first MOSFET devices were fabricated by Kahng and Atalla in 1960 at Bell Laboratories. The breakthrough occured finding a method of producing silicon dioxide with a low density of surface states which had previously screened the action of a gate on the semiconductor material. Perhaps the biggest breakthrough for the technology was the invention of the complemetary MOS (CMOS) architecture by Wanless and Sah in 1963. The CMOS architecture of using both n- and p-type transistors together, consumes power (apart from leakage currents) only when the state of the circuit is switched or active (see Sect. 10.1). Hence it is now possible to fabricate millions of MOSFETs on single silicon chips without requiring enormous efforts to cool the system.

As the MOSFET is the most important device in present semiconductor manufacture, more time and space will be devoted to it. We will first derive the basic capacitor equations of the device before deriving the equations explaining transistor action. Modern short-channel transistors will be investigated before limitations and problems will be discussed.

5.3.1 The MOS Capacitor

The basic operation for the MOS transistor structure relies on a simple capacitor structure created by the semiconductor substrate (almost always silicon), an insulator (typically silicon dioxide) and a metallic gate. The gate in most modern MOSFET transistors is made of poly-Si which is doped heavily enough to be metallic but metal gate transistors may also be fabricated and are typically used for rf applications due to the low gate resistance. The operation can be easily explained by considering the energy band diagrams for the system (Fig. 5.13).

Figure 5.13 shows the band structure as different voltages are applied to the gate for both n- and p-type silicon substrates. Without any voltages applied, the built-in potentials of the constituent layers will determine the band bending in the system as electrons will move to try and get the system to equilibrium. If we consider first the p-type substrate, by applying a negative gate voltage compared to the substrate, the Fermi level in the gate is pulled up and a quantum well is formed in the valence band and holes from the p-type substrate are accumulated at the interface of the Si/SiO_2. This is therefore termed an accumulation layer. By applying a positive bias to the gate, the bands in the semiconductor are pulled down depleting the number of holes near the Si/SiO_2 interface. A quantum well is formed in the conduction band and with sufficient positive bias applied to the gate, electrons form a two dimensional electron gas at the Si/SiO_2 interface. This is termed an inversion layer since the polarity of the carriers in the system has been inverted. One may use similar analysis for the n-type substrate, reversing the polarity of the carriers for each case.

Before deriving the appropriate equation, a number of parameters must first be defined. In all cases below, a p-type silicon substrate will be used so that the inversion layer forms an electron gas. The situation for a n-type substrate may be easily derived by changing the appropriate polarities in the equations. Using the conventions as shown in Fig. 5.14, the following parameters are defined:-

- E_i is the Fermi level for intrinsic semiconductors
- ψ_s is the surface potential
- ψ_B is the potential in the bulk of the semiconductor
- $q\psi_B = E_i - E_F$

The number of electrons and holes at the surface of the semiconductor i.e. at the Si/SiO_2 interface is

$$n_s = n_i \exp\left[\frac{q(\psi_s - \psi_B)}{k_B T}\right] \tag{5.99}$$

$$p_s = n_i \exp\left[\frac{q(\psi_B - \psi_s)}{k_B T}\right] \tag{5.100}$$

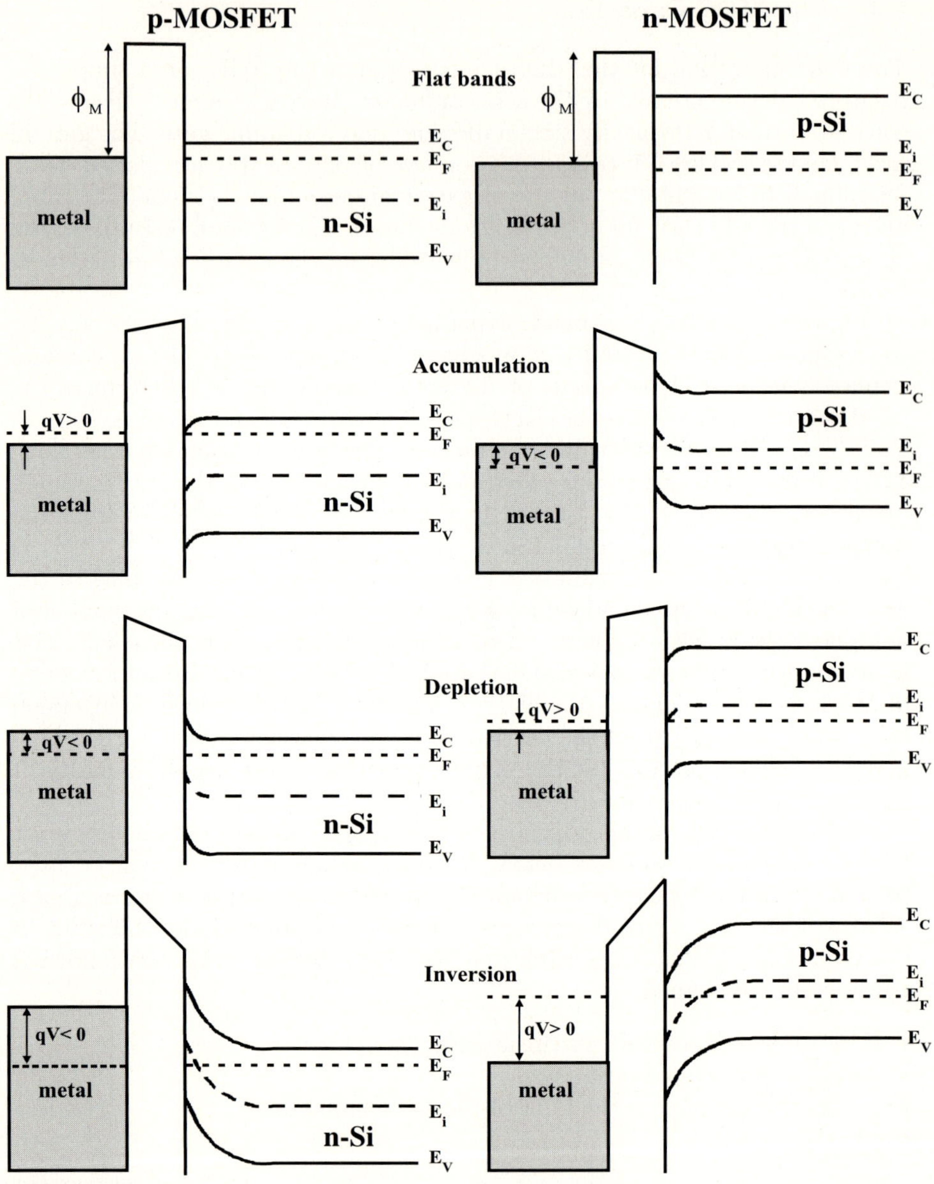

Fig. 5.13. Schematic diagrams of the operation of MOSFETs. Voltage V refers to the difference to the flat band voltage V_{FB}

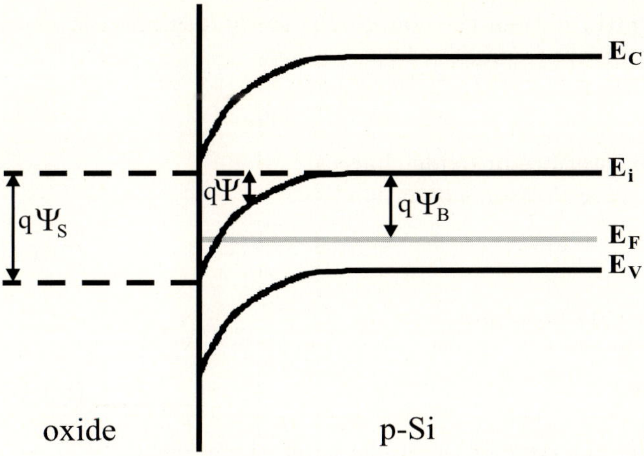

[scale=0.85]

Fig. 5.14. The definitions of the parameters used in a MOSFET

so we can define the different regimes through the surface and bulk potentials in the system as

- $\psi_s < 0 \longleftrightarrow$ accumulation of holes
- $\psi_s = 0 \longleftrightarrow$ flat-band condition
- $\psi_B > \psi_s > 0 \longleftrightarrow$ depletion of holes
- $\psi_B = \psi_s \longleftrightarrow$ mid-gap; $n = p = n_i$
- $\psi_B < \psi_s \longleftrightarrow$ inversion; E_i crosses E_F in the semiconductor
- $\psi_s \geq 2\psi_B \longleftrightarrow$ strong inversion

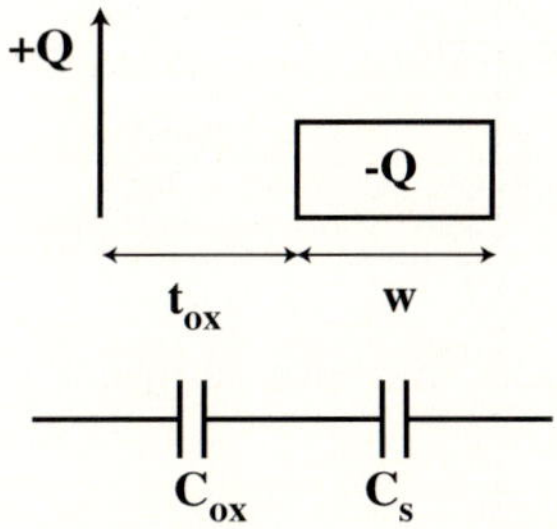

Fig. 5.15. The charge distribution in a MOSFET where the depletion layer is approximated by a lump of charge Q with width w. The oxide thickness is T_{ox}. Below is the equivalent small signal capacitance of the system which consists of the oxide capacitance, C_{ox} and the depletion capacitance C_s

The first important derivation is the capacitance of the depletion layer. Using the idea of image charges from electrostatics, if the gate has a charge $+Q$, then the depletion layer must have a charge $-Q$ (Fig. 5.15). If the depletion layer has a width, w then the charge due to the depletion layer in the semiconductor may be approximated by

$$-Q = -qwN_A \tag{5.101}$$

there we have neglected interface or oxide charges.

The next step is to solve Poisson's equation

$$\nabla.F = \frac{\rho}{\varepsilon_0 \varepsilon_r} \tag{5.102}$$

which on addition of (5.101) becomes

$$-\varepsilon_0 \varepsilon_s \frac{\mathrm{d}^2\psi}{\mathrm{d}z^2} = -qN_A \tag{5.103}$$

If the z-origin is taken at the edge of the undepleted silicon where
$\frac{\mathrm{d}\psi}{\mathrm{d}z} = 0$ and
$\psi = 0$ then the surface potential is given by

$$\psi_s = \frac{qN_A w^2}{2\varepsilon_0 \varepsilon_s} \tag{5.104}$$

Using the expression (5.101) and solving for Q gives

$$Q = \sqrt{2\varepsilon_0 \varepsilon_s q N_A \psi_s} \tag{5.105}$$

Now the small signal a.c. capacitance, C_s of the depletion layer is given by

$$C_s = \frac{dQ}{d\psi_s} = \frac{\varepsilon_0 \varepsilon_s}{w} \tag{5.106}$$

The total capacitance of the system is C_s and C_{ox} in series (Fig. 5.15) and with the applied voltage given by

$$V - V_{FB} = \psi_s + \frac{Q}{C_{ox}} \tag{5.107}$$

where V_{FB} is the flatband voltage. One may solve and find (for simplicity V_{FB} is set to zero)

$$\frac{C}{C_{ox}} = \left(1 + \frac{2V\varepsilon_{ox}^2 \varepsilon_0}{eN_A \varepsilon_s t_{ox}^2}\right)^{-1/2} \tag{5.108}$$

where t_{ox} is the gate oxide thickness.

Strong inversion is considered to occur when the induced carrier density in the inversion layer is equal and opposite to the bulk carrier density. It is at this point that there is sufficient charge in the inversion layer that it effectively screens out the effects of the gate voltage on the bulk of the semiconductor i.e. the depletion layer.

$$n_s = |N_A| = n_i \exp\left[\frac{q\psi_B}{k_B T}\right] \tag{5.109}$$

It is typically defined as $\psi_s = 2\psi_B$. The threshold voltage for strong inversion is therefore given by

$$V_T = 2\psi_B + \frac{Q_B}{C_{ox}} \tag{5.110}$$

Since the the inversion layer is now screening, the depletion layer width is effectively constant and at a maximum value. The bulk charge in the depletion layer is given by

$$Q_B = qN_A w_m \tag{5.111}$$

Here, w_m is the maximum width of the depletion region and is given by

$$w_m = \sqrt{\frac{4\varepsilon_s\varepsilon_0\psi_B}{qN_A}} \tag{5.112}$$

and the threshold voltage is given by

$$V_T = 2\psi_B + \frac{\sqrt{4\varepsilon_s\varepsilon_0\psi_B qN_A}}{C_{ox}} \tag{5.113}$$

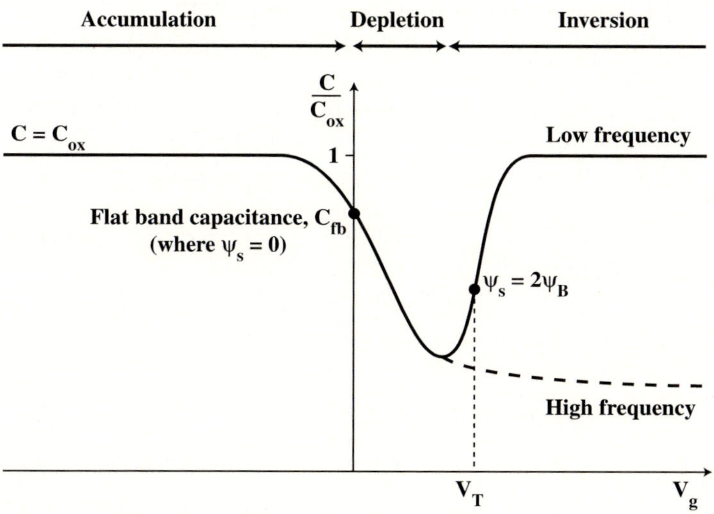

Fig. 5.16. The normalised capacitance against gate voltage, V_g for a MOS capacitor with the important regimes and points highlighted

Figure 5.16 plots out the typical capacitance from a MOS device. In accumulation, the substrate is conducting up to the oxide and so only the oxide capacitance is measured. this provides a convenient method of determining the oxide thickness. As the gate voltage is made more positive, the depletion layer is formed and the total capacitance of the system decreases. If the gate voltage is further increased then the inversion layer is formed. What is actually observed in the capacitance will depend, however, on the measurement frequency. At low measurement frequencies the capacity increases again to the value of oxid capacity. At higher frequencies the inversion charge modulation cannot follow the voltage modulation. The capacitance stays at the minimum value.

5.3.2 Carrier Transport in the MOS Transistor

In this section the current voltage properties due to the electron transport of the MOSFET will be derived. In Sect. 7.4, depletion mode devices will be discussed. In the following section, only enhancement mode devices will be discussed as these are technologically the most important. In addition, this section will only discuss the long channel behaviour of the devices. Later, problems and discrepancies with the results due to short channel effects (Sect. 5.3.6) will be reviewed.

The device can be considered to consist of four different regimes or regions as depicted in Fig. 5.17. In all cases a finite source-drain bias, V_{ds} is applied. The first region is where no bias is applied to the gate and hence no current can flow between the source and drain since the structure is n-p-n. This corresponds to the situation deplicted in region 1 of Fig. 5.17.

Below saturation in the linear regime (region 2 in Fig. 5.17) there is a continuous sheet of electrons in the inversion layer from source to drain. Current depends on Q_n is the inversion charge per unit area in the inversion layer. The carrier mobility of the electrons in the channel is μ_n. Thus the current per unit width is given by

$$\mu_n \left|Q_n\right| \frac{\mathrm{d}V}{\mathrm{d}x} \tag{5.114}$$

and therefore the potential gradient along the channel is

$$\frac{\mathrm{d}V}{\mathrm{d}x} \tag{5.115}$$

Using $I\mathrm{d}R = \mathrm{d}V$, the resistance of unit width is given by

$$\mathrm{d}R = \frac{\mathrm{d}x}{\mu \left|Q_n\right|} \tag{5.116}$$

In the saturation regime (region 4 in Fig. 5.17) the source end of the channel behaves as in the linear regime described above. At the drain end of the channel, however, the channel is depleted so carriers would not be available in equilibrium. Carriers entering from the source end are subject to an electric field that accelerates them towards the drain. The mobility of the

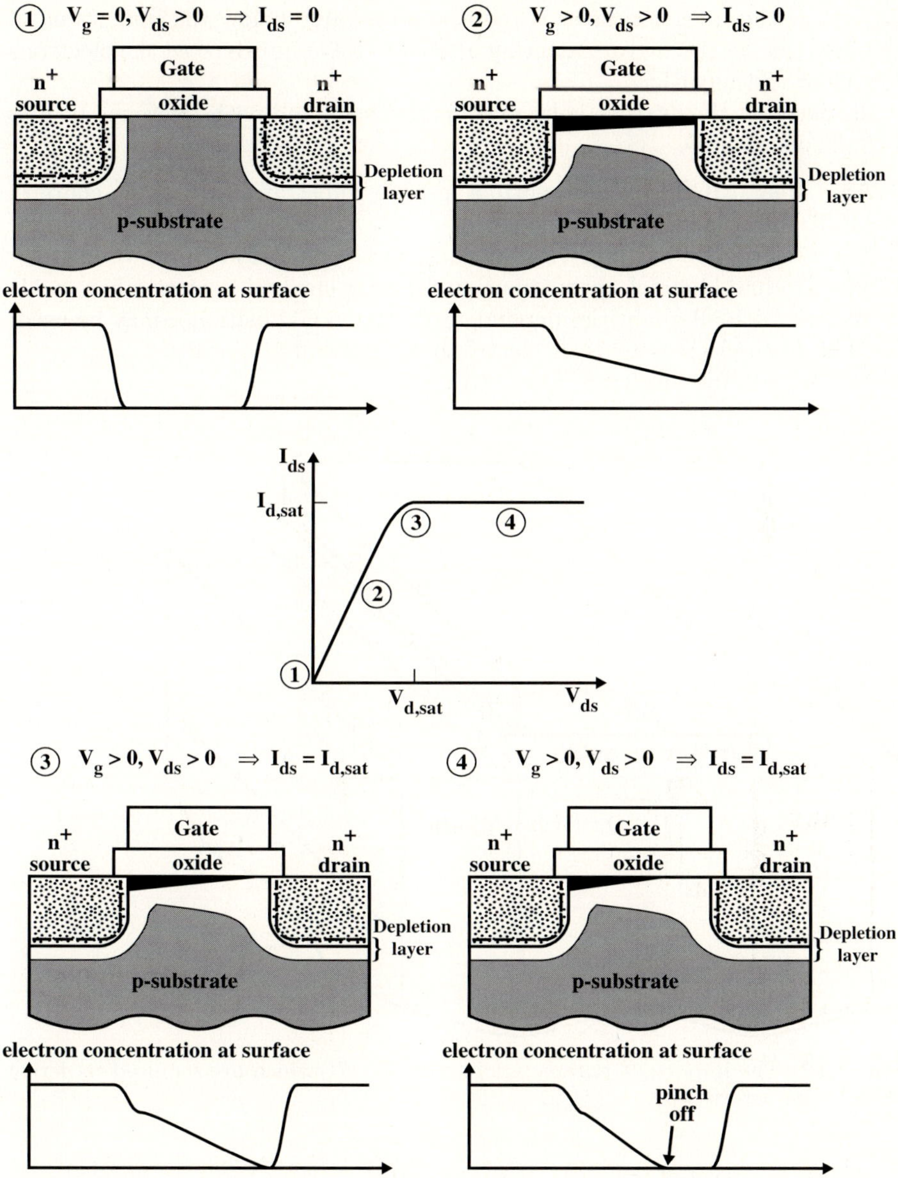

Fig. 5.17. The current voltage properties of a MOSFET along with the charge distribution along the device. Four different regimes are shown corresponding to 1. no applied voltage, 2. the linear regime, 3. pinch-off and 4. the saturation regime. The black region under the oxide corresponds to the inversion layer

carriers in the inversion layer is reduced to a value below the bulk value of mobility due to the extra scattering at the Si/SiO_2 interface as the electrons flow close to the surface.

In deriving the current voltage characteristics and with reference to Fig. 5.18, let us assume:-

1. An "ideal MOS" system with a p-type substrate (no oxide charges, $V_{FB}=0$).
2. The carrier mobility is μ_n in a channel of length L and width W.
3. The gradual channel approximation for long channel devices where a 1D model with all quantities dependent on the axial position x may be used. The axial electric fields $\ll$ electric fields normal to surface.
4. The source and substrate are at $0V$.

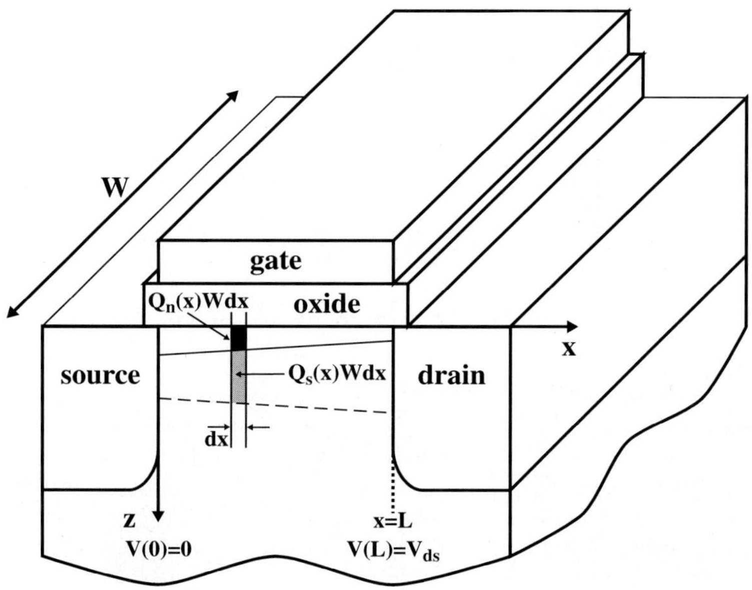

Fig. 5.18. The important parameters in a MOSFET which are required to derive the I-V characteristics

The source end of the channel has conditions equivalent to a MOS capacitor at strong inversion in equilibrium. Therefore

$$\psi_s(0) = 2\psi_B \tag{5.117}$$

$$w(0) = w_m \tag{5.118}$$

At other points in the channel, there is a resistive voltage drop of $V(x)$ in the channel so that

$$\psi_s(x) = 2\psi_B + V(x) \tag{5.119}$$
$$w(x) > w_m \tag{5.120}$$

The depletion charge per unit area is given by substituting 5.119 into 5.105 to give

$$Q_s(x) = -\sqrt{2\varepsilon_0\varepsilon_s N_A q\,(2\psi_B + V(x))} \tag{5.121}$$

but the inversion charge per unit area is

$$Q_n(x) = -(V_g - \psi_s(x))\,C_{ox} - Q_s(x) \tag{5.122}$$

and the total charge in the semiconductor is

$$Q(x) = Q_n(x) + Q_s(x) \tag{5.123}$$
$$= -(V_g - \psi_s(x))\,C_{ox} \tag{5.124}$$

Putting all these equations together produces

$$Q_n(x) = -\,C_{ox}(V_g - 2\psi_B - V(x)) + \sqrt{2\varepsilon_0\varepsilon_s N_A q\,(2\psi_B + V(x))} \tag{5.125}$$

The resistance of the channel may be derived by considering an element

$$\mathrm{d}R = \frac{\mathrm{d}V}{I_{ds}} \tag{5.126}$$

where the element has a length $\mathrm{d}x$ (Fig. 5.18) and the current between the source and the drain, I_{ds} corresponding to the applied bias, V_{ds} is

$$I_{ds} = W\,|Q_n(x)|\,\mu_n \frac{\mathrm{d}V}{\mathrm{d}x} \tag{5.127}$$

I_{ds} must be constant along the channel. Substituting in 5.125 and solving, this is reduced to

$$I_{ds}L = W\mu_n \Big[C_{ox}V_{ds}(V_g - 2\psi_B - {}^1\!/_2 V_{ds}) - \frac{2}{3}\sqrt{2\varepsilon_0\varepsilon_s N_A q}\left([V_{ds} + 2\psi_B]^{3/2} - [2\psi_B]^{3/2}\right)\Big] \tag{5.128}$$

Expanding $[V_{ds} + 2\psi_B]^{3/2}$ using a binomial expansion, the current can be approximated for small V_{ds} with respect to ψ_B and V_g by

$$I_{ds} \approx \frac{W\mu_n C_{ox}}{L}\left[V_{ds}\left(V_g - 2\psi_B\right) - \frac{V_{ds}}{2C_{ox}\psi_B}\sqrt{2\varepsilon_0\varepsilon_s N_A q}\left[2\psi_B\right]^{3/2}\right] \tag{5.129}$$

but rearranging (5.113) produces

$$V_T = \frac{1}{C_{ox}}\sqrt{2\varepsilon_0\varepsilon_s N_A q 2\psi_B} + 2\psi_B \tag{5.130}$$

and substituting into (5.129) produces the compact and standard form

$$I_{ds} \approx \frac{W\mu_n C_{ox}}{L}\left(V_g - V_T\right)V_{ds} \tag{5.131}$$

Equation (5.131) clearly shows the linear relationship between the source-drain current and voltages as shown in Fig. 5.17.

Before moving onto the saturation regime, a few important device parameters will be derived. The first is the channel conductance which is given by

$$g_D = \frac{1}{r_d} = \frac{\partial I_{ds}}{\partial V_{ds}} = \frac{W\mu_n C_{ox}}{L}\left(V_g - V_T\right) \tag{5.132}$$

The second important parameter in FET devices is the transconductance which is defined as

$$g_m = \frac{\partial I_{ds}}{\partial V_g} = \frac{W\mu_n C_{ox}}{L}V_{ds} \tag{5.133}$$

The transconductance is extremely important for many applications as it defines the ability of a MOS transistor to amplify the gate signal. The important result from (5.133) is that the transcoductance is inversely proportional to the gate length of the transistor.

We will now derive the appropriate equations for a MOSFET in the saturation regime (region 4 of Fig. 5.17). The electron density in the channel near the drain falls to zero which is termed pinched-off (region 3 of Fig. 5.17). If further bias is applied then the pinch-off point moves towards the source as shown in region 4 of Fig. 5.17. The current flowing from source to drain has therefore reached a maximum or has saturated. At pinch-off in saturation we have $Q_n(x) \rightarrow 0$. At the onset of saturation (the pinch-off point) $Q_n(L) \rightarrow 0$ with the voltage $V_{ds,sat}$, current $I_{ds,sat}$ and

$$C_{ox}\left(V_g - 2\psi_B - V_{ds,sat}\right) = \sqrt{2\varepsilon_0\varepsilon_s N_A q}\sqrt{V_{ds,sat} + 2\psi_B} \tag{5.134}$$

Equation (5.134) can be solved to find $V_{ds,sat}$ giving

$$V_{ds,sat} = V_g - 2\psi_B + \left(\frac{\varepsilon_0\varepsilon_s N_A q}{C_{ox}^2}\right)\left(1 - \left[1 + \frac{2V_g C_{ox}^2}{\varepsilon_0\varepsilon_s N_A q}\right]^{1/2}\right) \tag{5.135}$$

Through similar arguments the current at saturation can eventually be shown to be

$$I_{ds,sat} \approx \frac{W\mu C_{ox}}{2L}\left(V_g - V_T\right)^2 \tag{5.136}$$

and the corresponding transconductance

$$g_{m,sat} = \frac{\partial I_{ds}}{\partial V_g} \approx \frac{W\mu C_{ox}}{L}\left(V_g - V_T\right) \tag{5.137}$$

5.3.3 Threshold Voltage Control

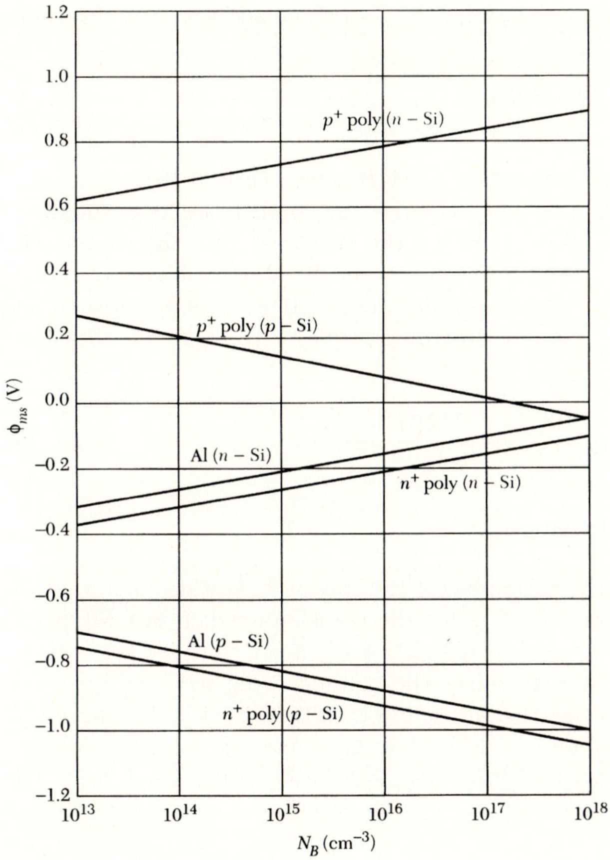

Fig. 5.19. The work function of different MOSFET gate materials as a function of silicon substrate doping (©Wiley)

The threshold voltage for a long channel MOSFET was derived in the previous section (equation 5.130). This derivation has ignored a number of factors that change the threshold voltage from the value in real devices. In particular the oxide may have some trapped charge in addition to the gate material having a different work function from the silicon substrate (Fig. 5.19). The charge in the oxide is normally divided into fixed charge, Q_f, mobile charge, Q_m (predominantly sodium or potassium which are fast diffusers in SiO_2), oxide trapped charge, Q_{ot} and interface trapped charge, Q_{it}. The interface trapped charge and fixed charge densities are typically $10^{14}\,\mathrm{m}^{-2}$ in modern MOSFETs. The fixed charge is believed to be related to broken Si bonds near the interface and cannot be charged or discharged over a wide range of surface potentials. The oxide trapped charge is related to defects created by energetic photons or electrons such as X-rays or injected hot electrons and can be removed by low temperature annealling ($\sim$400 °C). To take account of these effects a flat band voltage, V_{FB} must be added to the threshold voltage given by

$$V_{FB} = \phi_{ms} - \frac{Q_f + Q_m + Q_{ot} + Q_{it}}{C_{ox}} \tag{5.138}$$

where ϕ_{ms} is the metal-semiconductor work function (Fig. 5.19).

A second effect which may be used in some application is a substrate bias. If a reverse bias is applied between the substrate and the source, the depletion region is widened and therefore a larger threshold voltage is required to accommodate the larger charge in the 2DEG. If a voltage between the substrate and the source is applied of magnitude, V_{BS} then the threshold voltage becomes

$$V_T \simeq V_{FB} + 2\psi_B + \frac{\sqrt{2\varepsilon_0\varepsilon_r q N_A\left(2\psi_B + V_{BS}\right)}}{C_{ox}} \tag{5.139}$$

5.3.4 The Subthreshold Region

The subthreshold region, the region below the threshold voltage, is important as it determines the off current, I_{off} for the transistor which in CMOS will be a significant factor in the power dissipation of the circuit (see Sect. 10.1). The subthreshold slope also determines the voltage swing required to switch a transistor on and off. In this region the electron transport is dominated by diffusion rather than drift and is derived in the same way as the collector current in a bipolar transistor with homogenous base doping (Sect. 5.2). The MOSFET should be viewed along the Si/SiO_2 interface as a n-p-n transistor so that

$$I_{ds} = -qLWD_n\frac{\partial n}{\partial x} = -qLWD_n\frac{n\left(0\right) - n\left(L\right)}{L} \tag{5.140}$$

where $n\left(0\right)$ is the electron density at the source and $n\left(L\right)$ at the drain. The electron densities are given by (3.167) and for the present case

$$n(0) = n_i \exp\left[\frac{q(\psi_s - \psi_B)}{k_B T}\right] \quad (5.141)$$

$$n(L) = n_i \exp\left[\frac{q(\psi_s - \psi_B - V_{ds})}{k_B T}\right] \quad (5.142)$$

for a source-drain applied bias of V_{ds} with ψ_s the surface potential at the source. Substituting (5.141) and (5.142) into (5.140) gives

$$I_{ds} = qWD_n n_i \exp\left[\frac{-q\psi_B}{k_B T}\right]\left(1 - \exp\left[\frac{-qV_{ds}}{k_B T}\right]\right)\exp\left[\frac{q\psi_s}{k_B T}\right] \quad (5.143)$$

The surface potential can be approximated by $V_g - V_T$. Therefore when $V_g < V_T$ the source-drain current decreases exponentially as

$$I_{ds}\,(\text{subthreshold}) \sim \exp\left[\frac{q(V_g - V_T)}{k_B T}\right] \quad (5.144)$$

It is the slope of the subthreshold current against gate voltage which is one of the important MOSFET parameters and is defined as

$$S = \left(\frac{\mathrm{d}(\log_{10} I_{ds})}{\mathrm{d}V_g}\right)^{-1} \quad (5.145)$$

It is typically between 70 and 100 mV/decade. Typical subthreshold slopes are plotted in Fig. 5.24 for both long and short channel devices. These will be discussed in more detail when short channel effects are discussed in Sect. 5.3.6. The slope is increased if the interface trap density in the oxide is increased but apart from that and a small variance with the substrate bulk doping concentration, the subthreshold slope is relatively insensitive to most paramaters in a MOSFET. The slope is therefore only a function of temperature which has significant implications when the devices are scaled as will be discussed in the next section.

5.3.5 MOSFET Scaling

The basic idea behind scaling is that by reducing the dimensions of the transistor, the speed of a circuit may be increased while simultaneously increasing the density of transistors and reducing the power consumption of each transistor. The density of transistors in a circuit is limited by the rate at which heat can be removed from transistors. Therefore the lower the power level, the larger the density of devices which can be operated in a circuit. There are a number of different possibilities for scaling the transistor. The most important are constant field scaling and generalised scaling.

In constant field scaling, the device voltages and dimensions are scaled by a constant factor, κ, so that the electric field in the device remains constant (Fig. 5.20 and Fig. 5.21. The doping concentration in the substrate must be increased by a factor κ to keep Poisson's equation constant with respect to

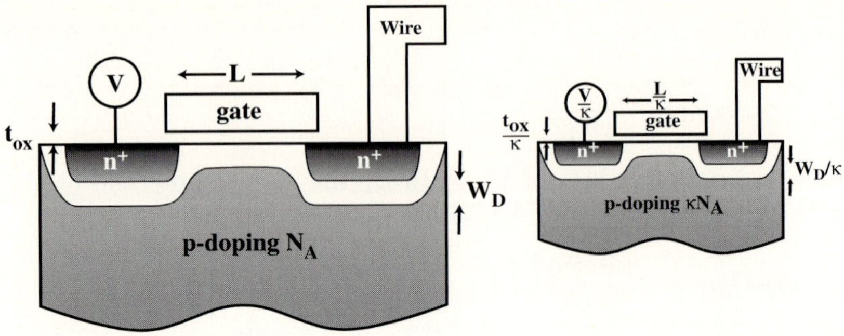

Fig. 5.20. The principle of constant electric field scaling of MOSFET device and circuit parameters

the scaling. All the capacitances in the device scale down by κ since they are proportional to the area and inversely proportional to the oxide thickness. The source-drain current per unit width is unchanged as the device is scaled down since the constant vertical electric field will result in the same mobility and carrier velocity. Therefore, since the width is scaled down by κ, the source-drain current is also scaled down by κ provided the operating voltage and threshold voltages are also scaled by a similar factor. The diffusion current per unit width, however, scales up by κ since the current is inversely proportional to the channel length, L. This has serious implications for the subthreshold behaviour at small gate lengths as will be discussed in Sect. 5.3.6. Since both the voltage and the current scale down by κ, the resistance of the channel remains constant as the device is scaled.

The main reason that scaling is so successful is due to the scaling of the circuit performance. The circuit delay time, neglecting parasitic resistances and capacitances along with interconnect delays, is proportional to the RC time constant which reduces by a factor κ. The power dissipation which is equal to the current-voltage product is reduced by κ^2. Therefore the power-delay product has been reduced by a factor of κ^3 and the power per unit area remains unchanged by scaling.

A number of problems occur with the constant electric field scaling. The built-in potentials in the device do not change with scaling as they are related to the bandgap of the semiconductor. The subthreshold slope is predominantly determined by the thermodynamics of the Boltzmann distribution of carriers and therefore cannot be scaled. In addition, there is a reluctance in the semiconductor industry to change the supply voltage between single generations of scaling as all other components in the system must then be changed. The actual electric field across the oxide has also been increasing rather than remaining constant. One method of circumventing these problems is not to scale the electric field as quickly as the device dimensions, a technique called generalised scaling.

Parameter	Constant Electric Field Scaling	Generalised Scaling
Device dimensions (L, W, t_{ox})	$1/\kappa$	$1/\kappa$
Doping concentration (N_A, N_D)	κ	$\alpha\kappa$
Voltage (V)	$1/\kappa$	α/κ
Electric field (F)	1	α
Carrier velocity (v)	1	1
Depletion-layer width (W_D)	$1/\kappa$	$1/\kappa$
Capacitance ($C=\varepsilon A/t$)	$1/\kappa$	$1/\kappa$
Current (I)	$1/\kappa$	α/κ
Channel resistance (R)	1	1
Circuit delay time ($\tau \sim CV/I$)	$1/\kappa$	$1/\kappa$
Power dissipation per circuit ($P \sim IV$)	$1/\kappa^2$	α^2/κ^2
Power-delay product per circuit ($P\tau$)	$1/\kappa^3$	α^2/κ^3
Circuit density ($\propto 1/A$)	κ^2	κ^2
Power density (P/A)	1	α^2

Fig. 5.21. The scaling of MOSFET device and circuit parameters

In general scaling, it is assumed that the electric field is being scaled by a factor α while the physical dimensions of the device are again scaled by κ. The changes to the major device and circuit parameters are shown in Fig. 5.21. The major and most serious change compared to constant field scaling is the increase in power dissipation, power-delay product and power density by α^2 if the devices are operated in velocity saturation. This requires more heat to be removed from the chip during operation.

As might be expected the gate length of a transistor cannot be scaled to smaller sizes indefinately. While physically the smallest possible size might be considered to be a channel of one atom in length, a number of significant problems are encountered at much longer channel lengths. A number of these are related to parameters which do not scale in the transistor or become impossible to control at the smallest gate lengths. These problems will be discussed in the next section.

5.3.6 Short Channel MOSFETs

From Sect. 5.3.5 it is clear that improved performance results from scaling down the device dimensions. Since a number of the device parameters such as bandgap, the built-in potentials and the subthreshold slope do not scale, a number of problems can occur if the length of the channel becomes too short. At least in theory, the channel length can be reduced indefinitely but in reality, a number of problems prevent scaling below specific lengths without modifications to the design of the MOSFET.

The short channel effect is normally defined as the decrease of the MOSFET threshold voltage as the channel length is reduced at a much faster level than that predicted from scaling rules as described in Sect. 5.3.5. The effects start to occur when the channel length becomes comparable to the depletion width in the vertical direction and the source-drain potential has a strong effect on the band bending over a substantial portion of the device. It is clear that if all parameters are kept constant in a device but the channel length is reduced then the electric field in the channel increases substantially in both the lateral and vertical directions of the transistor.

The threshold voltage calculated in (5.130) and (5.139) has assumed that only the gate effects the threshold voltage. In a short channel device, the doping in the source and drain can deplete out a significant fraction of the channel length and in doing so the threshold voltage is reduced. This is called threshold voltage roll-off and is shown in Fig. 5.22. As the source-drain voltage is varied, a substantial difference exists in the threshold voltage as the gate length is reduced.

A different method of considering short channel effects was originally derived by Troutman. He envisaged the enhancement mode transistor as a n-p-n device where the p-layer forms a barrier when the transistor is in the off state (similar to the model used to calculate the subthreshold current in Sect. 5.3.4. As the channel length is reduced then the barrier between the two n^+ Ohmic contacts is reduced, the so called drain-induced barrier lowering (DIBL). This creates a substantial increase in the subthreshold current and also reduces the threshold voltage compared to a long channel device. When a high drain voltage is appled to the short channel device, a further lowering of the barrier height occurs resulting in a larger decrease in threshold voltage.

Figure 5.24 shows subthreshold slopes for both long and short channel MOSFETs. For long channel devices the subthreshold slope is independent of drain voltages ($\geq 2k_BT/q$) as expected from (5.144). For the short channel devices the subthreshold slope is shifted to lower voltages as the source-drain voltage is increased. At even shorter gate lengths the substhreshold slope increases substantially with increasing source-drain voltage as the surface potential is controlled more by the drain than by the gate. Eventually the

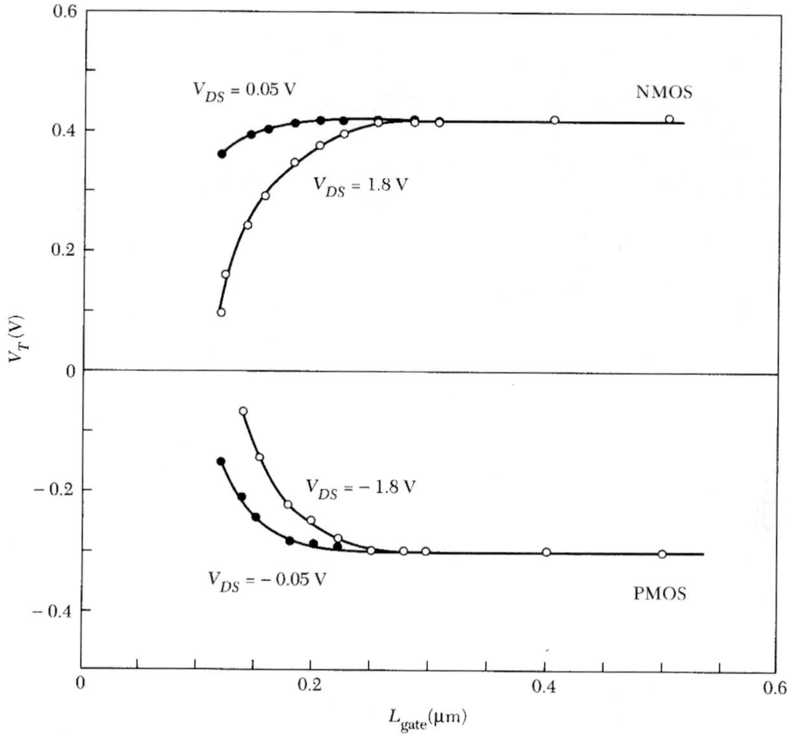

Fig. 5.22. The threshold voltage roll-off for 0.15 μm n- and p-channel MOSFETs (©IEEE)

device reaches a state where the gate has no control over the channel, the punch-through regime.

As was discussed in Chap. 3, the electrons in an inversion layer occupy discrete subbands in the triangular potential well formed between the oxide and the silicon layer. If classical electrostatics are considered then the electron density is at a maximum at the oxide / silicon interface but quantum mechanically, the electron wavefunction peaks at a small distance away from the interface. The energy levels of the nth subband are given by

$$E_n = \left(\frac{\hbar^2}{2m^*}\right)^{1/3} \left[\frac{3\pi qF}{2}\left(n+\tfrac{3}{4}\right)\right]^{2/3} \tag{5.146}$$

and the distance from the oxide surface to the centre of the wavefunction to the oxide surface is approximately

$$z = \frac{2E_n}{3qF} \tag{5.147}$$

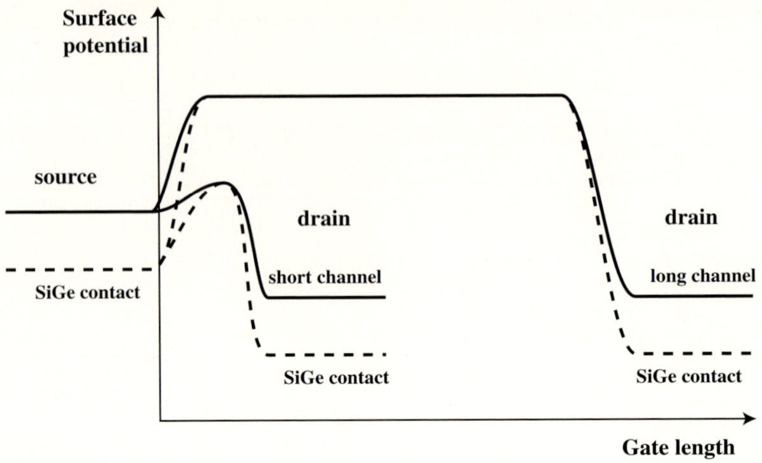

Fig. 5.23. The drain induced barrier lowering (DIBL) for two different gate length transistors. The vertical scale represents the surface potential. As the gate length is reduced, the effective barrier between the two contacts is reduced making it more difficult to switch the transistor off. By reducing the bandgap of the contacts with the addition of some Ge can lower the DIBL effect

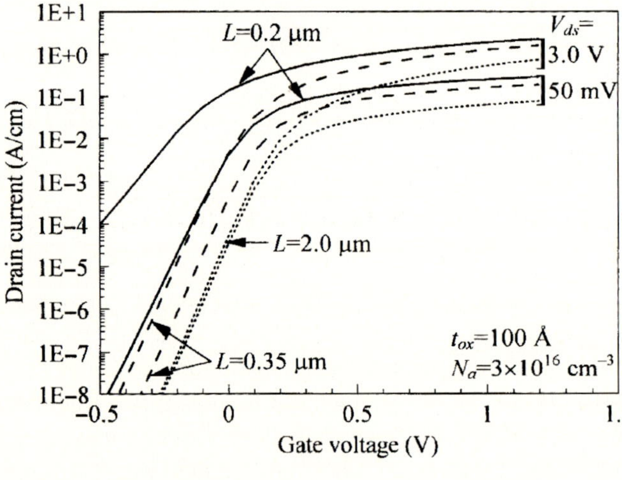

Fig. 5.24. The subthreshold slope of long and short channel MOSFETs showing short channel effects (©CUP)

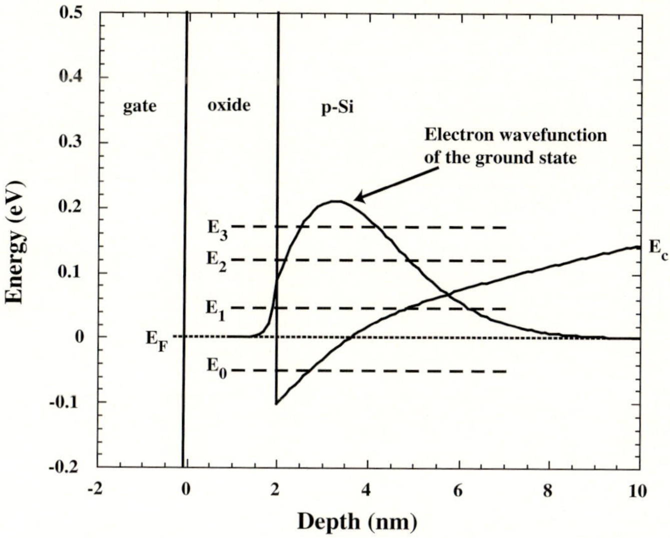

Fig. 5.25. The wavefunction of the ground state and quantised energy levels in a MOSFET with a 2 nm gate oxide. The gate voltage is 0.75 V. The width of the wavefunction is comparable to the thickness of the oxide for thin gate oxides

The wavefunction of an electron in the lowest subband of a MOSFET with a 2 nm gate oxide is shown in Fig. 5.25. The charge density in the device therefore is substantially different when the wavefunction is taken into account (Fig. 5.26). This distance can become significant when the oxide thickness is comparable in size.

The quantum mechnical properties of the electrons affect the operation of the MOSFET in a number of ways. The threshold voltage is increased at high electric fields since an increased amount of band bending is required to populate the first subband which lies above the conduction band edge. Once the inversion layer does form, a higher gate voltage is required to produce a given amount of charge in the inversion layer compared to the classical case.

The doping density of the substrate is increased for short channel devices to reduce DIBL and short channel effects. Too high a doping density of course will reduce the mobility in the channel from additional Coulombic scattering. One problem that becomes significant as the size of transistors is decreased is that the number of dopants in the channel becomes small and the statistical variation in numbers between devices can become significant. As an example take a $10^{24}\,\mathrm{m}^{-3}$ doped silicon substrate with a $0.1\,\mu$m gate length and width transistor. The mean number of dopants in the depletion region is 350 and the standard deviation is the square root of this which is 18.7. This corresponds to about a 5% variation between transistors which is quite significant.

One method which is employed to reduce the statistical dopant fluctuations and also help to keep a high mobility in the channel is the retrograde

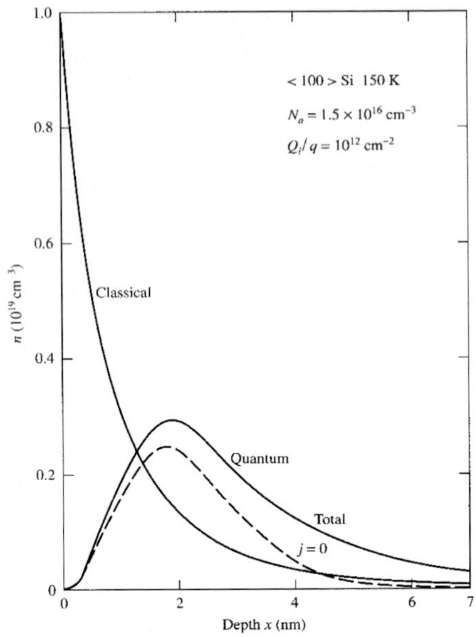

Fig. 5.26. The classical and the quantum mechanical charge density in a MOSFET as a function of depth. The dashed curve shows the electron density distribution for the lowest subband (©AIP)

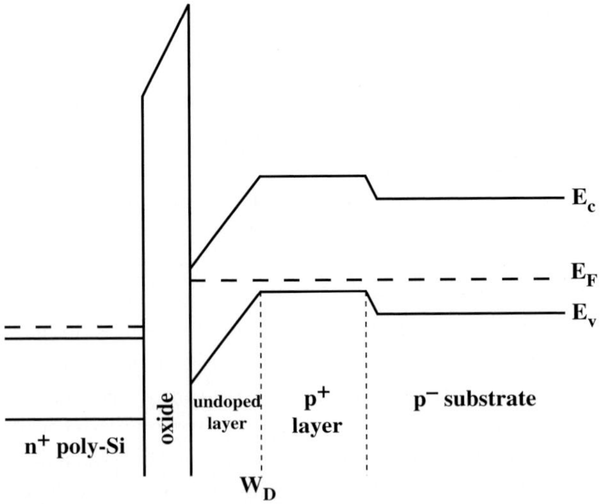

Fig. 5.27. The energy band structure of a retrograde doping profile in the substrate of a nMOSFET at threshold. The undoped region is designed to be the thickness of the maximum depletion width

doping profile (Fig. 5.27). This profile when designed correctly can eliminate all dopant fluctuation effects on the threshold voltage and is presently used in most production MOSFET designs.

5.3.7 MOSFET Device Performance

Each year the gate length of MOSFETs is reduced. Therefore any device results in this part of the book will be out of date before the publication of the book. It is informative to at least show some typical performance data as this should only have to be scaled to gain the performance for a different technology node. Fig. 5.28 shows the current-voltage plots for both a nMOS and (upside down) pMOS at the 0.13 μm technology node. It should be clear that there is a difference in the on current, I_{on} of about a factor of 2.5 which is related to the holes having a mobility lower than the eletrons by the same factor. The subthreshold plots for these transistors are shown in Fig. 5.29 where the I_{off} current of 3 nA per μm is marked for a source-drain voltage of 1.5 V.

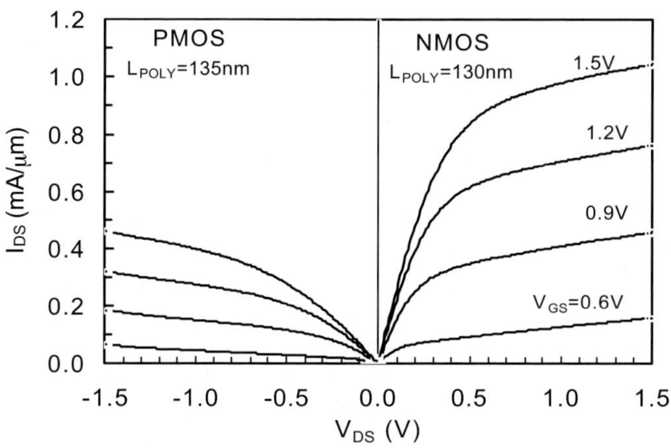

Fig. 5.28. IV characteristics from a 0.13 μm CMOS technology produced by Intel with a 1.5 nm thick oxide (©Intel)

5.3.8 Silicon On Insulator (SOI)

Silicon-on-Insulator or SOI is now in production for a number of different products including microprocessors for portable devices and low power application specific integrated circuits (ASICs). A good example is the IBM PowerPC^{TM} microprocessor many designs of which are now manufactured using a full SOI substrate to reduce the power dissipation for portable applications. There are a number of different types of SOI devices which can

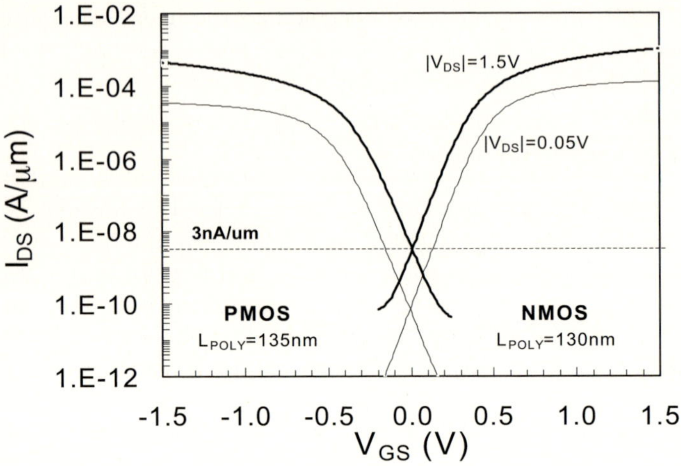

Fig. 5.29. Subthreshold characteristics from a 0.13 μm CMOS technology produced by Intel with a 1.5 nm thick oxide (©Intel). I_{off} is marked with the 3 nA μm^{-1} line where the subthreshold slopes of the n- and p-MOS transistors intersect at V=0 V

be fabricated and are aimed between two different opposite extremes (Fig. 5.30). The first is the reduction of the substrate capacitance by the incorporation of the buried oxide layer. This can either be transferred into a circuit improvement in speed of around 30% or more commonly is used to reduce the power dissipation of a circuit by around 30% compared to a bulk -Si substrate wafer. The main application for such devices has been low power portable devices such as mobile phones, laptop computers and personal digital assistants (PDAs). The second extreme is related to the inability to switch the channel of a MOSFET off as the channel length approaches sub-20 nm or so dimensions. This is predominantly related to the high doping required in both contacts and the substrate and at some gate length, the scaling relations discussed in Sect. 5.5.3 break down. By producing a partially or fully-depleted thin Si channel on top of oxide, either a back gate, double gate (Fig. 5.30(c)) or wrap-around gate (termed frequently as a FinFET)(Fig. 5.30(d)) can be fabricated which allows the channel to be completely depleted by the gates, thereby switching the transistor off.

There are two main techniques for the formation of SOI wafers. The first is termed Separation by IMplantion of OXygen (SIMOX) and involves the implantation of oxygen to the required depth in a bulk Si substrate before the substrate is heated to ~ 1250 °C. At this temperature SiO_2 is formed and most of the implantation damage is annealled out. The second is produced by wafer Bonding and Etch back SOI (BESOI). Here two wafers have thermal oxides grown on the surface and are then bonded together and annealled at high temperature (>1100 °C). One of the wafers is then thinned by either or a combination of chemical mechanical polishing and chemical etching. A

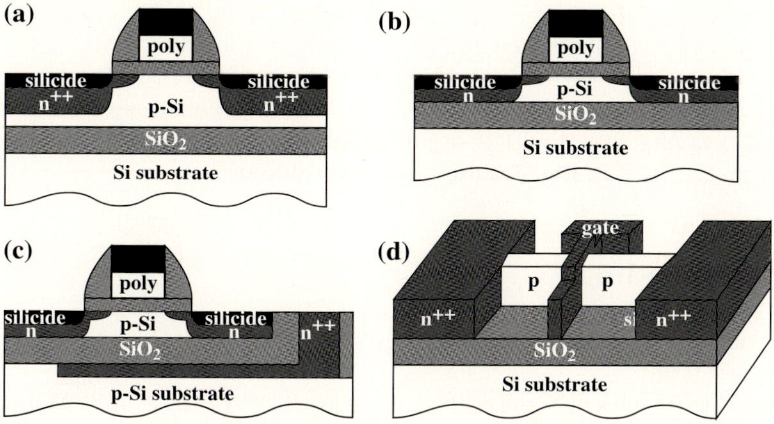

Fig. 5.30. (a) A partially depleted SOI MOSFET where the Si body above the oxide is only partially depleted by the gate electrode. (b) A full depleted SOI MOSFET where the gate electrode can fully deplete the Si body above the oxide. (c) one design of double gate MOSFET using SOI. (d) A wrap around gate or FinFET transistor

thin SiGe layer is frequently used as an etch stop layer allowing fast etches to stop accurately close to the required thickness for the thin top silicon layer. There is also a technique called Smart-Cut^{TM} which implants hydrogen into the wafer. On annealling the hydrogen layer forms bubbles and the wafer can then be easily fractured along these bubbles before being polished.

The general SOI transistor has smaller parasitic capacitances, smaller source/drain leakage and are less immune to soft errors caused by alpha particles. SOI transistors also allow higher speed and lower power consumption than bulk Si devices. In particular the buried oxide eliminates many leakage paths. The major problem with SOI apart from the higher substrate cost is the poor thermal conductivity of the SiO_2 layer that can produce thermal problems. There are two main type of standard MOSFETs which can be produced on SOI substrate. The first is a partially depleted body device (Fig. 5.30(a)). In this device the silicon layer above the oxide is thick and cannot be fully depleted by the transistor gate. One potential problem that can occur with this transistor is the "body" of silicon between the two contacts can become charged which may change the threshold voltage and change the digital on and off state voltages. This is termed the "body effect" and may be removed by arrange an earth contact to the body silicon.

The second type of standard SOI transistor is the fully depleted transistor (Fig. 5.30(b)). Here the body is so thin that there is also depletion from the bottom side due to the potential from the p-substrate. The threshold implant for the body is set so that the top gate can fully deplete the silicon body between the contacts when the transistor is switched off. As transistor gates are reduced below about 50 nm, it becomes substantially more difficult to switch the transistor off as the high doping levels between the two contacts cannot be achieved to deplete the channel. The fully-depleted SOI MOSFET can solve this problem by the additional depletion from the bottom interface. As gate-lengths are reduced below about 20 nm, this problem of switching a transistor off becomes worse and two gates such as Fig. 5.30(c) or a wrap around gate or FinFET as Fig. 5.30(d) may be required.

5.4 Further Reading

1. Y. Taur and T.H.Ning, Fundamentals of Modern VLSI Devices, CUP, Cambridge (1998)
2. S.M. Sze, The Physics of Semiconductor Devices, 2nd Edition, John Wiley and Sons, New York (1981)
3. S.M. Sze, Modern Semiconductor Device Physics, John Wiley and Sons, New York (1998)
4. C.Y. Chang and S.M. Sze, ULSI Devices, John Wiley and Sons, New York (2000)
5. S.M. Sze, Semiconductor Devices: Physics and Technology, 2nd Edition, John Wiley and Sons, New York (2002)

6. Heterostructure Bipolar Transistors - HBTs

The first major application for which SiGe heterolayers has been used is the heterostructure bipolar transistor or HBT. The original idea on the HBT actually appears in the original patent for the transistor. Herb Kroemer latter published a paper with more details about wide gap emitters along with the idea of grading the bandgap of the base to accelerate the carriers across the base. In the wide band gap emitter concept the top layer (emitter) consists of a material with wider band gap (e.g. GaAlAs) than in the underlying base (e.g. GaAs) and collector layers (table 6.1). A direct transfer of this concept to Si based electronics failed because of material problems. The most promising candidates, amorphous silicon (a-Si) and silicon carbide (SiC) suffered from emitter resistance and interface quality problems, respectively. The corresponding principle to the wide band gap emitter is the small band gap base (table 6.1) which contains two heterointerfaces and hence is also called double heterostructure bipolar transistor (DHBT). Silicon germanium is an appropriate material for the DHBT concept. Mainly it is used as pseudomorphic SiGe (strained SiGe, unstrained Si).

Table 6.1. Layer structure and material choice for the wide band gap emitter HBT and the small band gap base DHBT. The material A is with the wider band gap than the material B. The usual emitter configuration is assumed. Heterointerfaces are marked with HI.

Layer	HBT (wide gap emitter)		DHBT (small gap base)	
emitter	A		A	
	———	HI	———	HI1
base	B		B	
			———	HI2
collector	B		A	
substrate	B		A	

The first patent on a SiGe/Si small band gap base HBT was filed in 1977 by E. Kasper and P. Russer. The first SiGe HBTs had to wait until 1987 when the epitaxial growth became good enough to attempt transistors. Numerous other papers and groups followed. The first circuits were being sold in the market place in 1999, a little over a decade after the first devices were fabricated.

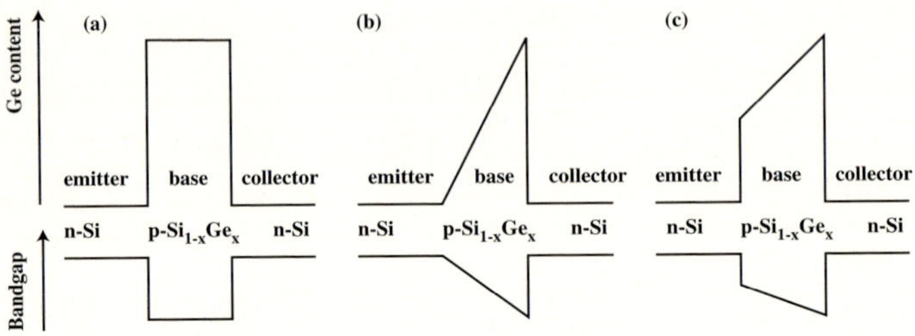

Fig. 6.1. A schematic diagram of the Ge profile in different types of HBTs and the resulting bandgap energy. (a) a box profile, (b) a graded or triangular profile and (c) a trapezoidal profile which is a combination of (a) and (b)

There are three main types of base design in SiGe HBTs as shown in Fig. 6.1: (a) box like (b) graded base and (c) trapezoidal base (a combination of the first two). Strictly speaking, a HBT is engineered to have a larger bandgap in the base at the emitter-base junction. Therefore only the box and trapezoidal profiles are true HBTs and the linearly graded profile is a graded base transistor. Each design has its own merits and problems. The precise design that will be used will depend on the application and the other components in the circuits. It is worthwhile deriving the important parameters of the HBT designs before comparing the advantages and disadvantages. We will only derive the box and linear graded parameters. This will demonstrate the two extremes and the trapezoidal profile results are a mixture between the two.

Figure 6.2 shows the band structure for a HBT. For the HBT, the diagram shows that the barrier for electrons travelling from the emitter to the collector is reduced. The additional valence band discontinuity between the base and the emitter reduces hole injection from the base into the emitter. One would therefore expect the collector current and hence the gain of the transistor to be much higher than a standard silicon bipolar transistor.

Figure 6.3 shows the band structure for a linearly graded base transistor. The linear grade in the base accelerates the electrons across the base but the emitter base junction remains identical to a standard silicon homojunction transistor. It is clear that the electric field accelerating carriers across the base will reduce the base transit time and hence increase the cut-off frequency of the transistor. The effect of the grading on the other parameters is not as obvious from the band structure and will require solutions of the appropriate equations.

Most of the important parameters in HBTs may be calculated by simply replacing the effective intrinsic carrier concentration in the base from the bulk silicon expression (Sect. 5.2) to one which accounts for the SiGe in the

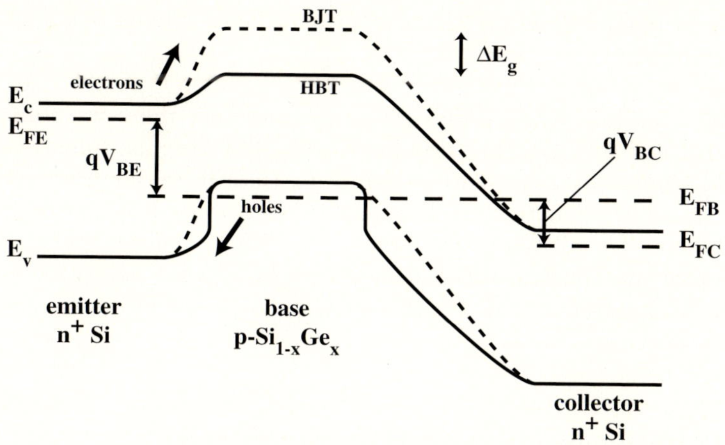

Fig. 6.2. A schematic diagram of the band structure for a Si bipolar transistor (BJT - dotted lines) and a box Ge profile HBT

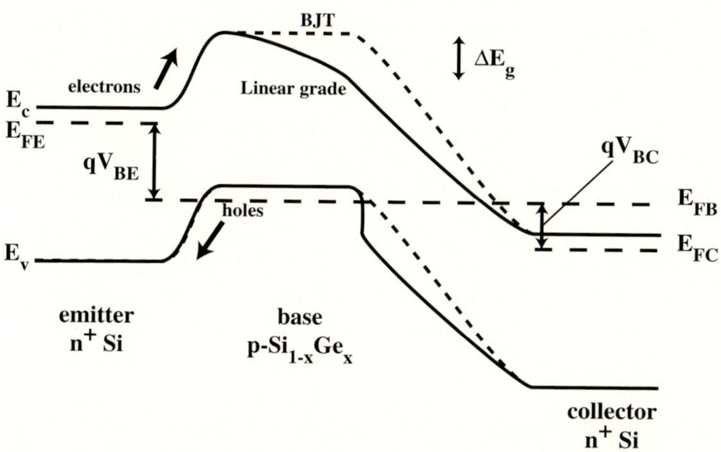

Fig. 6.3. A schematic diagram of the band structure for a Si bipolar transistor (BJT - dotted lines) and a linearly graded Ge-base profile transistor

base and then solving the formulae. The effective intrinsic carrier density in the base is given by (neglecting differences in the effective densities of states between Si and SiGe)

$$n_{ieB}^2\,(\mathrm{SiGe}) = n_{ieB}^2\,(\mathrm{Si}) \exp\left[\frac{\Delta E_g}{k_B T}\right] \qquad \text{box profile} \tag{6.1}$$

$$n_{ieB}^2\,(\mathrm{SiGe}) = n_{ieB}^2\,(\mathrm{Si}) \exp\left[\frac{\Delta E_g}{k_B T}\frac{z}{W_B}\right] \qquad \text{linear grade} \tag{6.2}$$

where ΔE_g is the bandgap reduction for the HBT and the maximum bandgap reduction for the linear graded base.

The basic advantage of the HBT is given by the different barrier for electron and hole injection which allows the device designer more freedom. A SiGe HBT with the same doping profile as a Si BJT (bipolar junction transistor) would exhibit a higher current gain, β_0, than its Si counterpart alone. We will, therefore, call this Ge box profile HBT design as the high current gain HBT in the following section. Of significantly more practical importance is producing a transistor design where the current gain can be traded off for the high speed. On this high speed design we will first discuss the SiGe HBT properties.

6.1 Trade-off between current gain and speed

From the introduction the reader will understand that the SiGe-HBT which has the same doping and widths as the Si-BJT is superior to the latter in current gain. But the success of the SiGe-HBT is based on a trade off between current gain and speed and the understanding of the real HBT is often blurred by this trade off. We start with explaining the speed disadvantages of the Si-BJT and continue with analysing the fundamental steps in the mentioned trade off.

The fundamental speed limit of the Si-BJT is given by its base design. In order to get a reasonable current gain β_0 the base doping $N_A^{(B)}$ has to be roughly two orders of magnitude lower than the emitter doping $N_D^{(E)}$. The limited base doping levels of about $10^{18}/cm^3$ with the thin base thickness (around 100nm) result in a low base conducitivity measured as a high base sheet resitivity R_{ShB} of around 4kΩ/sq. A high base resistance R_B influences directly the maximum oscillation frequency f_{max} but also f_T measurements indirectly because with high base resistances the base signals are reduced which leads to current crowding at the emitter edges except in emitter stripes with very low stripe width b_E. The progress in Si-BJT speed was therefore linked to the shrinking of lateral dimensions allowing higher base sheet resistivities to obtain the same base resistances .

With SiGe-HBTs the base doping $N_A^{(B)}$ is not more limited by the obtainable current gain because gain is produced by the E-B heterointerface. The trade off follows the scheme given below

- **Step1:** Increase of the base doping level $N_A^{(B)}$ by typically an order of magnitude. The upper limit of the increase is given by $exp(\Delta E_g/k_BT)$ but practically often the increase is limited by the outdiffusion of boron in following process steps. Directly opposed high doping levels in a p-n junction lead to tunneling currents which are suppressed by a thin low doped emitter (LDE) between base and usual high doped emitter (HDE).

This LDE is frequently unmentioned and hidden in the specific process sequence of the producer.

- **Step2:** Reduction of the base thickness w_B by a factor two to five. The corresponding decrease of the base transit time τ_B (mainly following a w_B^2 law) increase both, f_T and f_{max}. The reduction of τ_B lets the collector transit time τ_C as dominating time constant in many HBT designs.
- **Step 3:** Reduction of the collector thickness w_{BC} decreases the collector transit time τ_C. The gain in τ_C is partly compensated by the inherent $(R_E+R_C)C_{BC}$ time constant which increases with w_{BC} ($C_{BC}\sim 1/w_{BC}$). The optimum design depends on the technological parameters emitter and collector resistance, and base area A_{BC}. But, the thinner collector reduces the collector-base breakdown voltage BV_{CB0} and the collector-emitter breakdown voltage BV_{CE0}.
- **Step 4:** Increase of the collector doping $N_D^{(C)}$.
  The collector width w_{BC} is depleted under operation conditions (the undepleted collector width would otherwise increase the collector resistance). The thinner collector width allows higher collector doping levels to be depleted which in turn shifts the modified Kirk effect to higher current levels. Higher current levels reduce the input time delay ($\sim 1/I_C$). The Kirk effect limits ultimately the useful current density by an effective base extension into the collector above a critical current value. In a DHBT the base extension is additionally influenced by the second interface, therefore it is termed modified Kirk effect.

Table 6.2. HBT layer design steps for trade off β vs f_T (see text for more details).

No	step	advantage	disadvantage
1	$N_A^{(B)}\uparrow$	$R_{ShB}\downarrow$,$f_{max}\uparrow$	tunneling circuits
1a	LDE	avoids tunneling	
2	$w_B\downarrow$	$\tau_B \rightarrow f_T\uparrow$, $f_{max}\uparrow$	
3	$w_{BC}\downarrow$	$\tau_B \rightarrow f_T\uparrow$, $f_{max}\uparrow$	$\tau_{RC}\uparrow$, $BV_{CE0}\downarrow$
4	$N_D^{(C)}\uparrow$	$J_{crit}\uparrow$, $f_T\uparrow$, $f_{max}\uparrow$	

6.2 The High Speed SiGe HBT

In order to simplify the considerations but without loss of generality we assume an emitter up finger geometry (Fig. 6.4) with abrupt and flat doping and Ge profiles (Fig. 6.5). The emitter finger with area A_E (finger length l_E times finger width b_E) defines the current injection $I_E = A_E \times J_E$. For high speed operation the current density $J_E \cong J_C$ is driven to high values near the onset of the modified Kirk effect. The corresponding power consumption P in the transistor heats the transistor above the ambient temperature (self heating). Transistors with smaller finger width, b_E, may be driven to higher current densities for the same junction temperature. On one side the base contact is typically self-adjusted so that the sum of the base finger width, b_B, and emitter finger width, b_E, roughly define the collector finger width, $b_C \cong b_E + b_B$. Another frequently used design has the base contacts on both sides of the emitter which results in a smaller base resistance, R_B, but higher collector-base junction capacitance because then the collector finger width is roughly given by $b_C \cong b_E + 2b_E$. The collector is contacted via a highly doped subcollector (typical sheet resistivities of 10 to 25 $\Omega\,\mathrm{cm}^2$) which is fabricated as a buried layer before the epitaxy of the n-collector. The collector contact is separated from the base contact by a distance s to reduce capacitive coupling between the input signals at the base and the output signals at the collector in the common emitter circuit.

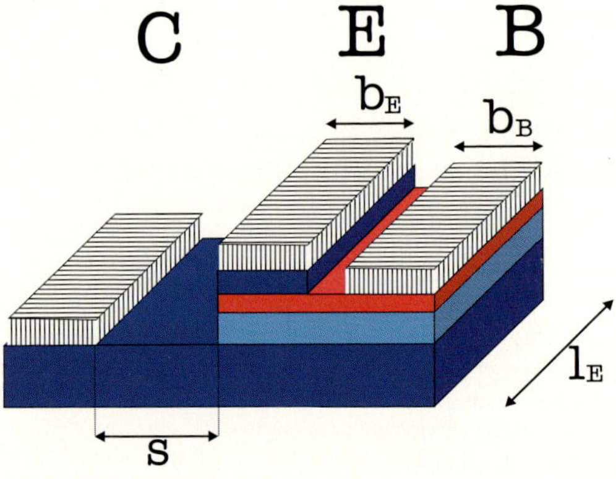

Fig. 6.4. A sketch of the SiGe-HBT design for an emitter (E) up geometry with a base contact on one side (B). The n-collector (C) is contacted via a highly doped n^+-subcollector. Regions with n-doping are blue coloured (n^+ emitter and n^+ subcollector darkblue; n-collector light blue) and the p-SiGe region is coloured red

The vertical structure consists of a two stage emitter, the p-type SiGe base and the collector-subcollector sequence. The high doped emitter (HDE) provides the reserve of electrons important for injection with high current gain, β_0, whereas the emitter-base junction capacity C_{BE} and the amount of the trap assisted tunnelling current (excess current) are determined by the low doped emitter (LDE). The p-doped SiGe base is cladded on both sides by intrinsic or n-type SiGe spacers to avoid out diffusion of the p-doping into the Si-layers which has very detrimental effects on the HBT function. The n-collector is optimally designed to be depleted under the operating voltage V_{CE}. A thicker collector would increase the collector resistance R_C

and a thinner collector would increase the base-collector capacitance C_{BC}. The sub-collector functions as in a BJT with the sheet resistance R_{Sh} as most important property. For the chosen flat profile the sheet resistance is given by

$$(R_{Sh})^{-1} = q \int_z n(z)\mu(z)\mathrm{d}z = qN_{D,SubC}W_{SubC}\mu_p \tag{6.3}$$

where $N_{D,SubC}$ is the donor concentration in the subcollector with width, W_{SubC}.

Table 6.3. Fundamental differences in the designs of a high speed Si-BJT and a SiGe HBT.

	LDE	SiGe	typical base doping	width	typical collector doping
Si-BJT	no	no	$10^{18}\,\mathrm{cm}^{-3}$	100 nm	$3 \times 10^{16}\,\mathrm{cm}^{-3}$
SiGe-HBT	yes	yes	$10^{19}\,\mathrm{cm}^{-3}$	30 nm	$3 \times 10^{17}\,\mathrm{cm}^{-3}$

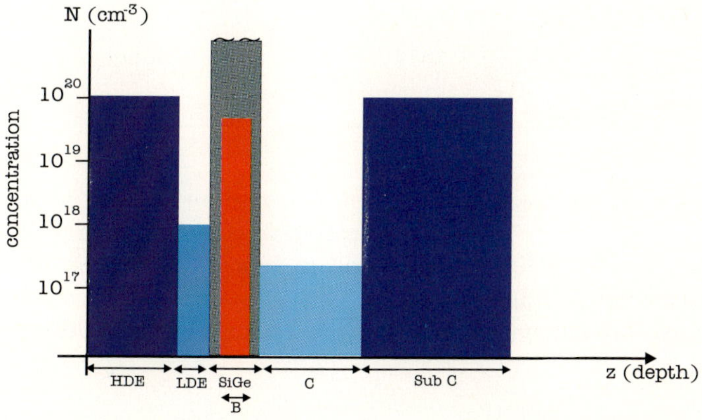

Fig. 6.5. A schematic profile of the doping and Ge concentration (drawn in the z-direction below the emitter contact). The following doping regions are shown: High doped emitter (HDE), low doped emitter (LDE), SiGe layer (SiGe), p-doped SiGe base (B), collector (C) and subcollector (Sub C). The Ge concentration is decreased by a factor 10 to fit into the figure

This general doping structure is similar to that of a Si BJT with the characteristic differences summarised in Table 6.3. The most important difference concerns the base structure. The base contains a SiGe alloy, its p-doping is much higher and the thickness is considerably thinner. The higher doping

and lower thickness of the base follows from the trade off between the current gain and speed. The changes in the emitter and collector are consequences caused by the requirements to avoid an increase in tunnelling across the base-collector junction and to reduce equivalently the collector transit time.
The actual thicknesses and concentration levels are dictated by the requirements for current gain, speed, breakdown voltages and by the technological constraints from growth and device processing. As an example we choose a high speed HBT with low breakdown voltage BV_{CE} (collector-emitter voltage) and a moderate speed HBT with higher breakdown voltage.

Table 6.4. Layer structures of two different HBT transistors (HBT1 high speed f_T=220 GHz, BV_{CE}=1.8 V, HBT2 moderate speed f_T=60 GHz, BV$_{CE}$=6 V).

Structure	No.	thickness (nm)	doping ($10^{17}\,\mathrm{cm}^{-3}$)	Ge content (%)
Subcollector	1,2	4000	n, 50	-
Collector	1	50	n, 4	-
	2	250	n, 0.4	
Base	1	12	p, 80	28
	2	27	p, 50	18
LDE	1	20	n, 4	-
	2	100	n, 10	
HDE	1,2	200	n, 200	-

As shown by these examples the thicknesses of the base and collector are shrunk to get the higher speed. The collector thickness shrinkage results in lower breakdown voltages, BV$_{CE}$. As general in bipolar transistors the collector-emitter breakdown voltage is a factor of 2-3 lower than the corresponding collector-base breakdown voltage, BV$_{CB}$ because of the feedback from impact ionised holes.

For the analysis of the high frequency behaviour of the HBT an investigation of the current dependence of the transit frequency is the most important step. For this we have to rearrange the single components of the emitter collector transit time, τ_{EC}, (5.86). Assuming that τ_{E2} may be neglected and using (5.94) and the saturation velocity, v_{sat} (see Fig. 3.31), we obtain

$$\frac{1}{2\pi f_T} = \tau_{EC}$$

$$= \frac{W_B^2}{2D_n} + \frac{W_B}{v_e} + \frac{W_{BC}}{2v_{sat}} + C_{BC} \cdot (R_E + R_C) + \frac{k_B T}{qI_C}(C_{BE} + C_{BC}) \quad (6.4)$$

where τ_B and τ_{hc} are current independant, τ_C is split into a current dependant part and a RC load time $\tau_{RC} = C_{BC}(R_E + R_C)$. From (6.4) one would expect a linear plot of τ_{EC} vs $1/I_C$ (Fig. 6.6). Indeed the linear fit is very good up to high current densities where deviations by the modified Kirk effect lead to a steep increase in the transit time. The Kirk effect in a bipolar transistor is caused by the base extension under high current densities. In a HBT the BC heterojunction strongly influences the current flow. We name this the modified Kirk effect because of the similar results as in the original description. In the given example the fixed inner transit time is roughly 0.6 ps with about 0.2 ps for τ_B, τ_{hc} and τ_{RC}, respectively. From the slope of the linear fit a capacitance $C_{BE} + C_{BC}$ of 12.2 fF/μm^2 is extracted from which, in the given geometry, one third is attributed to C_{BC}. The attainable minimum transit time, τ_{min}, is limited to values above the fixed inner transit time because of the modified Kirk effect which limits the current density to about a critical current density of

$$J_{crit} = qN_{DC}v_{sat} \tag{6.5}$$

The critical current density J_{crit} is $6.4 \times 10^5\,\mathrm{A\,cm^{-2}}$ for the collector doping N_{DC} of $4 \times 10^{17}\,\mathrm{cm^{-3}}$.

The transit frequency f_T is not dependant on the base resistance R_B because for the current gain measurement a base current modulation is given to the inner transistor. For the power gain the maximum oscillation frequency f_{max} is the corresponding frequency limit which is strongly influenced by the loading of the capacitance, C_{BC}, via the base resistance, R_B. Rewriting (5.98) gives

$$\frac{f_{max}}{f_T} = \frac{1}{2}\sqrt{\frac{\tau_{EC}}{R_B C_{BC}}} \tag{6.6}$$

which means a high transit frequency, f_T, and a low $R_B C_{BC}$ time constant are important for a high oscillation frequency. Let us give a simplified assessment of the order of magnitude for the time constant $R_B C_{BC}$. For this we assume the geometry of Fig. 6.4, a base contact with $b_B = b_E$ and a simplified model of the base resistance R_B which contains an area proportional to the external base conductance

$$\frac{1}{R_{B_e}} = \frac{(b_E \cdot l_E)}{\rho_{CB}} \tag{6.7}$$

and an inner base resistance, R_{B_i}

$$R_{B_i} = R_{ShB} \cdot \frac{b_E}{6 l_E} \tag{6.8}$$

which is proportional to the base sheet resistance, R_{ShB}. The collector-base capacitance, C_{BC}, is given by

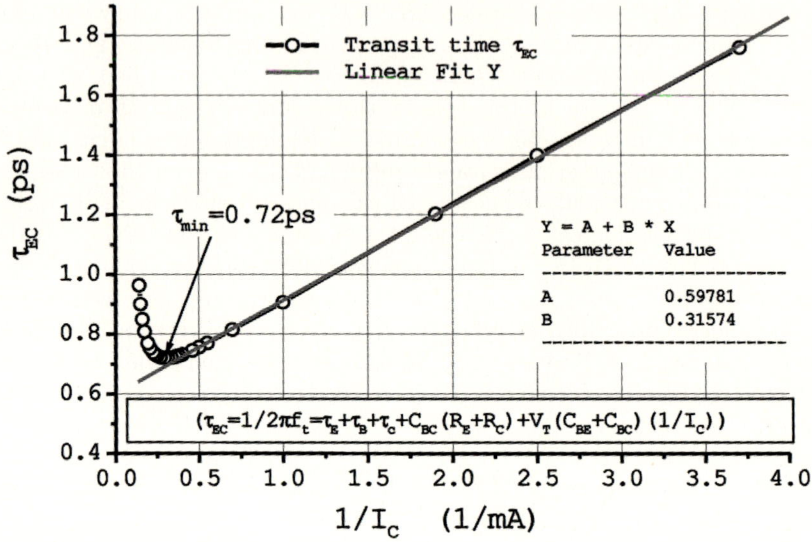

Fig. 6.6. The inverse transit frequence $1/2\pi f_T = \tau_{EC}$ versus inverse current $1/I_C$. Numerically calculated (BLAZE) transit frequency for a $1\,\mu$m$\times 1\,\mu$m emitter finger with the structure given in table 6.4 (high speed transistor)

$$C_{BC} = 2 \cdot (b_E \cdot l_E)\frac{\varepsilon_0 \varepsilon_{Si}}{W_{BC}} \tag{6.9}$$

In this simplified model the transit time constant $R_B \cdot C_{BC}$ is obtained by

$$R_B \cdot C_{BC} = (2 \cdot \rho_{CB} + \frac{1}{3} R_{ShB} \cdot b_E^2) \cdot (\frac{\varepsilon_0 \varepsilon_{Si}}{W_{BC}}) \tag{6.10}$$

This time constant depends both on the vertical structure of the device which determine W_{BC} and the base sheet resistance, R_{ShB}, but also on the lateral structure which defines b_E and ρ_{CB} (remark: from (6.7) ρ_{CB} contains not only the specific contact resistance of the base but also the other area dependant parts of the external base resistance). The ultimate f_{max} which may be obtained is given by neglecting the external base resistance contributions ($\rho_C \to 0$).

$$f_{max} < \frac{1}{b_E}\sqrt{f_T \cdot w_{BC} \cdot \frac{3}{4\pi \cdot R_{ShB} \cdot \varepsilon_0 \varepsilon_{Si}}} \tag{6.11}$$

From the device technology viewpoint a small emitter finger ($b_E \downarrow$) is important to obtain high f_{max} values.

6.3 The Linear Graded Profile

The saturated collector current density requires (5.68) to be modified to

$$J_{C0}\,(\mathrm{SiGe}) = \frac{q}{\int_0^{W_B} \frac{p_p(z)}{D_{nB}(\mathrm{SiGe}) n_{ieB}^2(\mathrm{SiGe})} \mathrm{d}z} \tag{6.12}$$

Let us assume that the doping profile, p_p is near constant throughout the base in both transistors and approximately equal to the doping density in the base N_B. Therefore (6.12) simplifies to

$$J_{C0}\,(\mathrm{SiGe}) \approx \frac{qD_{nB}n_{iB}^2\,(\mathrm{Si})}{N_B W_B} \exp\left[\frac{\Delta E_g}{k_B T}\right] \qquad \text{box profile} \tag{6.13}$$

for the HBT and the linear graded transistor becomes

$$J_{C0}\,(\mathrm{SiGe}) \approx \frac{qD_{nB}n_{ieB}^2\,(\mathrm{Si})}{N_B} \frac{1}{\int_0^{W_B} \exp\left(-\frac{\Delta E_g}{k_B T}\frac{z}{W_B}\right) \mathrm{d}z} \qquad \text{linear grade}$$

$$= \frac{qD_{nB}n_{ieB}^2\,(\mathrm{Si})}{N_B W_B} \frac{\frac{\Delta E_g(\mathrm{SiGe})}{k_B T}}{1 - \exp\left(-\frac{\Delta E_g(\mathrm{SiGe})}{k_B T}\right)} \tag{6.14}$$

Since the base currents are approximately similar between the SiGe and Si base transistors, it is the collector currents which produce the major changes. Therefore comparing the collector currents between the Si and SiGe transistors we find

$$\frac{I_C\,(\mathrm{SiGe})}{I_C\,(\mathrm{Si})} = \exp\left[\frac{\Delta E_g}{k_B T}\right] \qquad \text{box profile} \tag{6.15}$$

$$\frac{I_C\,(\mathrm{SiGe})}{I_C\,(\mathrm{Si})} = \frac{\frac{\Delta E_g}{k_B T}}{1 - \exp\left[-\frac{\Delta E_g}{k_B T}\right]} \qquad \text{linear grade} \tag{6.16}$$

It is therefore the collector currents which have the biggest effect on the gain. The gain for a box profile transistor is modified using (5.75) and (5.79) to produce

$$\frac{\beta\,(\mathrm{SiGe})}{\beta\,(\mathrm{Si})} = \exp\left[\frac{\Delta E_g}{k_B T}\right] \qquad \text{box profile} \tag{6.17}$$

$$\frac{\beta\,(\mathrm{SiGe})}{\beta\,(\mathrm{Si})} = \frac{J_{C0}\,(\mathrm{SiGe})}{J_{C0}\,(\mathrm{Si})} = \frac{\frac{\Delta E_g(\mathrm{SiGe})}{k_B T}}{1 - \exp\left(-\frac{\Delta E_g(\mathrm{SiGe})}{k_B T}\right)} \qquad \text{linear grade} \tag{6.18}$$

To calculate the Early voltage, (5.74) must be used along with substituting the intrinsic electron concentrations in the SiGe bases from (6.1) and (6.2). Therefore

$$V_A\,(\mathrm{SiGe}) \simeq \frac{qD_{nB}n_{ieB}^2\,(\mathrm{Si})}{C_{BC}} \int\limits_0^{W_B} \frac{N_B\,(z)}{D_{nB}\,(z)\,n_{ieB}^2\,(\mathrm{SiGe,z})} \mathrm{d}z \tag{6.19}$$

must be solved for the two cases to give

$$V_A\,(\mathrm{SiGe}) \simeq V_A\,(\mathrm{Si}) \qquad \text{box profile} \tag{6.20}$$

$$V_A\,(\mathrm{SiGe}) \simeq \frac{qN_BW_B}{C_{BC}}\frac{k_BT}{\Delta E_g}\left[\exp\left(\frac{\Delta E_g}{k_BT}\right)-1\right]A_{BC} \qquad \text{linear grade} \tag{6.21}$$

Therefore there appears to be no change to the Early voltage in an HBT while the linear graded transistor has a nearly exponential change. In reality, however, the Early voltage in typical box profile HBTs will be greater than a normal silicon bipolar transistor because V_A is proportional to base doping which can be increased in a box profile HBT without trading off other parameters. The small bandgap has no role to play in modifying the Early voltage of a HBT.

To find the base transit time, (5.92) must be rewritten as

$$\tau_B\,(\mathrm{SiGe}) \approx \int\limits_0^{W_B} \frac{n_{ieB}^2(\mathrm{SiGe},z)}{N_B\,(z)} \int\limits_x^{W_B} \frac{N_B\,(z')\,\mathrm{d}z'\mathrm{d}z}{D_{nB}\,(z')\,n_{ieB}^2(\mathrm{SiGe},z')} \tag{6.22}$$

Substituting (6.1) and (6.2) produces the following results

$$\tau_B\,(\mathrm{SiGe}) = \tau_B\,(\mathrm{Si}) \qquad \text{box profile} \tag{6.23}$$

$$\tau_B\,(\mathrm{SiGe}) \approx \frac{W_B^2}{D_{nB}}\frac{k_BT}{\Delta E_g}\left(1-\frac{k_BT}{\Delta E_g}\left[1-\exp\left(\frac{-\Delta E_g}{k_BT}\right)\right]\right) \qquad \text{linear grade} \tag{6.24}$$

Since from (5.91) the emitter transit time is inversely proportional to the gain of the transistor ((6.17) and (6.18)), this value will be reduced in both types of transistor. Therefore the cut-off frequency will be increased in both the box profile HBT and the linear graded base transistors compared to standard silicon bipolar devices. For the box profile HBT, the improvement in f_T is predominantly due to the exponential increase in gain with decreasing bandgap reducing τ_E. For the linearly graded base, the in-built electric field from the Ge profile reduces τ_B while the nearly linear increase in gain with reducing bandgap will reduce τ_E thereby increasing f_T. The comparisons made above are given with the same doping profiles for the Si and SiGe transistors. In real designs the high current gain of the HBT is traded against higher base doping N_B (lower base resistance) and smaller base width W_B (lower base transit time).

Table 6.5. A comparison of performace gains for HBTs and linearly graded SiGe base transistors (same dopant profile as a Si bipolar function transistor)

Parameter	High gain SiGe HBT	Linear graded SiGe base
$\frac{n_{iB}^2(\text{SiGe})}{n_{iB}^2(\text{Si})}$	$\exp\left[\frac{\Delta E_g}{k_B T}\right]$	$\exp\left[\frac{\Delta E_g}{k_B T}\frac{z}{W_B}\right]$
$\frac{J_C(\text{SiGe})}{J_C(\text{Si})}$	$\exp\left[\frac{\Delta E_g}{k_B T}\right]$	$\frac{\Delta E_g}{k_B T\left(1-\exp\left[\frac{-\Delta E_g}{k_B T}\right]\right)}$
$\frac{\beta(\text{SiGe})}{\beta(\text{Si})}$	$\exp\left[\frac{\Delta E_g}{k_B T}\right]$	$\frac{\Delta E_g}{k_B T\left(1-\exp\left[\frac{-\Delta E_g}{k_B T}\right]\right)}$
$\frac{V_A(\text{SiGe})}{V_A(\text{Si})}$	1	$\frac{k_B T}{\Delta E_g}\left(\exp\left[\frac{\Delta E_g}{k_B T}\right]-1\right)$
$\frac{\beta(\text{SiGe})V_A(\text{SiGe})}{\beta(\text{Si})V_A(\text{Si})}$	$\exp\left[\frac{\Delta E_g}{k_B T}\right]$	$\exp\left[\frac{\Delta E_g}{k_B T}\right]$
$\frac{\tau_B(\text{SiGe})}{\tau_B(\text{Si})}$	1	$\frac{2k_B T}{\Delta E_g}\left(1-\frac{k_B T}{\Delta E_g}\left(1-\exp\left[\frac{\Delta E_g}{k_B T}\right]\right)\right)$
$\frac{\tau_E(\text{SiGe})}{\tau_E(\text{Si})}$	$\sim\exp\left[\frac{\Delta E_g}{k_B T}\right]$	$\sim\frac{k_B T}{\Delta E_g}\left(1-\exp\left[\frac{\Delta E_g}{k_B T}\right]\right)$

The great advantage of the addition of Ge into the base of either the box profile HBT or the linear graded base transistor is the additional flexibility in device performance to optimise circuit performance. For digital bipolar circuits, the most important parameters are the circuit switching speed and the power dissipation. Therefore minimising the base resistance, the base-collector capacitances and all parasitic capacitances in the device or circuit increases the performance more than optimising other parameters. Analogue circuits will also benefit from optimisation of the important parameters in the digital circuit. The main parameters to optimise are the cut-off frequency, the maximum frequency of oscillations, the base resistance and the Early voltage. In both cases the reduction of the base resistance becomes important. This requires increasing the doping density in the base. The easiest way to achieve this is to trade off the extra gain that a SiGe base transistor gives for increased doping density in the base and therefore lower base resistance. Table 6.5 shows a summary of the performance of SiGe box profile HBTs and linearly graded base transistors compared to standard silicon bipolar devices.

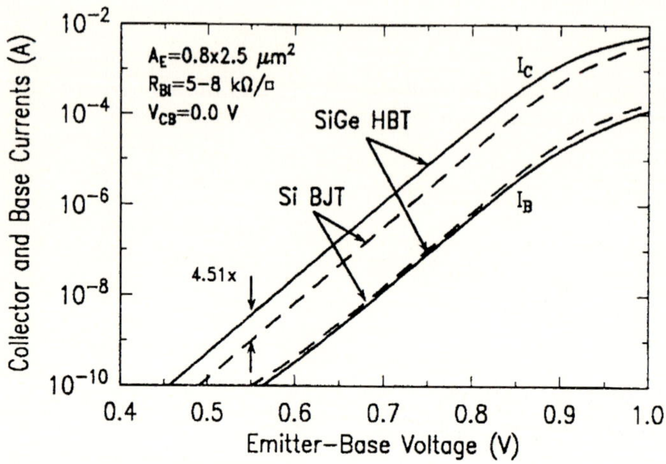

Fig. 6.7. A Gummel plot showing the typical difference between a Si bipolar and a SiGe HBT device. The collector current for the SiGe linear graded base is approximately 4.5 times that for the Si bipolar. (©IEEE)

6.4 SiGe HBT Device Performance

In this section some experimental data from both Si bipolar and SiGe HBT transistors will be reviewed. Most of the data is from IBM using a 0.5 μm BiCMOS process using 200 mm wafers and UHVCVD growth of the SiGe base. The SiGe base profile is a linear grade from 0 to 15% over 100 nm. While these are not the most aggressive or highest performance devices around they have been compared with comparable Si bipolar devices produced using the same process and therefore illustrate the performance differences between Si and SiGe transistors very nicely. Box shaped transistors with their higher current gain differ in several parameters (emitter doping, base doping, base width) from their silicon counterparts and the influence of one parameter is not as easily observed.

Figure 6.7 shows a Gummel plot for both the Si and SiGe devices with all other major common parameters constant (at least as best is possible for the different types of bipolar transistors). While the base currents are almost identical, the increase in the SiGe collector current is very obvious. The common emitter configuration characteristics of a similar device with a different area is shown in Fig. 6.8.

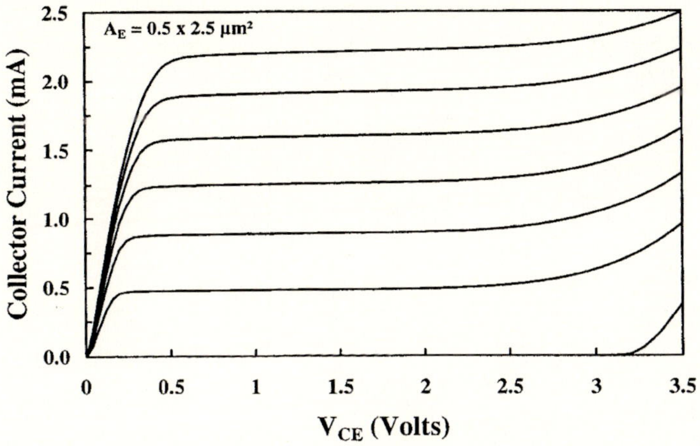

Fig. 6.8. Common emitter characteristics of a 0.5 × 2.5 μm^2 npn linearly graded base SiGe bipolar transistor manufactured by IBM. I_B=0 to 30 μA in 5 μA steps. (©IEEE)

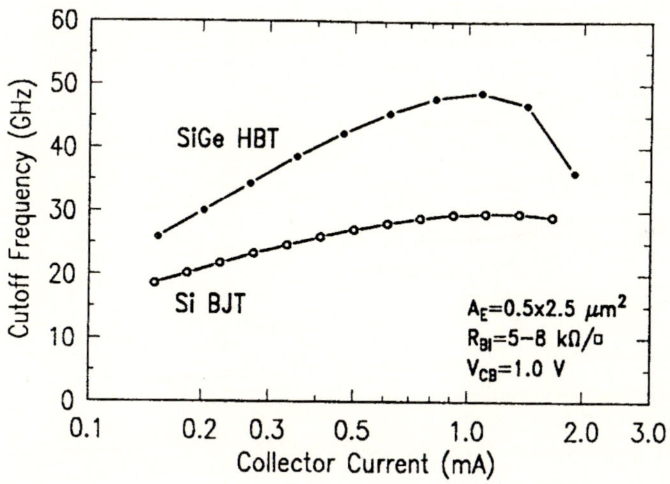

Fig. 6.9. A comparison between the cut-off frequencies for both Si and SiGe linearly graded base bipolar transistors fabricated in the same lot of wafers. (©IEEE)

One of the major reasons for the SiGe base bipolar transistors is the increase in speed. Fig 6.9 shows the cut-off frequency for both the Si and SiGe transistors. The SiGe device has a significantly larger peak cut-off frequency. This is of course very useful for applications which depend on having the highest speed possible. What is more important for many other applications is that a given speed can be reached by the SiGe device at a much lower collector current. Since power = current × voltage, this results in much lower

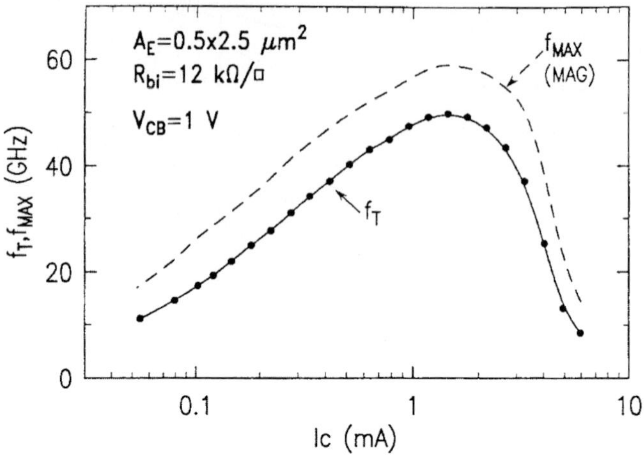

Fig. 6.10. The cut-off frequency and f_{max} versus collector current for a linear graded base SiGe bipolar transistor. (©IEEE)

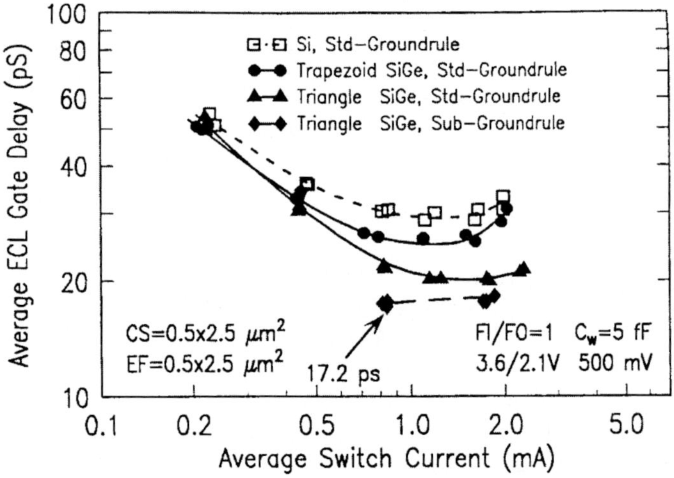

Fig. 6.11. The ECL ring oscillator gate delay for a Si bipolar transistor and variations in the Ge profile of SiGe base bipolar transistors. (©IEEE)

power consumption. Particularly for portable solutions such as mobile phones, it is this particular characteristic of the SiGe base bipolar transistor which is most useful. For high speed applications, both f_T and f_{max} are important. Both are plotted for a SiGe transistor in Fig. 6.10. This is a typical design for a transistor where f_{max} is approximately 10 to 20% higher than f_T.

All the figures above have been related to single devices but for applications it is the circuit performance which matters. There are of course many different types of circuit that can be built but so that general comparisons

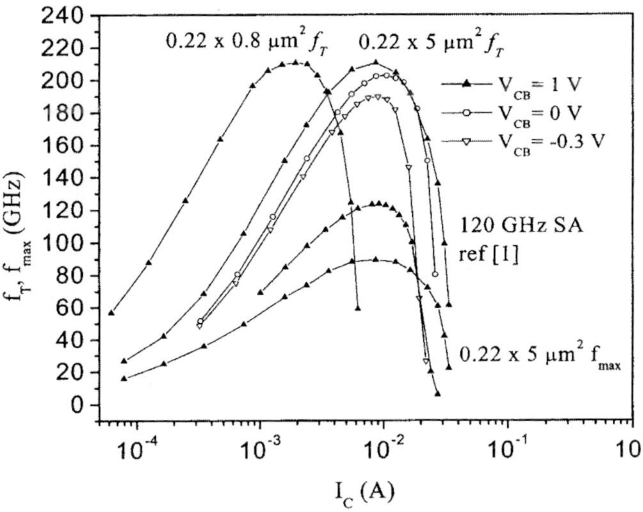

Fig. 6.12. f_T characteristics for an IBM linear graded base SiGe bipolar transistor at various V_{CB} for a 0.22 × 5 μm² non-self aligned device (©IEEE).

can be made between technologies, there are a few standard circuit configurations which are used. One of the major test circuits for a bipolar technology is the emitter common logic (ECL) ring oscillator gate delay which is shown for both Si and SiGe technology in Fig. 6.11. At about 1 mA switching current, the best SiGe base design has a average gate delay almost half that of the Si bipolar technology. The other way to compare the technology is to take a fixed gate delay for which the SiGe technology requires half the switching current of the Si bipolar circuit. This will translate into a SiGe linear base transistor power dissipation which is approximately half that of the Si bipolar. The combination of bipolar devices with complementary MOS (CMOS) logic is termed BiCMOS. The availability of high speed HBTs with good βV_A product pushed strongly the introduction of BiCMOS circuits.

The results above have shown some of the early production devices from IBM but of course research has been continuing. In Fig. 6.12 one of the results from the literature is shown where the cut-off frequency has been increased to 210 GHz using a linear graded base SiGe technology with the base Ge content varied between 0 and 25%. The device is not self-aligned and so the f_{max} value is relatively low in this particular device. It does show, however, that bipolar transistors with SiGe in the base can have operating parameters far above 200 GHz, an area which was previously only believed to be achievable using III-V technology.

As mentioned before a clear analysis of the frequency behaviour is given by a plot of τ_{EC}=$(2\pi f_T)^{-1}$ versus $(1/I_C)$. For the three typical SiGe-HBT generations this analysis is done in Fig. 6.13. From the slope of the curves

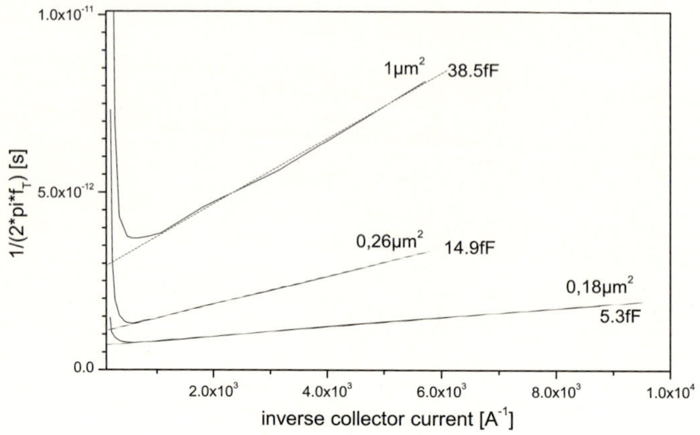

Fig. 6.13. Analysis of the frequency behaviour by a plot of τ_{EC}=$(2\pi f_T)^{-1}$ versus $(1/I_C)$. For the three typical SiGe-HBT gemerations

one can immediately conclude about the capacity $C_{BE} + C_{BC}$ of the different transistors. The slope is given by $V_t(C_{BE} + C_{BC})$yielding 38.5 fF, 14.9fF and 5.3fF for the $1\mu m^2$ (f_T=45 GHz), $0.26\mu m^2$(f_T=120 GHz) and $0.18\mu m^2$(f_T=210 GHz) transistor, respectively. From exrapolation of the slope toward $(1/I_C)$=0 the main contributions (τ_B+τ_C+τ_{RC}) to the current independent part of the time constant τ_{EC}may be assessed to be 3.0ps, 1.1ps and 0.6ps, respectively. This demonstrates nicely the reduction of the time constants in the subpicosecond regime obtained by decrease in base and collector thickness. The onset of the Kirk effect (I_C=2mA for all three transistors with decreasing area) demonstrates the increasing critical current density caused by the increased collector doping from $10^{17}/cm^3$ to nearly $10^{18}/cm^3$.

6.5 Further Reading

1. Y. Taur and T.H.Ning, Fundamentals of Modern VLSI Devices, CUP, Cambridge (1998)
2. C.Y. Chang and S.M. Sze, ULSI Devices, John Wiley and Sons, New York (2000)
3. D.J. Paul, Semiconductor Science and Technology **19**, R59 (2004)
4. J.D. Cressler and G. Niu, Silicon-Germanium Heterojunction Bipolar Transistors, Artech House, Norwood (2003)
5. J. Eberhardt, E. Kasper, Solid-State Electronics 45, 2097 (2001)
6. E. Kasper et al., Solid-State Electronics 48, 837 (2004)

7. Hetero Field Effect Transistors (HFETs)

The classical approach to heterostructure field effect transistors (HFET) is given both in III/V-materials and SiGe/Si by using modulation doped structures (modulation doped FET-MODFET). These structures have proven carrier confinement in quantum dots with improved mobility in two dimensional transport. These devices are also called HEMT (high electron mobility transistor) or TEGFET (two dimensional electron gas FET). Silicon microelectronics driven are the more recent approaches to provide HFETs with insulating gates as in a MOSFET and to combine n-channel and p-channel devices as in a CMOS logic. Further on, one hast to consider the specific band ordering in the SiGe/Si system to understand the differences to the approaches in III/V-systems. Let us first remember the band ordering in SiGe/Si as function of the strain situation (Fig.7.1). In the technologically simplest case (a) a pseudomorphic, thin layer SiGe is compressively strained on a Si substrate. Then electrons are more or less not influenced by the heterointerface (the conduction band offset ΔE_C is below $\pm$20meV, the sign of the offset is under scientific debate since more than a decade). The holes are confined to the low band gap side SiGe ($\Delta E_V \cong 0.7x$ eV, x Ge content). The electronic nature of this interface (SiGe strained) is between type I and type II. With reversed strain, that means a tensile silicon layer grown on a virtual substrate with unstrained SiGe layer on top, the electronic nature of the interface switch to a type II (case (B) in Fig. 7.1) where the electrons are confined to the large band gap semiconductor (Si) and the holes to the small band gap SiGe.

With this knowledge in mind let us now discuss the possible routes for a complementary, insulated gate HFET logic on silicon (Tab.7.1).

The easiest integration is promised by the first route via pseudomorphic SiGe. The principle scheme is shown in Fig. 7.2. The n-MOS is made in Si as usual. The SiGe p-MOS is made in areas where selectively a thin SiGe/Si layer stack is deposited on top of the Si-substrate. The Ge-content is low (around x=0.15), the improvement in the p-channel stems from the valence band splitting (heavy, light holes) and the better carrier transport in the 2D hole gas.

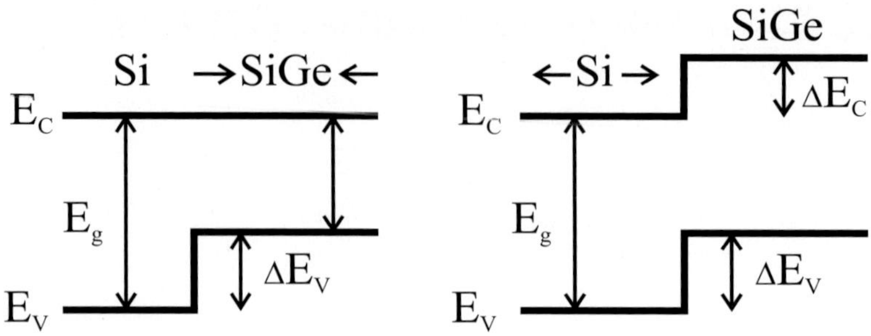

Fig. 7.1. Band ordering of (a) compressively strained SiGe on unstrained Si, (b) tensile strained Si on unstrained SiGe

Table 7.1. Main routes to hetero (CMOS) FET circuits.

Route	p-channel	n-channel	Remarks
Pseudomorphic SiGe	strained SiGe (compression)	unstrained Si (conventional n-MOS)	⊕ Easy integration p-SiGe and n-Si ⊖ only p-channel improvement (10-30%)
strained Si	strained Si (tension) on SiGe, holes are only confined by strong electric fields to a Si channel	strained Si (tension), electrons jump into Si (type II)	⊕ moderate p and n-channel improvements ⊖ virtual substrates, parallel p-channel (SiGe)
double channel	strained SiGe or Ge (compression), buried p-channel	strained Si (tension), surface n-channel	⊕ Ultimate, symmetric p- and n-channel, High improvements ⊖ High Ge content, high strain, virtual substrate, processing with low temperature budget

In the strained Si route a tensile strained surface layer is provided for both, holes and electrons. The electrons like to jump into strained Si, the holes have to be bound by strong vertical gate fields and tend to create parasitic parallel channels in underlying SiGe layers. The standard approach to generate tensile strain in Si is by a virtual substrate (Fig.7.3) composed of a relaxed SiGe layer on a Si-substrate.
The strain can be alternatively transferred from a virtual substrate to a SOI-wafer (silicon on insulator) or generated by nanostructures (e.g. Ge islands or

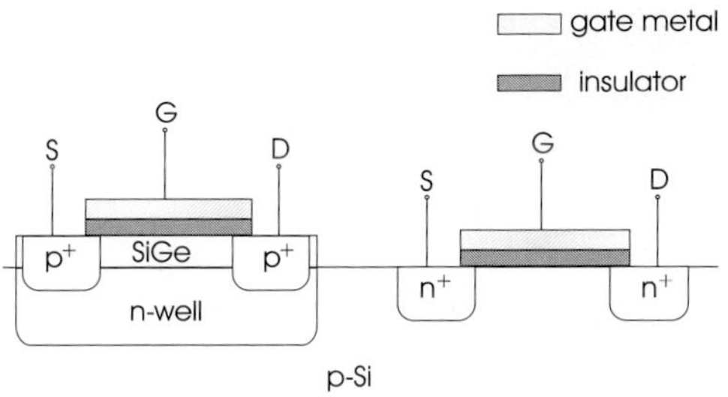

Fig. 7.2. Scheme of the pseudomorphic SiGe HFET (CMOS) route

implants). An unique solution would be provided by a double channel configuration, especially when high Ge content alloys or pure Ge for the p-channel is foreseen.

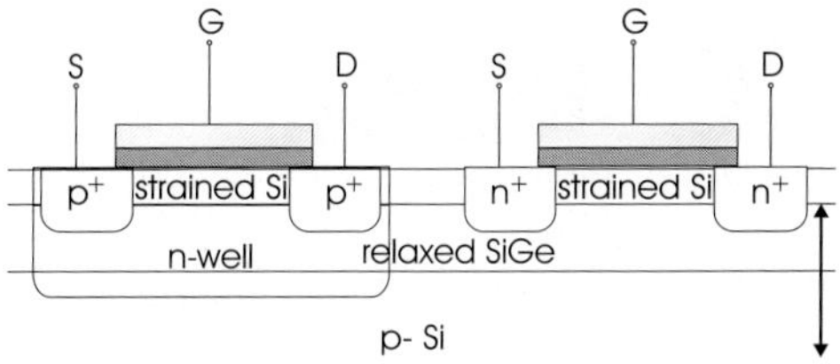

Fig. 7.3. Scheme of the strained Si CMOS route

All know semiconductors (with bandgaps large enough for room temperature operation) suffer from hole mobilities much lower than those of the electrons, only Ge has hole mobilities (μ_p in undoped Ge is $1900cm^2/Vs$) equally high than the mobilities of electrons in Si (μ_n in undoped Si is $1450cm^2/Vs$). This heterostructure couple offers the ultimate symmetric CMOS structure with equally high electron and hole mobilities. The strain in such structure (Fig.7.4) can even further enhance the mobility. The electrons (type II interface) move in the tensile strained Si, the holes in the compressively strained Ge. The technological barriers for realisation of these high performance double channel CMOS are large and require considerable research effort to overcome. Main obstacles are growth of high quality virtual substrates with Ge-content 0.5 or more, the high strain in sub 10nm layers, the processing of Ge and strained Si/Ge and the combination with high-k dielectric gate materials.

In the following the vertical gate arrangement, the strained Si-MOS and the MODFET results are discussed in more detail.

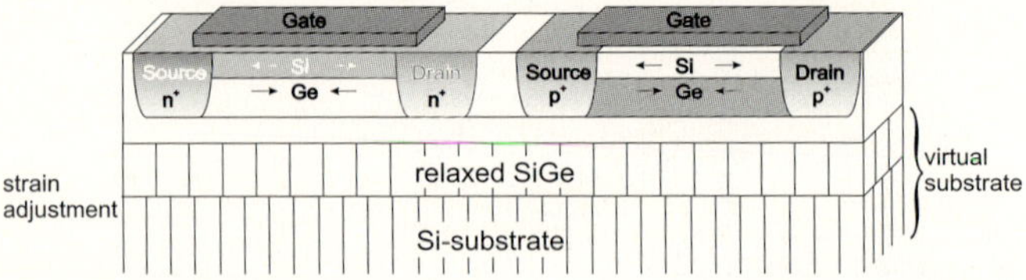

Fig. 7.4. Double channel HFET (CMOS). Scheme of the layer structure

7.1 Vertical Heterojunction MOSFETs

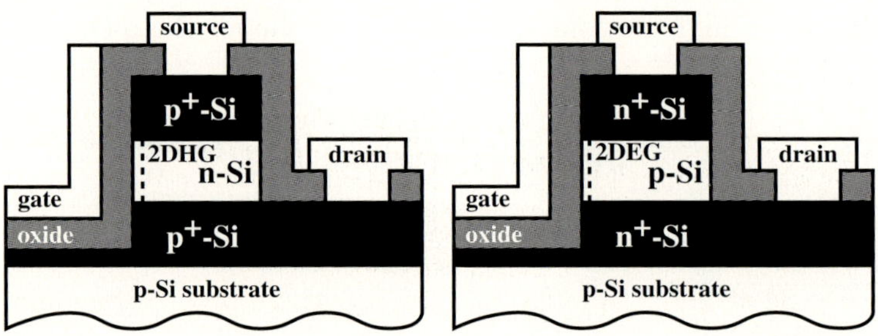

Fig. 7.5. An example of a p-VMOS on the left and a n-VMOS on the right

Vertical sidewall MOSFET designs were very popular when it was thought that optical lithography could not be used much below 0.25 μm. The great advantage of a vertical FET is that the vertical dimensions of a FET can be easily controlled down to sub-10 nm. Already gate oxides on MOSFETS are below 1.5 nm in production while the lateral gate-lengths are 60 nm demonstrating how the vertical dimensions of standard MOSFETs are typically 100 times smaller than the lateral lithographically defined dimensions. A typical device layout is shown in Fig. 7.5. The figures also demonstrate the major disadvantage of the vertical MOSFET concept. Since the gate is not self-aligned with the channel, the parasitic capacitances which include the gate to source and gate to drain capacitances are significantly higher than a planer FET.

The addition of SiGe into the vertical structure also has the potential to allow some bandgap engineering of the channel. In particular, it is possible to reduce DIBL by having contacts with a smaller bandgap (Fig. 7.6(a)) and also provide a built in electric field to accelerate the carriers across the channel, especially for a hole device such as demonstrated in Fig. 7.6(b). As optical lithography has progressed with ArF and KrF sources along with phase shifting technology, the vertical FET has fallen out of favour as the lateral self-aligned MOSFET has significantly better performance.
In order to overcome the non self-aligned gate structures several suggestions were made the most promising of them are using double gate transistors on

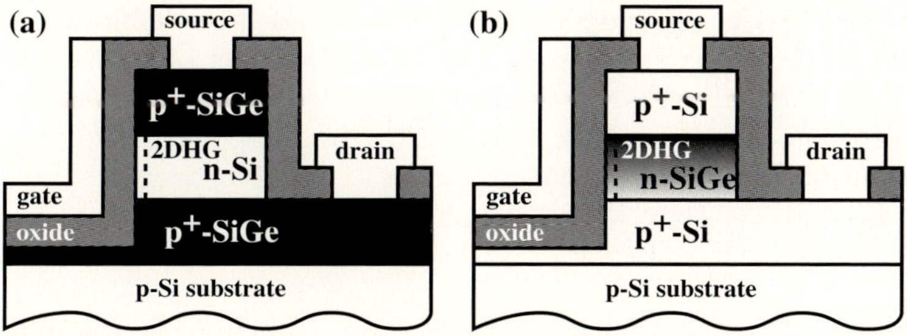

Fig. 7.6. Two examples of vertical MOSFETs where in (a) the lowered bandgap in the Ohmic contact regions reduces DIBL while a graded SiGe channel in (b) produces an inbuilt electric field to accelerate carriers along the channel

fully depleted SOI. Figure 7.7 shows the structure called FinFET (see also chapter 5.3.8). In a SOI structure (silicon on insulator on a Si substrate) all of the undoped top silicon layer except the small transistor finger (the fin) is removed by etching. A double gate surrounds the fin. Both ends of the fin are contacted by source and drain. Parasitic overlap capacities are strongly reduced by the etching step. Many authors believe such structures (vertical, double gated, fully depleted channel) will dominate the CMOS technology below 45nm gate length.

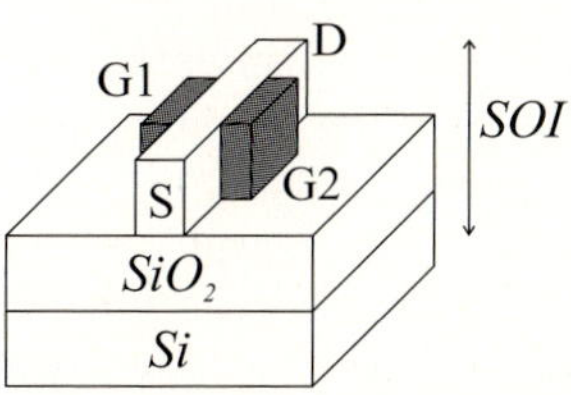

Fig. 7.7. FinFET- a double gate (G1, G2) vertical transistor on fully depleted SOI. The top layer (either strain Si or SiGe) on the SOI (silicon on oxide) structure is etched outside the transistor finger between Source (S) and Drain (D)

7.2 Strained-Si CMOS

CMOS devices are now being aggressively scaled to gate-lengths below 100 nm and predictions suggest that the scaling is likely to continue for at least until 2014. A number of problems, however, are being found as the MOSFET gate-lengths are reduced. In particular the gate oxide thickness in state-of-the-art production devices is now below 2 nm and thinner oxides increase the off-state current through the increase in quantum mechanical tunnelling of charge through the gate insulator. A secondary effect of the reduction of the gate insulator thickness is the reduction of electron and hole mobilities in the inversion layers of CMOS transistors. Therefore a number of technology solutions are being pursued to find methods of circumventing these problems. The real problem can be discovered if we return to the equation for the saturation current of a MOSFET (7.1)

$$I_{ds,sat} \approx \frac{W \mu_n C_{ox}}{2L_g} (V_g - V_T)^2 \tag{7.1}$$

To increase the on current in digital electronics you can decrease the gate length, decrease the oxide thickness (or increase the gate insulator capacitance with new materials), increase the gate overdrive voltage or increase the mobility. Increasing the gate overdrive increases the power dissipation and most microprocessors are already at a density that the power cannot be increased. As threshold voltage is almost fixed at the lowest gate lengths and the oxide thickness is already near the limit, the only other parameter that can be improved when scaling the gate length is the mobility.

One of the leading contenders for improving the mobilities of the inversion layer carriers is the use of strained-Si technology. A number of different schemes are being researched to produce appropriate strain in the n- and p-channel devices but most include SiGe technology. Many of the main microelectronic companies are involved in SiGe technology research at some level.

The growth of a $Si_{1-x}Ge_x$ heterolayer on top of a silicon or a relaxed $Si_{1-y}Ge_y$ buffer layer or virtual substrate results in a compressively strained SiGe channel for $x > y$. By growing a strain relaxation buffer of $Si_{1-y}Ge_y$ followed by a tensile strained-Si layer results in a structure which from a processing point of view, looks very similar to a silicon wafer and can be processed in a fashion much closer to a standard CMOS process. This is the basis of strained-Si CMOS. The tensile strain splits the conduction band valleys with the Δ_2 valleys being lowered in energy and the Δ_4 valleys being increased in energy to such an extent that only the lower Δ_2 valleys have any significant population of carriers. A quantum well is formed with a conduction band discontinuity of $\sim 0.6y\,\text{eV}$ for a strained-Si grown on top of a relaxed $Si_{1-y}Ge_y$ buffer and this combined with the high effective mass in the vertical direction confines the electrons in the tensile strained-Si surface layer. The reduction of intervalley scattering has demonstrated significant increases in the n-MOSFET mobility both at room and low temperatures. Strained-Si on insulator has also been used to increase the mobility enhancements.

For holes the situation is very different. For both compressive or tensile strain, the light-hole and heavy-hole bands are split but only by a small amount so both have significant populations of carriers especially with the high electric fields produced in short-channel CMOS devices. The major change is the reduction in the density of states effective hole mass for both compressive or tensile strain. For the light hole mass, tensile strain reduces the mass value but compressive strain increases the mass. The reduction in

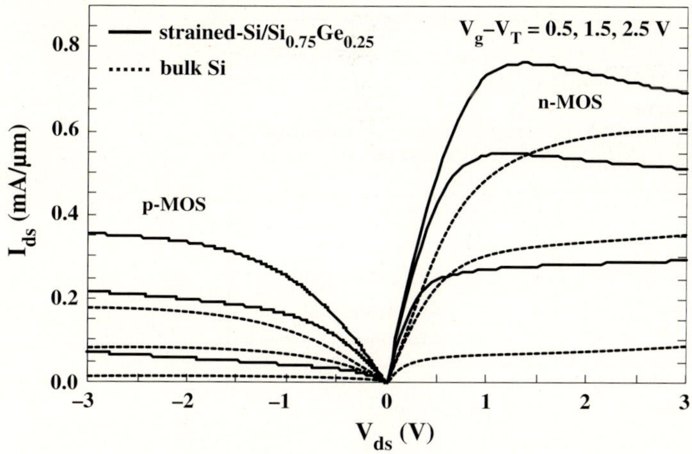

Fig. 7.8. The drain current versus gate overdrive ($|V_g - V_T|$) of 0.5, 1.5 and 2.5 V for 0.3 μm gate length by 5 μm wide MOSFETs. The dashed lines are the bulk Si control devices and the solid lines are strained-Si devices fabricated on relaxed $Si_{0.75}Ge_{0.25}$ virtual substrates

the heavy hole mass, however, is significantly higher for compressive strain. A second issue is that a tensile strained-Si layer grown on a relaxed $Si_{1-y}Ge_y$ buffer is higher in energy to holes than the relaxed substrate. This combined with the lower effective mass in the vertical direction results in a larger spread of the wavefunction into the substrate compared to electrons. It can result in a parasitic channel of holes in the relaxed $Si_{1-y}Ge_y$ buffer especially if high Ge contents in the substrate are used to improve the mobility since then only a thin strained-Si channel can be grown under the critical thickness. Therefore the use of a buried, compressively strained-$Si_{1-x}Ge_x$ quantum well may have advantages in improving the hole mobility in such devices by the use of the lower effective mass and by confining the holes away from the Si/SiO_2 interface. The mobility in the strained-Si surface p-MOSFET has been limited to less than 30% for standard virtual substrates with Ge contents of 20% and below with silicon-on-insulator devices required for any significant mobility improvement.

Figure 7.8 shows the current voltage characteristics for identically processed CMOS bulk Si devices along with strained-Si MOSFETs on relaxed $Si_{0.75}Ge_{0.25}$ virtual substrates for a number of gate overdrive voltages. The higher on-currents for the strained-Si devices are clearly shown for these transistors with 80% improvements in the n-MOS current and 160% improvement in the p-MOS current. The decrease in the strained-Si nMOS current at the highest gate overdrive and source-drain bias current is due to self-heating. The thermal conductivity of $Si_{0.75}Ge_{0.25}$ is approximately 18 times lower than silicon and therefore at high voltages and currents the dissipated power can-

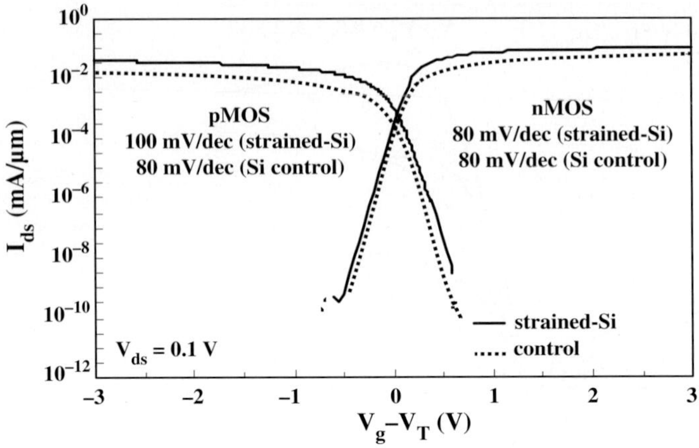

Fig. 7.9. The subthreshold plots for 0.3 μm gate length by 5 μm wide MOSFETs at $V_{ds} = 0.1$ V. The dashed lines are the bulk Si control devices and the solid lines are strained-Si devices fabricated on relaxed $Si_{0.75}Ge_{0.25}$ virtual substrates

not be completely removed from the transistor lead to reduced performance. This is the major problem of strained-Si technology.

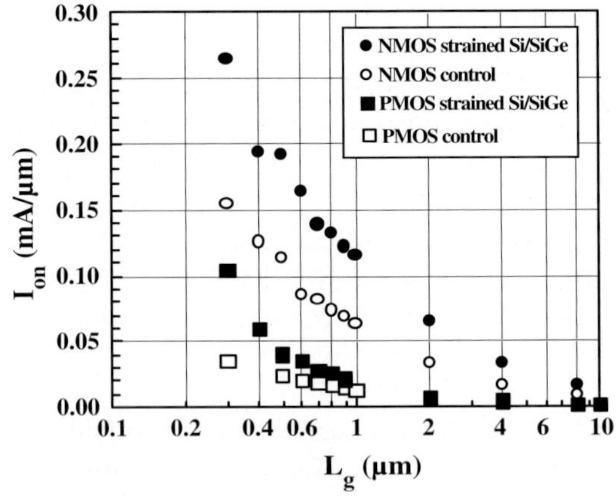

Fig. 7.10. The on-current as a function of gate length for 5 μm wide MOSFETs at V_{ds} $|V_g - V_T| = 1.0$ V for both bulks Si control devices and strained-Si fabricated on relaxed $Si_{0.75}Ge_{0.25}$ virtual substrates

Figure 7.9 shows the subthreshold slopes and off-currents for both the stained-Si and control Si devices. Very little of the subthreshold performance is traded off for the improvement in on-current with subthreshold slopes of

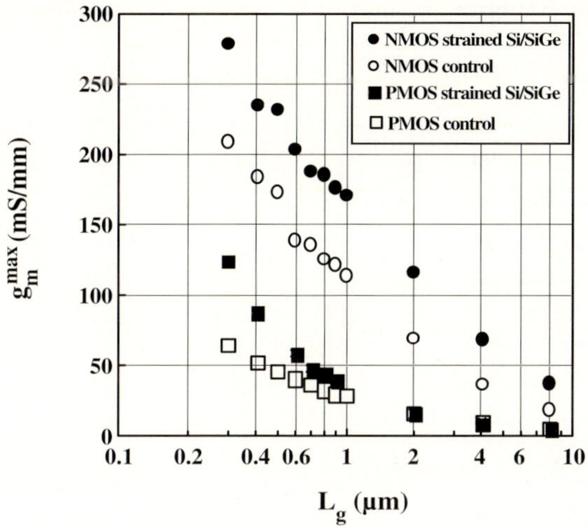

Fig. 7.11. The maximum transconductance as a function of gate length for 5 μm wide transistors for both bulks Si control devices and strained-Si fabricated on relaxed $Si_{0.75}Ge_{0.25}$ virtual substrates

typically 80 mV/dec being produced. The only way of improving the subthreshold slope significantly is by cooling the sample as it is dominated by k_BT thermal broadening. The on-current as a function of gate length is shown in Fig. 7.10 for a gate overdrive of 1 V and a source-drain bias of 1 V. The performance improvements with the strained-Si technology over the silicon control can be easily observed. These performance improvements as a function of gate length can also be observed in the maximum transconductance as demonstrated in Fig. 7.11.

While from a theoretical point of view, the higher the strain the larger the splitting of the valleys, the transistor performance improvements do not keep increasing as the Ge content in the virtual substrate and therefore the strain in the silicon layer increases (Fig. 7.12). As the Ge content in the virtual substrate is increased then the top strained-Si layer thickness must be decreased since the critical thickness is decreasing. This produces more confinement for the electrons in the strained-Si n-MOS devices and eventually when the strained-Si layer is below about 5 nm, significant interface roughness scattering will reduce the improvements by reducing the mobility of the electrons and holes. A secondary problem is that the present virtual substrate growth technology produces significantly rougher surfaces as the Ge content is increased. While chemical mechanical polishing can be used, this actually creates a vicinal surface as the cross hatch is cut across at a constant height. Therefore any oxide is grown on a vicinal surface which increases the interface roughness scattering and reduces the mobility. Also any thermal treatment of the structure at the growth temperature will result in the cross hatch pattern returning to the surface as the structure attempts to get to the equilibrium state.

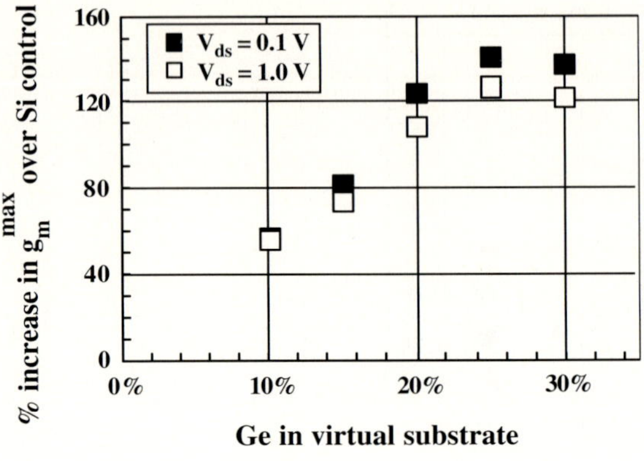

Fig. 7.12. The percentage increase in the maximum transconductance for 0.3 μm gate length by 5 μm wide strained-Si on relaxed $Si_{1-x}Ge_x$ virtual substrates n-MOSFETs as a function of Ge content in the virtual substrate

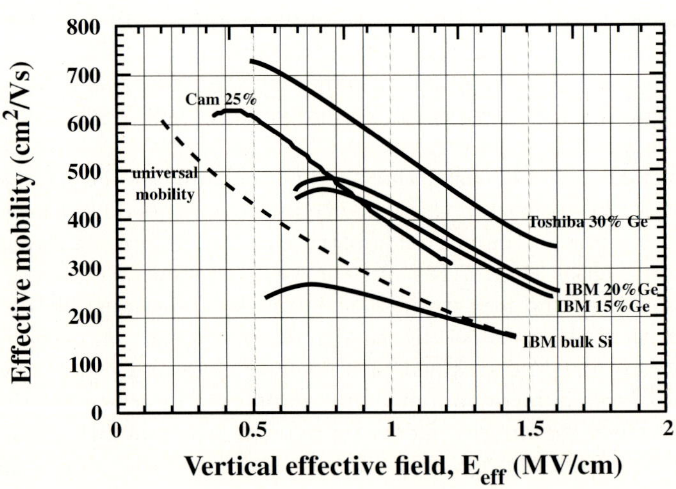

Fig. 7.13. The mobility extracted from 100 μm gate length n-MOS devices with a small applied bias of 10 mV to reduce electron or hole heating in the channel. Data from IBM and Toshiba strained-Si devices have also been included along with the strained-Si on $Si_{0.75}Ge_{0.25}$ from the previous figures. The Si universal curve is the maximum possible mobility for a given electrical field with appropriate doping to switch off a MOSFET

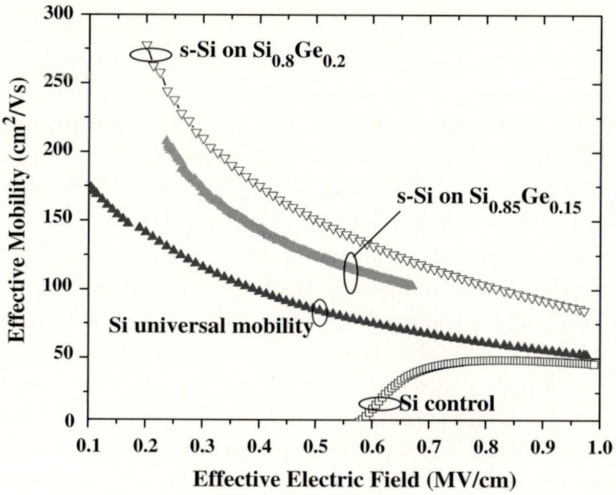

Fig. 7.14. The mobility extracted from 100 μm gate length p-MOS devices with a small applied bias of -10 mV to reduce electron or hole heating in the channel. The top curve is strained-Si on $Si_{0.8}Ge_{0.2}$ virtual substrates, the next down is strained-Si on $Si_{0.85}Ge_{0.15}$ virtual substrates. The Si universal curve is the maximum possible mobility for a given electrical field with appropriate doping to switch off a MOSFET and the Si control is a standard p-MOSFET processed identically to the strained-Si devices

Effective mobilities μ_{eff} as functions of vertical effective electric fields E_{Eff} can be calculated for long channel devices (L= 100 μm) according to the expressions

$$\mu_{eff} = \frac{L}{W} g_d (V_g) Q_{inv} \tag{7.2}$$

and

$$E_{eff} = \frac{1}{\epsilon_{Si}} (Q_b + \eta Q_{inv}) \tag{7.3}$$

where $\eta = 1/3$ for holes and Q_{inv} and Q_b are the inversion layer (i.e. channel) and the bulk (depletion) charge densities, respectively. The drain conductance $g_d\,(V_g)$ can be obtained from $-I_d(V_g)/V_{ds}$ measured at low drain bias ($V_{ds} =$ 10 mV). The charge densities are easiest to obtain by computing from split C-V measurements using the following equations:

$$Q_{inv} = \int_{V_g}^{\infty} C_{gc} \mathrm{d}V_g \tag{7.4}$$

and

$$Q_b = \int_{V_g}^{V_{fb}} C_{gb} \mathrm{d}V_g \tag{7.5}$$

where C_{gc} is the gate-to-channel and C_{gb} the gate-to-body capacitance (per unit area). The flat-band voltage V_{fb} which limits the integration in (7.5) is determined from the high-frequency MOS capacitance measurements, and the overlap capacitance needs to be subtracted from C_{gc} before performing the integration in (7.4). The $\mu_{eff} - E_{eff}$ characteristics for Si control and strained-Si devices are plotted in Figs. 7.13 and 7.14 for n- and p-channel devices respectively. A number of results from the literature have also been included with the n-channel devices for comparison. N-channel devices have demonstrated the highest mobility enhancements over the silicon control devices with values of over 100%. The p-channel devices demonstrate smaller mobility enhancements of 50 to 70% for the strained-Si on $Si_{0.85}Ge_{0.15}$ and 85 to 100% for the $Si_{0.8}Ge_{0.2}$ strained-Si devices over the Si universal mobility.

Strained-Si CMOS is a technology which is now a major part of the International Technology Roadmap for Semiconductors. It may be the case in the future that every transistor manufactured in microprocessors and silicon chips will have SiGe in them.

7.3 Metal-Gated MOSFETs

All III-V high frequency transistors use metal gates with high conductivity to improve the performance of the devices. The disadvantage of the technology in GaAs MODFETs (Sect. 7.4) is that the Ohmic contacts are not self-aligned to the gate. In MOSFETs, poly-Si gates are used since they can be doped by implantation while the Ohmic contacts are also being implanted. Such self-aligned contacts reduce parasitic capacitances in the transistor thereby improving performance. Poly-Si has a significantly lower conductivity than silicide which has a lower conductivity than metal. To improve the high frequency performance of MOSFETs, one technique is to replace the poly-Si gate with metal. A number of different schemes exist for metal gate MOSFETs. Potentially metal gate replacement techniques allow both a self-aligned structure and a metal gate to be integrated into a single transistor.

The International Technology Roadmap for Semiconductors predicts that metal gate technology is required for short channels lengths below about 30 nm. If the conductivity of the gate remains constant and the gate is reduced in size then the resistance of the gate increases. Therefore a change from silicided poly-Si to metal gates will reduce the gate resistance.

7.4 Modulation Doped Field Effect Transistors (MODFETs)

The first modulation-doped material was produced at Bell Laboratories in 1987 by Horst Stoermer and colleagues. The basic concept is to remove the donor impurity ions from the carrier electrons in the system, thereby reducing ionised impurity scattering and increasing mobility. The system therefore requires a quantum well, a doped region and a spacer between the two regions. Theoretically, as long as the donor energy is above the conduction band edge in the quantum well, the electrons will diffuse into the well (Fig. 7.15). In real systems as the spacer layer is increased, the Coulombic interaction with the carriers in the quantum well is reduced but the number of carriers which may populate the well is also reduced. Hence the heterolayers have to be optimised with the spacer thickness being one of the major variables.

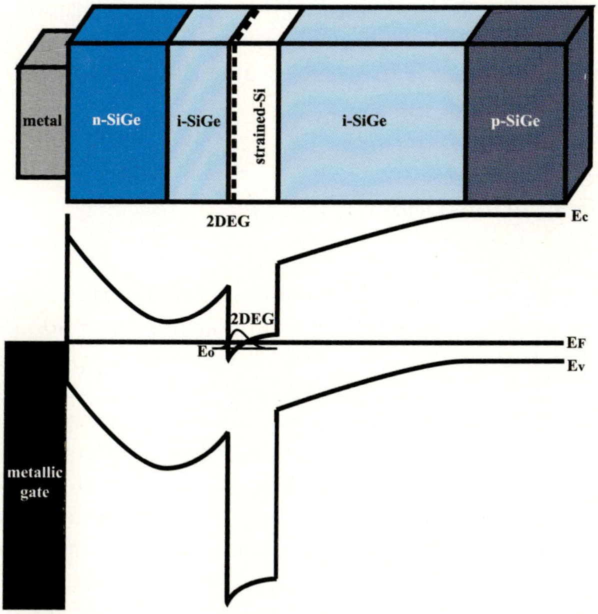

Fig. 7.15. A schematic diagram of a n-MODFET using SiGe technology. The lower figure shows the appropriate band structure with a quantum well and the electrons in the system forming a two dimensional electron gas in the well

In the GaAs system, MODFETs are generally called high electron mobility transistors (HEMTs) although the term two-dimensional electron gas field-effect transistor (TEGFET) is also used. They are one of the fasterfield effect transistors along with having one of the lowest noise properties of any transistor. This low noise is the result of high mobility coupled with low interface roughness scattering as the carriers in the system are at a smooth heterointerface rather than a rough oxide interface. They are therefore ideal devices for high speed analogue devices especially in the rf market with typical applications such as low noise amplifiers for mobile phones, satellite telecommunications and automotic radar systems. The first modulation-doped SiGe material was grown by John Bean and co-workers at Bell Laboratories in 1982 while the first n-type devices was grown by Erich Kasper and co-workers at Daimler-Benz in 1986.

MODFETs are generally depletion mode devices. This is fine for discrete devices such as those found in the front ends of rf systems since the power consumed can be easily dissipated. For high integrtation densities, however, they consume much higher powers than enhancement mode devices mainly through the leakage currents between Ohmic contacts being enhanced by the donor (or acceptor) doping layer in the channel and gate leakage since Schottky barriers are normally used rather than oxide gates. The integration densities of MODFETs are therefore limited compared to enhancement mode FETs. It is possible to design MODFET devices which work in enhancement mode but again leakage and power levels are still greater than comparable fully enhancement mode devices.

7.4.1 Low Temperature Properties of Two Dimensional Modulation-Doped Electron and Hole Gases

While the low temperature properties of modulation doped material are not that useful for building circuits, by going to low temperatures the phonon scattering in the system which typically dominates the room temperature properties can be removed, thereby allowing extraction of other scattering mechanisms in the system. In particular this allows the roughness of heterointerfaces and the background impurity concentrations in the material to be minimised. In addition, by sweeping a magnetic field, Schubnikov de Haas oscillations can be found in the longitudinal resistivity (ρ_{xx}) and quantum Hall platuea in the transverse resistivity (ρ_{xy}) as Landau levels formed in the system and swept through the chemical potential. The temperature dependence of the Shubnikov de Haas oscillations also allows the extract of the effective mass of the electrons or holes in the modulation doped sample. Figure 7.16 shows the longitudinal and transverse magnetoresistivity from a modulation doped strained-Si quantum well grown on a relaxed $Si_{0.77}Ge_{0.23}$ virtual substrate grown by DERA and measured at Cambridge. The mobility is $361\,000\,cm^2/Vs$ with a carrier density of 3.4×10^{11} cm^{-2} at 55 mK. This sample also shows a number of fractional quantum Hall plateau in the transverse resistivity which demonstrate that the material is of extremely high quality. When Landau levels are split, high quality 2DEGs in silicon will also have spin and valley degeneracies split by the application of a magnetic field which can be resolved in Fig. 7.17. At lower magnetic fields these splittings which are weaker than the Landau level splittings will disappear.

To optimise the low temperature mobility when modulation doping is being used, the spacer thickness must be carefully chosen. Larger spacers reduce the remote ionised impurity scattering since the electrons are taken further away from the donors which created them, thereby reducing the Coulomb

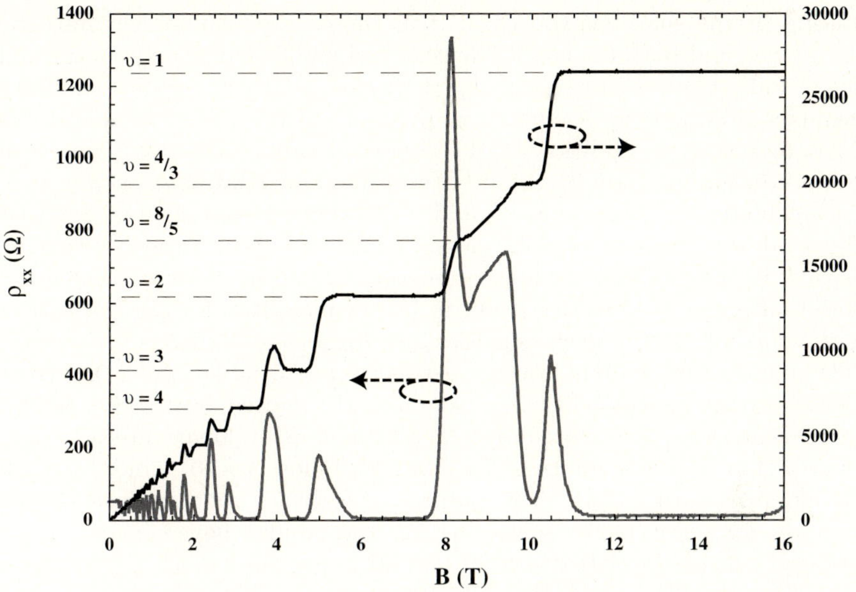

Fig. 7.16. The magnetoresistivity of a modulation doped Si/SiGe 2DEG grown by DERA with a mobility of $361\,000\,\mathrm{cm}^2/\mathrm{Vs}$ and a carrier density of $3.4\times10^{11}\,\mathrm{cm}^{-2}$ at 55 mK. The major integer and fractional quantum Hall plateau in the transverse magnetoresistance are shown

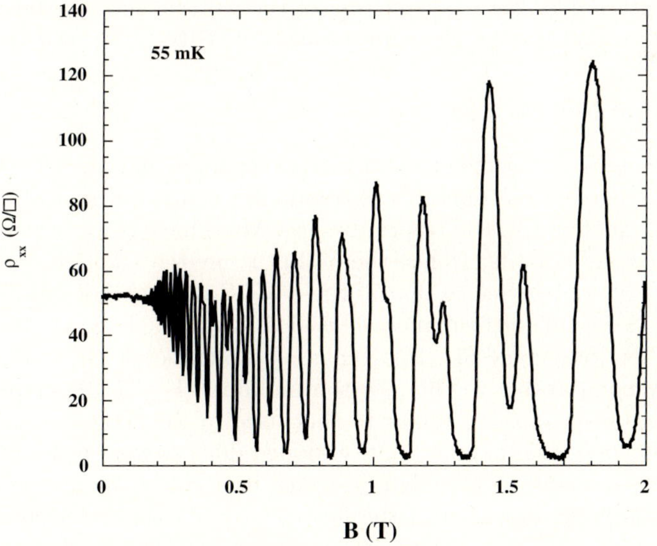

Fig. 7.17. An expanded version of the longitudinal resistivity of the 2DEG shown in Figure 7.16 showing both spin and valley splitting of the Landau levels

scattering. If the spacer is too thick then the carriers cannot get into the quantum well and both the carrier density and mobility in the quantum well typically falls. Figure 7.18 shows a plot of carrier density against mobility for samples produced by a number of different groups, all at temperatures of 4.2 K or below. In addition to the experimental results, two theoretical curves produced by Frank Stern and Steven Laux at IBM have been drawn. These calculate the mobility as a function of carrier density for two different background impurity density. The carrier density has been varied by varying the spacer in the modulation doped structure. It can be observed that lower background impurity densities result in higher mobilities for a given carrier density and that lower carrier densities give higher mobilities.

The maximum in mobility occurs between 3 and 5 $\times 10^{11}$ cm^{-2}. At higher carrier densities, the mobility is prodominantly limited by remote ionised impurity scattering. Therefore either larger amounts of dopant in the supply layer or a thin spacer is limiting the mobility. Below 3 $\times 10^{11}$ cm^{-2}, the fall off in mobility is the result of a mixture of interface roughness scattering and remote ionised impurity scattering. All the samples have Ge contents in the virtual substrates of between 20 and 30%. For the higher Ge contents, the interface roughness is higher which will reduce the mobility while the splitting of the valleys is also higher which should increase the mobility since the probability of electrons occupying the upper valleys is reduced. For low carrier densities, lower Ge contents in the virtual substrate are required for higher mobilities. The optimum mobility is therefore a compromise between remote ionised impurity scattering, the splitting of the valleys, the interface roughness in the system and the background impurity density.

7.4.2 Pseudomorphic MODFETs

One of the limiting factors in performance of CMOS is the p-channel MOSFET. The mobility and effective masses are about 2.5 times lower in the p-MOS than the n-channel and so the transistors in CMOS have to be scaled accordingly to balance the circuits. Hence the ability to match the size and performance of the p-channel device to that of the n-MOS would be of significant benefit to CMOS circuit performance. The ideal way of achieving this would be to have a pseudomorphic $Si_{1-x}Ge_x$ channel grown below the oxide with a large mobility comparable to the electron mobility (Fig. 7.19). This type of heterostructure layer is very similar to that used in the HBT.

The first pseudomorphic channel modulation-doped devices were produce by John Bean and co-workers at Bell Laboratories in 1982. A significant number of publications have appeared in the field trying to attempt such a heterostructure MOSFET device. Two major problems have been encountered in trying to realise a technology that may be integrated in a CMOS production line. The first is that to integrate strained layers into a CMOS line, the structures should be as close to compatible with conventional processing as possible unless the performance gain is so large that increased production

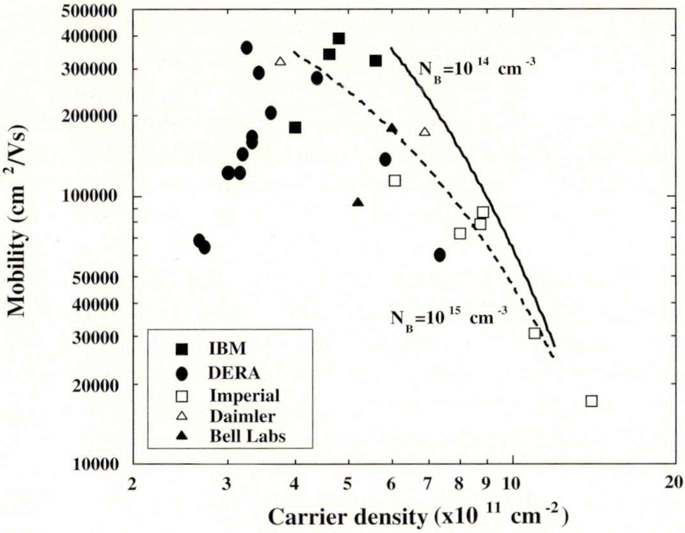

Fig. 7.18. The mobility below 4.2 K against carrier densities for modulation doped Si/SiGe 2DEG grown by a number of different groups. The maximum in mobility occurs for carrier densities between 3 and 5 $\times 10^{11}$ cm^{-2}. At higher carrier densities, the mobility is prodominantly limited by remote ionised impurity scattering. At low carrier densities, interface roughness scattering becomes prevalent

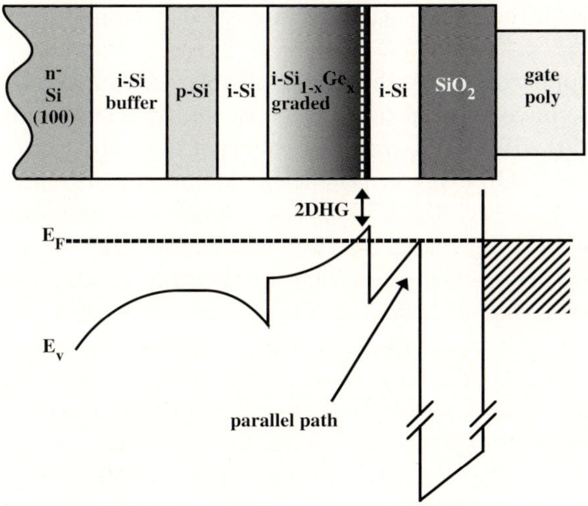

Fig. 7.19. A buried channel-SiGe MOSFET demonstrating where parallel conduction at the Si/SiO_2 interface can occur

costs can be justified. Any strained layer incorporated must be below the equilibrium or Matthews and Blakeslee critical thickness otherwise dislocations and defects will result either from growth or during process annealling stages reducing performance and yield. The high thermal budgets used in present CMOS production are not ideal for strained layers and may cause strain relaxation or diffusion. The second problem is that the valence band discontinuity of $Si_{1-x}Ge_x$ to Si is small especially at low Ge contents and with the Ge content being kept low to allow the $Si_{1-x}Ge_x$ layer to be stable to thermal treatments, parallel conduction can occur in the transistors producing a 2D hole gas (2DHG) at the Si/SiO_2 interface and the $Si/Si_{1-x}Ge_x$ interface (Fig. 7.19). The parallel conduction significantly reduces the performance of such devices. Finally, while such devices may have lower noise since the holes are situated at a heterointerface with a higher mobility, the transconductance will be lower than a comparable p-MOSFET or strained-Si p-MOSFET since the holes have been moved further away from the gate.

IBM demonstrated a modulation-doped heterostructure $Si_{1-x}Ge_x$ MOSFET which was processed using a CMOS fabrication process but performance was only 20% or so better than a conventional p-MOS transistor. It is cheaper to down scale the p-MOS gate-length than to incorporate a technology with epitaxy into a production line with such performance and so a number of companies have stopped research in this area. The best mobility achieved was about $220\,cm^2/Vs$ in the $Si_{1-x}Ge_x$ p-MOS at 300 K and a value closer to $500\,cm^2/Vs$ is probably required before the technology can compete on economic grounds.

Better mobilities have appeared for pseudomorphic $Si_{1-x}Ge_x$ channel MODFETs but many of these have parallel conduction or multiple subbands at 300 K and the true mobilities will be substantially different from the values quoted. As the gate-lengths in CMOS decrease, the thermal budgets also have to decrease to prevent diffusion especially of dopants (Ohmic contacts and threshold implants). A point may be reached where the extra confinement awarded by such a heterostructure transistor may be beneficial at small gate-lengths and the economics and reduced thermal budgets could then allow the technology to be implemented. The reason behind the relatively low mobilities in such structures are still not fully understood. Theoretical modelling suggests that higher mobilites should be realised at room temperature but these have not come to fruition in experiments. Even the low temperature mobilities are much lower than predicted with values up to $17\,000\,cm^2/Vs$ in the metallic regime at 300 mK. Alloy scattering looks the most promising explanation as the theories used are by nature very simplified and may underestimate the scattering but interface roughness and charge scattering along with other mechanisms cannot be completely overlooked. Ge clustering has been observed in experiments and this may modify scattering processes, especially alloy scattering, from the theories which assume a uniformly random Ge distribution.

7.4.3 Virtual Substrate MODFETs

From a performance view point, some of the most exciting $Si_{1-x}Ge_x$ FET results to date have been achieved with modulation doped field effect transistors (MODFETs) grown on virtual substrates. There are many different types of heterolayer design for MODFETs as shown in Fig. 7.20. The standard design used is a direct copy of most GaAs / AlGaAs HEMT designs with the doping above the quantum well and a metal Schottky gate is used (Fig. 7.20(a)). Schottky gating in Si technology is far poorer than oxide gates due to much higher leakage currents. This leakage may be reduced by placing the doping below the quantum well but it still remains orders of magnitude above an oxide gate (Fig. 7.20(b)). This design also has the advantage of reducing the channel to gate distance and thereby increasing the transconductance of the device. The disadvantage is that surface segregation of the dopant or memory effects during growth can either make doping under the channel impossible or very difficult. The carrier density may be increased, thereby maximising the conductance of the MODFET by placing doping on both sides of the quantum well (Fig. 7.20(c)). This has particular applications in a number of high power analogue applications. Finally all of the above designs may also have a gate oxide incorporated to reduce the gate leakage current (Fig. 7.20(d)).

For all the designs discussed above an equivalent p-MODFET design also exists (Fig. 7.21). Basically a high Ge content layer will give a quantum well in the valence band. The major consideration of the design of the structure is the quality of the strain relaxation buffer. While a pure Ge channel will have no alloy scattering (or at least only a tiny amount from the leakage of the wavefunction of the holes into the SiGe spacer) and therefore on paper appear the best for performance. Pure Ge channels require a high Ge content in the strain relaxation buffer, typically $Si_{0.4}Ge_{0.6}$ is required for a sensible quantum well thickness, then the surface roughness of the buffer can be quite severe and limit the performance. Khalid Ismail and colleagues at IBM has used $Si_{0.3}Ge_{0.7}$ quantum wells on $Si_{0.7}Ge_{0.3}$ strained relaxation buffers which are also the typical content used for the n-MODFET devices. This substantially reduces the surface roughness allowing short gate length transistors to be easily fabricated along with potentially allowing the p-MODFETs to be integrated with the n-MODFETs for complementary MODFET architectures.

7.4.4 Analytical Description of MODFET Operation

The operation of the MODFET is similar to that of the MOSFET as described in Sect. 5.3.2. Here we will describe the model of Drummond et al. which was originally designed for GaAs / AlGaAs MODFETs but is equally applicable to SiGe MODFETs. The reader should consider that in this original paper the energy terms had voltage units which was changed in this paper. This will allow the important operating parameters of a MODFET to

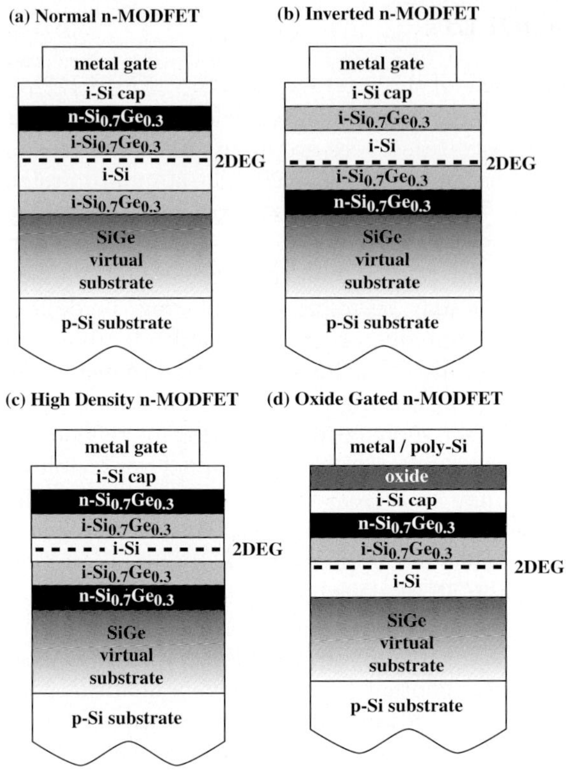

Fig. 7.20. Different designs for SiGe n-MODFETs. (a) Standard n-MODFET (b) the inverted MODFET with the doping under the channel (c) doping on both sides of the channel to increase the carrier density and (d) an oxide gated device

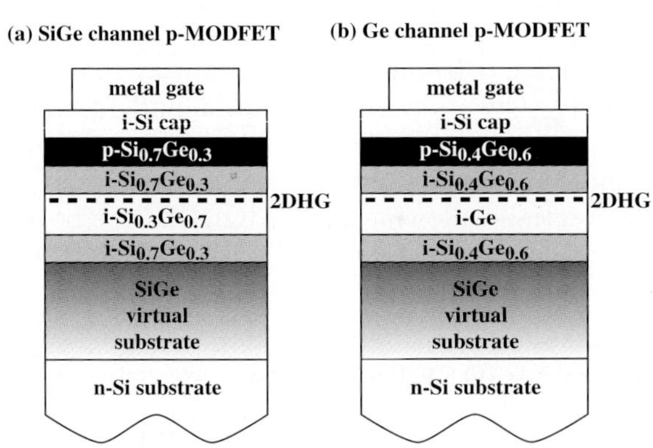

Fig. 7.21. Schematic diagrams of SiGe p-MODFETs with (a) a SiGe channel and (b) a pure Ge channel

be described in terms of the sheet charge density in the 2DEG as a function of mobility and carrier velocity. We will derived all the equations for electrons in an n-MODFET but these can easily be modified to those required for holes in a p-MODFET.

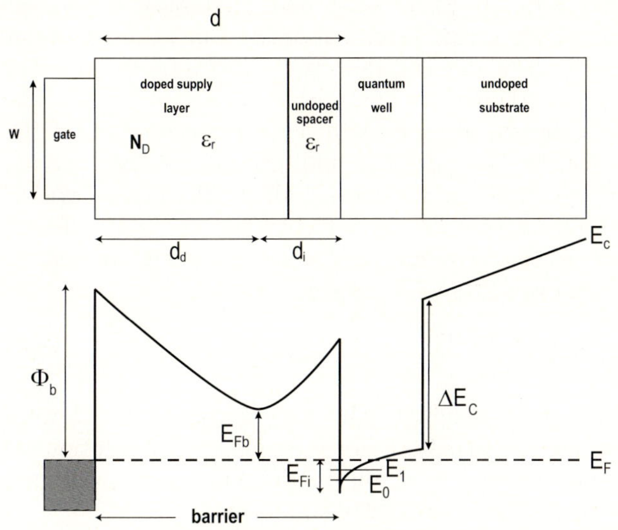

Fig. 7.22. A schematic diagram of the layered structure and conduction band with all the parameters required to model the properties of a MODFET

To start we impose the constraint that the amount of charge depleted from the donor supply layer must equal the charge accumulated at the heterointerface while the Fermi level is kept constant. The electron charge depleted from the donor supply layer is given by

$$n_s = \sqrt{\frac{2\varepsilon_r\varepsilon_0}{q^2}\left(\Delta E_c - E_{Fb} - E_{Fi}\right)N_D + N_D^2 d_i^2} - N_D d_i \tag{7.6}$$

where E_{Fb} is the separation between the conduction band in the barrier layer and the Fermi level, N_D is the doping density in the donor supply layer, ΔE_c is the conduction band discontinuity between the quantum well and the barrier, ϵ_r is the dielectric constant of the barrier material, E_{Fi} is the Fermi level with respect to the conduction band edge in the quantum well and d_i is the thickness of the undoped spacer between the quantum well and the doped supply layer. The electron charge stored at the heterointerface in the quantum well is given by

$$n_s = \int_{E_{z,0}}^{\infty} g_{2D}\left(E\right) f\left(E\right) \mathrm{d}E \tag{7.7}$$

$$n_s = g_{2D}(E)k_BT$$

$$\ln\left[\left(1+\exp\left\{\frac{E_{Fi}-E_0}{k_BT}\right\}\right)\left(1+\exp\left\{\frac{E_{Fi}-E_1}{k_BT}\right\}\right)\right] \tag{7.8}$$

where E_0 and E_1 are the lowest two subbands in the triangular quantum well as calculated in Sect. 3.1.3. It is assumed here that only the lowest energy states are filled or partly filled. More subbands can be used but it substantially complicates the problem and does not substantially improves the accuracy of the result. The 2D density of states is as given in (3.91). The application of a voltage to the surface gate will deplete some or all of the charge at the heterointerface in the quantum well. Simultaneously solving (7.6) and (7.8) with a known sheet carrier density allows the Fermi level to be found in the system. Alternatively the sheet charge density can be found if the Fermi level is known. With the addition of a gate bias with respect to the source, V_g, the additional condition 7.9 fixes n_s to

$$n_s = \frac{\varepsilon_r\varepsilon_0}{qd}\left[V_g - \frac{1}{q}\left(\phi_b - \frac{q^2N_Dd_d^2}{2\varepsilon_r\varepsilon_0} + E_{Fi} - \Delta E_c\right)\right] \tag{7.9}$$

where ϕ_b is the Schottky barrier height of the gate metal. d is the total distance between the gate and the 2DEG channel. Simultaneous solutions of (7.6) and (7.9) allow the carrier density in the 2DEG as a function of gate bias to be determined. (7.9) can be rewritten in the form

$$n_s = \frac{\varepsilon_r\varepsilon_0}{q(d+\Delta d)}\left[V_g - V_T\right] \tag{7.10}$$

where the threshold voltage is defined as

$$qV_T = \phi_b - \Delta E_c - \frac{q^2N_Dd_d^2}{2\varepsilon_r\varepsilon_0} + \Delta E(T) \tag{7.11}$$

and where $\Delta E(T)$ is a term for the temperature dependence of the system with Δd related to the linearity of the carrier density to E_{Fi} such that

$$E_{Fi} = \Delta E(T) + \frac{q^2\Delta d}{\varepsilon_r\varepsilon_0}n_s \tag{7.12}$$

These terms are determined from experiment and basically represent the deviations between experimental values and the model.

Let us now add a drain bias to the system between the source and drain of the FET. In the long channel limit or for small source-drain biases, the variation in the bias along the 2DEG channel can be added to the gate bias describing a channel potential $V(x)$ to produce

$$n_s = \frac{\varepsilon_r\varepsilon_0}{q(d+\Delta d)}\left[V_g - V_T - V(x)\right] \tag{7.13}$$

For small values of $V(x)$ it can be assumed that the mobility, μ, is constant along the channel and so for a transistor of gate width, W it is found that the source drain current is

$$I_{ds} = q\mu_n n_s W \frac{\mathrm{d}V(x)}{\mathrm{d}x} \tag{7.14}$$

$$I_{ds} = \mu_n W \frac{\varepsilon_r \varepsilon_0}{(d + \Delta d)} [V_g - V_T - V(x)] \frac{\mathrm{d}V(x)}{\mathrm{d}x} \tag{7.15}$$

By integrating between the source and drain for a constant drain current the source drain current for a gate of length L_g is given by

$$I_{ds} = \mu_n W \frac{\varepsilon_r \varepsilon_0}{(d + \Delta d) L_g} \left\{ (V_g - V_T) V_{ds} - \frac{V_{ds}^2}{2} \right\} \tag{7.16}$$

To obtain the saturation source-drain current, we will use firstly the two-piece model which represent an abrupt transition from the constant mobility regime to a velocity saturated regime. When the velocity of the electrons reaches saturation with a value, $v_{sat} = \frac{F}{\mu_n}$, the saturation current is given by

$$I_{ds,sat} = \mu_n W \frac{\varepsilon_r \varepsilon_0}{(d + \Delta d) L_g} \frac{\left[(V_g - V_T) V_{ds,sat} - \frac{V_{ds,sat}^2}{2} \right]}{1 + (\mu_n V_{ds,sat} / v_{sat} L_g)} \tag{7.17}$$

A more accurate method is to use a smooth transition between the constant mobility and the velocity saturated regime. The Si velocity field curve saturates at about 6 $\times 10^6$ V/m and can be described by the expression

$$v = \frac{\mu F(x)}{1 + \mu F(x) / v_{sat}} \tag{7.18}$$

where $F(x)$ is the electric field in the channel which is equal to $\frac{\mathrm{d}V(x)}{\mathrm{d}x}$. The field is not constant in the channel, however. Therefore the source-drain current

$$\begin{aligned} I_{ds} &= W \frac{\varepsilon_r \varepsilon_0}{(d + \Delta d)} (V_g - V_T - V(x))\, v(x) \\ &= \left[\mu_n \frac{\mathrm{d}V(x)}{\mathrm{d}x} \right] \left[1 + \frac{\mu \frac{\mathrm{d}V(x)}{\mathrm{d}x}}{v_{sat}} \right]^{-1} \frac{W \epsilon_r \epsilon_0}{d + \Delta d} (V_g - V_T - V(x)) \end{aligned} \tag{7.19}$$

where $v_{sat} = \mu F_{sat}$ with F_{sat} the electric field at which the velocity saturates. By integrating (7.19) from the source to the drain while remembering that the drain current must be constant, one obtains

$$I_{ds} = \frac{\mu_n W \frac{\varepsilon_r \varepsilon_0}{(d + \Delta d)} \left\{ (V_g - V_T) V_{ds} - \frac{1}{2} V_{ds}^2 \right\}}{L_g + \frac{\mu_n}{v_{sat}} V_{ds}} \tag{7.20}$$

It should be noted that if the saturation velocity approaches infinity, (7.20) reduces to (7.16) which is the long channel case. By using (7.19) and (7.20), the saturation source-drain current can be found as

$$I_{ds,sat} = \frac{2\left(V_g - V_T\right)^2 \mu_n W \frac{\varepsilon_r \varepsilon_0}{L_g (d+\Delta d)}}{\left\{1 + \sqrt{\left[1 + \frac{2\mu_n (V_g - V_T)}{v_{sat} L_g}\right]}\right\}^2} \tag{7.21}$$

We can now find the transconductance in the saturation regime as

$$\begin{aligned} g_{m,sat} &= \frac{\mathrm{d}I_{ds,sat}}{\mathrm{d}V_g} |_{V_{ds}=cons\tan t} \\ &= \frac{\left(V_g - V_T\right) \mu_n W \frac{\varepsilon_r \varepsilon_0}{L_g (d+\Delta d)}}{\left\{1 + \frac{2\mu_n (V_g - V_T)}{v_{sat} L_g} \sqrt{\left[1 + \frac{2\mu_n (V_g - V_T)}{v_{sat} L_g}\right]}\right\}} \end{aligned} \tag{7.22}$$

The maximum transconductance is obtained when the sheet charge density is completely undepleted under the gate which produces

$$g_{m,\max} = \frac{q\mu_n W n_s}{L_g} \left[1 + \left(\frac{q\mu n_s \left(d + \Delta d\right)}{\varepsilon_r \varepsilon_0 v_{sat} L_g}\right)^2\right]^{-1/2} \tag{7.23}$$

For short gate lengths which represents almost all present day MODFETs, (7.23) reduces to

$$g_{m,\max} = \frac{v_{sat} W \varepsilon_r \varepsilon_0}{(d + \Delta d)} \tag{7.24}$$

The transconductance from experimental measurements or the extrinsic transconductance is actually smaller than the intrinsic value given by (7.24) as the source resistance, R_s reduces the transconductance through negative feedback. The equivalent circuit for a MODFET is shown in Fig. 7.23 where the resistances, capacitances and inductances which determine the high frequency behaviour of a MODFET are shown.

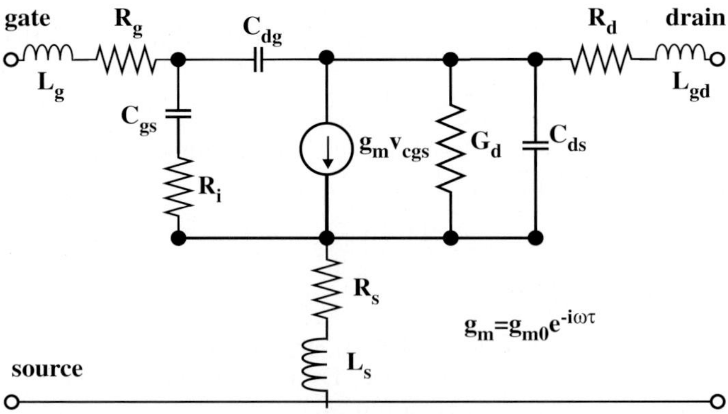

Fig. 7.23. The equivalent circuit model that is commonly used for the analysis of MODFETs and extraction of the important parameters

Therefore the extrinsic transconductance is

$$g_{m,\max}\big|_{ext} = \frac{g_{m,\max}}{1 + R_s g_{m,\max}} \tag{7.25}$$

The current gain cutoff frequency, defined as the frequency at which the current gain equals unity, since determined to be

$$f_T = \frac{g_m}{2\pi C_{gs}} \tag{7.26}$$

where C_{gs} is the gate-source capacitance since the feedback capacitance is negligible compared to the input capacitance. It is clear from (7.26) that the higher the saturation velocity and the smaller the gatelength, the higher f_T. The maximum oscillation frequency, defined as the frequency at which the power gain equals unity is given by

$$f_{\max} = \frac{f_T}{2\sqrt{(R_g + R_i + R_s)\,G_d + 2\pi f_T R_g C_{dg}}} \tag{7.27}$$

where the resistances and capacitances are those defined by the equivalent circuit model in Fig. 7.23.

From all the equations above, it is clear that to optimise the performance of MODFETs in circuits, either the saturation velocity must be increased or the gatelength of the transistor must be decreased. As the former is a fixed material parameter, it is the reduction of gatelength which primarily determines the transconductance and high speed performance. In addition, reduction of parasitic resistances and capacitances will increase the f_{max} of a transistor. These are the reasons that the gatelength of MODFETs in all material technologies is agressively scaled to lower gatelengths.

7.4.5 SiGe MODFET Performance

As a comparison of performance of both SiGe n and p-MODFETs, Fig. 7.24 shows the cut-off frequency, f_T as a function of gate-length for a number of different transistor families. Care has been taken only to plot $Si_{1-x}Ge_x$ devices with f_{max} comparable or higher than f_T. The lines correspond to theoretical scaling of the device geometry and for small gate-lengths the experimental points lie below these curves due to increased parasitics. The $Si_{1-x}Ge_x$ n-MODFET has switching times comparable to n-MODFETs in GaAs/AlGaAs devices and larger than GaAs MESFETs. The $Si_{1-x}Ge_x$ p-MODFET is even more impressive as the results are faster than any other p-channel transistors published. Mobilities at 300 K (77 K) of 2 830 cm^2/Vs (18 000 cm^2/Vs) at 2×10^{12} cm^{-2} (8×10^{11} cm^{-2}) for the n-MODFET and 1 300 cm^2/Vs (14 000 cm^2/Vs) at 1.5×10^{12} cm^{-2} (1.0×10^{12} cm^{-2}) for the p-MODFET have been demonstrated. Transconductances in the n-type for a 0.25 μm gate-length device at 300 K (77 K) was 330 mS/mm (600 mS/mm) and for a 0.1 μm p-MODFET 237 mS/mm. At low temperatures the results are more impressive with two dimensional electron gas (2DEG) mobilites up to 390 000 cm^2/Vs at 400 mK and two dimensional hole gases (2DHGs) reaching 55 000 cm^2/Vs at 4.2 K. The n-channel enhancements are due to reduced intervalley scattering with the strain splitting the valley degeneracy and the higher saturation velocity achievable at lower electric fields (Fig. 3.32). The p-channel enhancements result from the lower effective mass and Ge-like strain modified valance band structure. Initial modelling of circuit performance is also encouraging. n-MODFETs with loads of 200 fF were shown to exhibit 560 ps delays in NOR gates at 1.1 V compared to 1400 ps delays for the equivalent CMOS. The CMOS had to be run at 3.3 V to achieve the same delay, consuming nine times the power of the $Si_{1-x}Ge_x$ MODFETs. The possibility of high-speed complementary logic with such devices looks promising on paper.

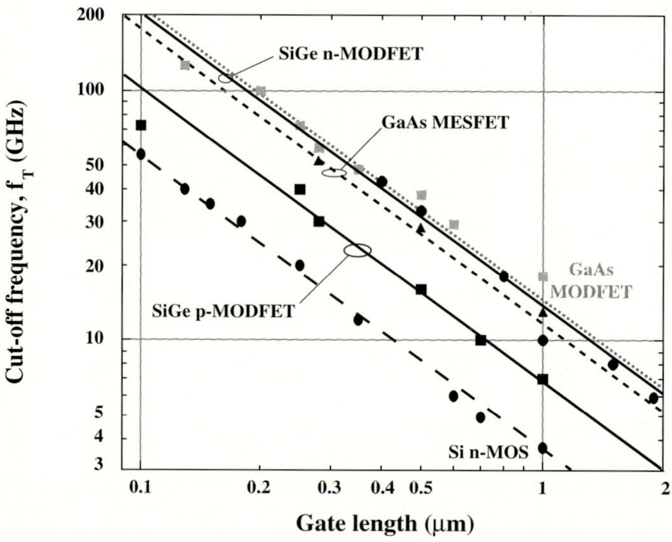

Fig. 7.24. The cut-off frequency, f_T as a function of the gatelength of transistors for n-MOSFETs, strained-SiGe p-MODFETs grown on virtual substrates, strained-Si n-MODFETs, GaAs n-MESFETs and GaAs n-MODFETs

7.5 Further Reading

1. M.J. Kelly, Low Dimensional Semiconductors: Materials, Physics, Technology, Devices" OUP, Oxford (1995)
2. J.H. Davies, The Physics of Low Dimensional Semiconductors, CUP, Cambridge (1998)

3. S.M. Sze, Semiconductor Devices: Physics and Technology, 2nd Edition, John Wiley and Sons, New York (2002)
4. L.D. Nguyen, L.E. Larson and U.K. Mishra, Proc. IEEE, **80**, 494 (1992)
5. D.J. Paul, Semiconductor Science and Technology **19**, R59 (2004)
6. L. Risch, Silicon Nanoelectronics: the next 20 years, in Silicon: Evolution and Future of a Technology, (P. Siffert, E. Krimmel, eds.), Springer, Berlin (2004)
7. T.J. Drummond, H. Morkoc, K. Lee, M. Shur, IEEE-EDL 3, 311 (1982)
8. H. Heinrich, Modulation-Doped Filed-Effect Transistors, IEEE Press, New York (1991)
9. E. Kasper, S. Heim, Appl. Surf. Sci 224, 3 (2004)
10. N. Arora, MOSFET models for VLSI circuit simulation, Springer-Verlag, Wien (1993)
11. S. Karmalkar and G. Ramesh, IEEE-ED 47, 11 (2002)

8. Tunneling Phenomena

8.1 Tunnel Diodes

One of the true quantum devices which has no classical analogue is the tunnel diode where the quantum mechanical tunnelling of electrons or holes is used to produce a negative differential resistance (NDR) region in the current-voltage characteristics. This NDR region allows the design of a number of circuits for applications including logic and memory. Tunnel diodes come in two flavours, the intra-band resonant tunnelling diode (RTD) and the inter-band tunnel diode normally termed an Esaki diode. Esaki diodes have been around since the 1960s while the first RTD was produced in GaAs by Leo Esaki and Daniel Tsu in 1974. In the III-V system, RTDs are now a mature technology with many logic and memory circuits demonstrated along with the fastest operation of any transit time device of 712 GHz. In the silicon system, however, the band structure is not ideal for producing RTDs. The first SiGe RTD was produced by Liu in 1988 and the first room temperature operation with extremely limited performance was in 1991. Since 1998, a number of major breakthroughs have been achieved in both the inter-band and the intra-band tunnel diodes which are now good enough to be able to demonstrate circuits at room temperature.

8.2 Resonant Tunnelling

Section 3.1 has shown the subband states in a quantum well along with tunnelling through a single barrier, and so the next system of interest is a double barrier system commonly known as a resonant tunnelling diode or RTD. The basic system for calculating the transmission coefficient for multiple barriers and specifically a double barrier system is shown in Sect. 8.2.1. The simplest way to calculate the transmission coefficient for the system is to consider an electron in the central quantum well which may escape to the left or the right (Fig. 8.1). The qualitative ideas behind resonant tunnelling will be described in detail which should allow a detailed description of the physics without complicated mathematical rigor. The operation of a RTD in many ways is similar to a Fabry-Perot resonator in optics.

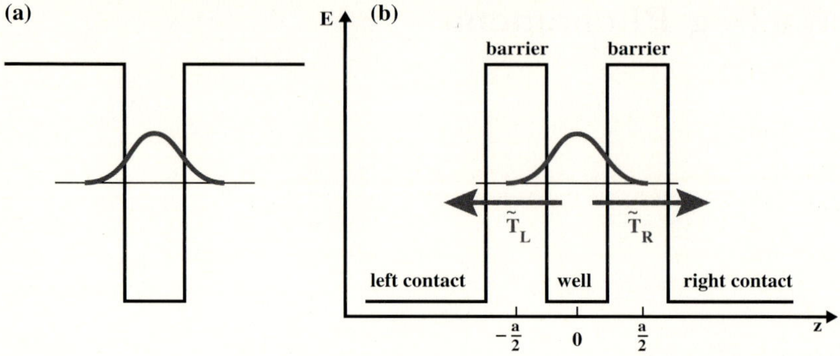

Fig. 8.1. (a) The wavefunction of an electron in the first bound state subband of a quantum well. (b) The same quantum well but with two barriers of finite thickness on each side where the bound state becomes a resonance or quasi-bound

Figure 8.2 shows a schematic of the energy band structure of a RTD with a quantum well between two barriers in the conduction band of a semiconductor heterostructure system such as Si / SiGe. On the right is shown the ideal transmission coefficient for electrons incident on the barriers with energy E which is derived below in Sect. 8.2.1. The transmission coefficient peaks every time an electron with energy equal to one of the subband energies in the quantum well is incident on the barriers. The current through the system therefore peaks every time the electrons can tunnel into an allowed subband state - basically the subband is a resonant state - hence the name resonant tunnelling diode.

The detailed current voltage properties of a RTD may be derived from the transmission properties and the band structure of the material. Let us define the transmission and reflection coefficients through the right barrier as $\tilde{T}_R$ and $\tilde{R}_R$ respectively and those through the left barrier as $\tilde{T}_L$ and $\tilde{R}_L$ (Fig. 8.1).

8.2.1 $\tilde{T}$-Matrices

Here the idea of $\tilde{T}$-matrices will be introduced. More detailed derivations can be found in the references at the end of the chapter. The great advantage of $\tilde{T}$-matrices is that once the transmission and reflection coefficients are known for a number of barriers, then the transmission through any combination of the barriers may be calculated using matrix algebra. For double or triple barrier systems, this is far easier to calculate than the complex integrals from more common techniques.

8.2.2 The Single Barrier

Let us generalise the tunnelling problem to any barrier where the incident wavefunction has amplitude A with wavevector k_1, the reflected wavefunc-

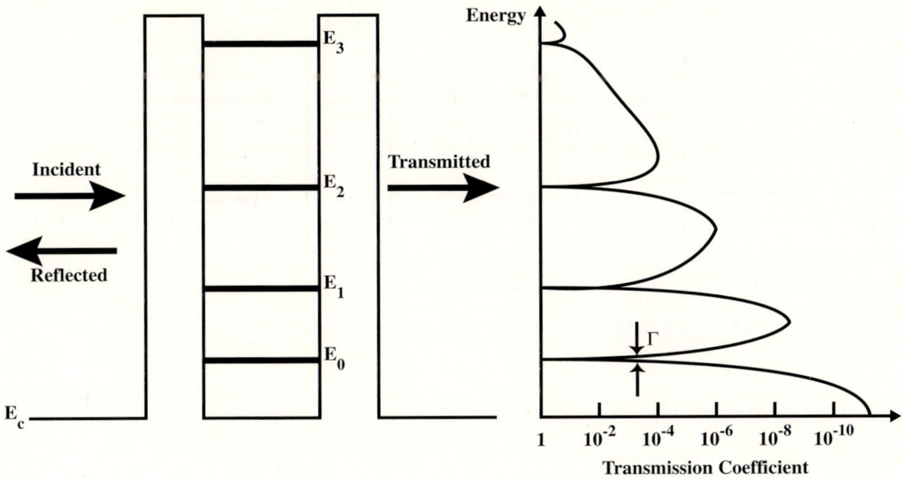

Fig. 8.2. A schematic diagram of the energy band structure of a RTD with a number of subbands in the quantum well defined between two barriers of e.g. SiGe. The diagram on the right shows the transmission coefficient with energy and the resonances which occur as the energy of an incident electron coincides with each of the energies of the subbands in the quantum well

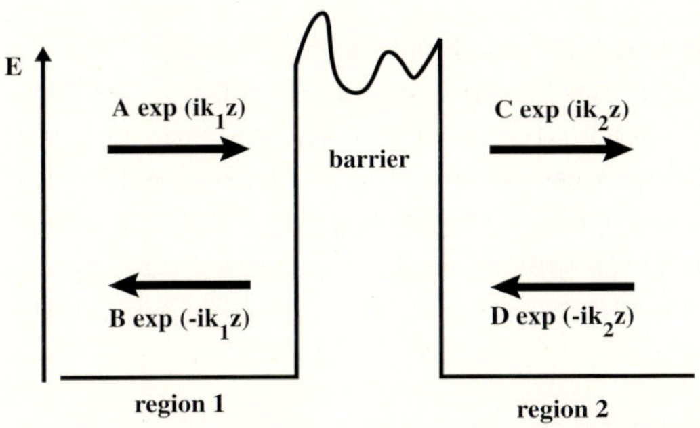

Fig. 8.3. A schematic showing a general transmission problem with a barrier of arbitrary height and shape. Each side of the barrier may have incoming and outgoing waves and the wavevectors on each side of the barriers may be different

tion has amplitude B and wavevector k_1, the transmitted wavefunction has amplitude C and wavefunction k_2 and a final incident wavefunction to the barrier from the right with amplitude D and wavefunction k_2 is also included to represent any reflections from anything to the right of the barrier (Fig. 8.3). Let us define the transmission coefficient for tunnelling from region 1 to region 2 as $\tilde{T}^{(21)}$ such that

$$\begin{pmatrix} C \\ D \end{pmatrix} = \tilde{T}^{(21)} \begin{pmatrix} A \\ B \end{pmatrix} = \begin{pmatrix} \tilde{T}_{11}^{(21)} & \tilde{T}_{12}^{(21)} \\ \tilde{T}_{21}^{(21)} & \tilde{T}_{22}^{(21)} \end{pmatrix} \begin{pmatrix} A \\ B \end{pmatrix} \tag{8.1}$$

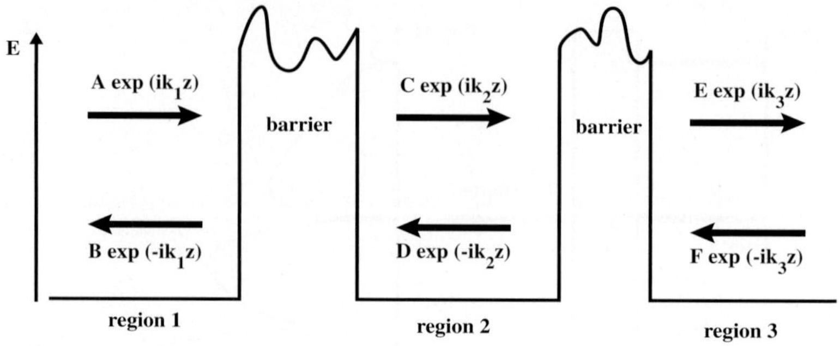

Fig. 8.4. A schematic diagram of a double barrier system

Expanding this to the two barriers of Fig. 8.4 such that

$$\begin{pmatrix} C \\ D \end{pmatrix} = \tilde{T}^{(21)} \begin{pmatrix} A \\ B \end{pmatrix} \qquad \begin{pmatrix} E \\ F \end{pmatrix} = \tilde{T}^{(32)} \begin{pmatrix} C \\ D \end{pmatrix} \tag{8.2}$$

which when combined produces

$$\begin{pmatrix} E \\ F \end{pmatrix} = \tilde{T}^{(32)} \tilde{T}^{(21)} \begin{pmatrix} A \\ B \end{pmatrix} \equiv \tilde{T}^{(31)} \begin{pmatrix} A \\ B \end{pmatrix} \tag{8.3}$$

The order of the superscripts should now be clear as $\tilde{T}^{(31)} = \tilde{T}^{(32)}\tilde{T}^{(21)}$. Therefore as long as one may define the $\tilde{T}$ matrices for all the barriers in a system, these may be multiplied together to find the total transmission and reflection in the system. The matrices must be written in the opposite order to the numbering of the barriers from left to right so that they act on the amplitudes of the waves in the correct sequence.To find the transmission and reflection coefficients, the waves on either side of a single barrier are related through

$$\begin{pmatrix} \tilde{t} \\ 0 \end{pmatrix} = \tilde{T} \begin{pmatrix} 1 \\ \tilde{r} \end{pmatrix} = \begin{pmatrix} \tilde{T}_{11} & \tilde{T}_{12} \\ \tilde{T}_{21} & \tilde{T}_{22} \end{pmatrix} \begin{pmatrix} 1 \\ \tilde{r} \end{pmatrix} \tag{8.4}$$

and so

$$\tilde{r} = -\frac{\tilde{T}_{21}}{\tilde{T}_{22}} \qquad \tilde{t} = \frac{\tilde{T}_{11}\tilde{T}_{22} - \tilde{T}_{12}\tilde{T}_{21}}{\tilde{T}_{22}} \tag{8.5}$$

Using the conservation of current and time reversal invariance, it can be shown that $\tilde{T}_{22} = \tilde{T}_{11}^*$, $\tilde{T}_{21} = \tilde{T}_{12}^*$ and $\left|\tilde{T}_{11}\right|^2 - \left|\tilde{T}_{12}\right|^2 = \det\left|\tilde{T}\right| = 1$. Hence the previous equations may be simplified to

$$\tilde{r} = -\frac{\tilde{T}_{12}^*}{\tilde{T}_{11}^*} \qquad \tilde{t} = \frac{1}{\tilde{T}_{11}^*} \tag{8.6}$$

Let us now re-examine the single barrier problem using the $\tilde{T}$-matrices.

$$\tilde{T}^{(31)} = \begin{pmatrix} e^{-ik_1 b/2} & 0 \\ 0 & e^{ik_1 b/2} \end{pmatrix} \tilde{T}(k_1, k_2) \begin{pmatrix} e^{ik_2 b/2} & 0 \\ 0 & e^{-ik_2 b/2} \end{pmatrix} \begin{pmatrix} e^{ik_2 b/2} & 0 \\ 0 & e^{-ik_2 b/2} \end{pmatrix} \tilde{T}(k_2, k_1) \begin{pmatrix} e^{-ik_1 b/2} & 0 \\ 0 & e^{ik_1 b/2} \end{pmatrix} \tag{8.7}$$

The middle pair of matrices can be multiplied to produce a matrix with $exp\,(\pm ik_2 b)$ on the diagonal, reflecting the change in phase while travelling between the barrier and the free space. Multiplying this out gives

$$\begin{aligned} \tilde{T}^{(31)} &= \frac{1}{2k_1 k_2} \begin{pmatrix} e^{-ik_1 b/2} & 0 \\ 0 & e^{ik_1 b/2} \end{pmatrix} \\ &\times \begin{pmatrix} 2k_1 k_2 \cos(k_2 b) + i\left(k_1^2 + k_2^2\right)\sin(k_2 b) & -i\left(k_1^2 - k_2^2\right)\sin(k_2 b) \\ i\left(k_1^2 - k_2^2\right)\sin(k_2 b) & 2k_1 k_2 \cos(k_2 b) - i\left(k_1^2 + k_2^2\right)\sin(k_2 b) \end{pmatrix} \\ &\times \begin{pmatrix} e^{-ik_1 b/2} & 0 \\ 0 & e^{ik_1 b/2} \end{pmatrix} \end{aligned} \tag{8.8}$$

After completing the multiplication it is found that

$$\tilde{T}_{21}^{(31)} = \frac{i\left(k_1^2 - k_2^2\right)\sin(k_2 b)}{2k_1 k_2} \tag{8.9}$$

$$\tilde{T}_{22}^{(31)} = \frac{2k_1 k_2 \cos(k_2 b) - i\left(k_1^2 + k_2^2\right)\sin(k_2 b)}{2k_1 k_2} \exp\left[ik_1 b\right] \tag{8.10}$$

To find the transmission coefficient we require to find the transmission amplitude

$$\tilde{t} = \frac{\tilde{T}_{11}\tilde{T}_{22} - \tilde{T}_{12}\tilde{T}_{21}}{\tilde{T}_{22}} = \frac{1}{\tilde{T}_{22}} = \frac{2k_1 k_2 \exp\left[-ik_1 b\right]}{2k_1 k_2 \cos(k_2 b) - i\left(k_1^2 + k_2^2\right)\sin(k_2 b)} \tag{8.11}$$

8.2.3 Double Barriers - The Resonant Tunnelling Diode

The next stage is to calculate the transmission coefficient for the double barrier system. The trick to make the algebra as simple as possible is to define the origin at the centre of the well so the structure is symmetric around the origin in the z-axis. In addition it is also easier to consider a trapped electron in the quantum well which can escape to the right with transmission amplitude $\tilde{t}_R$, relection amplitude $\tilde{r}_R$ and transmission coefficient $\tilde{T}_R$ or to the left with transmission amplitude $\tilde{t}_L$, relection amplitude $\tilde{r}_L$ and transmission coefficient $\tilde{T}_L$ (Fig. 8.1). The electron will travel to the right through the right barrier and to the left through the left barrier. The T-matix for an electron impinging from the left is

$$\tilde{T}_L = \begin{pmatrix} 1/\tilde{t}_L^* & r/\tilde{t}_L \\ \tilde{r}_L^*/\tilde{t}_L^* & 1/\tilde{t}_L \end{pmatrix} \tag{8.12}$$

Using the same technique as the single barrier, we now form the $\tilde{T}^{(31)}$ transmission coefficient for the double barrier

$$\begin{aligned} \tilde{T} &= \begin{pmatrix} \mathrm{e}^{-ika/2} & 0 \\ 0 & \mathrm{e}^{ika/2} \end{pmatrix} \begin{pmatrix} 1/\tilde{t}_R^* & -\tilde{r}_R^*/\tilde{t}_R^* \\ -\tilde{r}_R/\tilde{t}_R & 1/\tilde{t}_R \end{pmatrix} \\ &\times \begin{pmatrix} \mathrm{e}^{ika/2} & 0 \\ 0 & \mathrm{e}^{-ika/2} \end{pmatrix} \begin{pmatrix} \mathrm{e}^{ika/2} & 0 \\ 0 & \mathrm{e}^{-ika/2} \end{pmatrix} \\ &\times \begin{pmatrix} 1/\tilde{t}_L^* & \tilde{r}_L/\tilde{t}_L \\ \tilde{r}_L^*/\tilde{t}_L^* & 1/\tilde{t}_L \end{pmatrix} \begin{pmatrix} \mathrm{e}^{-ika/2} & 0 \\ 0 & \mathrm{e}^{ika/2} \end{pmatrix} \\ &= \begin{pmatrix} \frac{(1-\tilde{r}_L^*\tilde{r}_R^*\exp(-2ika))}{\tilde{t}_L^*\tilde{t}_R^*} & \frac{(\tilde{r}_L\exp(ika)-\tilde{r}_R^*\exp(-ika))}{\tilde{t}_L\tilde{t}_R^*} \\ \frac{(\tilde{r}_L^*\exp(-ika)-\tilde{r}_R\exp(ika))}{\tilde{t}_L^*\tilde{t}_R} & \frac{(1-\tilde{r}_L\tilde{r}_R\exp(2ika))}{\tilde{t}_L\tilde{t}_R} \end{pmatrix} \end{aligned} \tag{8.13}$$

The transmission amplitude follows from the bottom right entry

$$\tilde{t} = \frac{\tilde{t}_L\tilde{t}_R}{1-\tilde{r}_L\tilde{r}_R\exp[2ika]} \tag{8.14}$$

Rewriting the complex reflection amplitudes in the polar form $\tilde{r}_L = |\tilde{r}_L|\exp(i\vartheta_L)$ then

$$\begin{aligned} \tilde{T} = |\tilde{t}|^2 &= \frac{\tilde{T}_L\tilde{T}_R}{1+\tilde{R}_L\tilde{R}_R - 2\sqrt{\tilde{R}_L\tilde{R}_R}\cos[2ka+\vartheta_L+\vartheta_R]} \\ &= \frac{\tilde{T}_L\tilde{T}_R}{\left(1-\sqrt{\tilde{R}_L\tilde{R}_R}\right)^2 + 4\sqrt{\tilde{R}_L\tilde{R}_R}\sin^2\left(\frac{\phi}{2}\right)} \end{aligned} \tag{8.15}$$

where the phase $\phi = 2ka + \vartheta_L + \vartheta_R$. This is the transmission coefficient for a double barrier with distance a between the centres of the barriers.We are interested in finding the variation of $\tilde{T}$ with energy (i.e. $\tilde{T}(E)$) to allow the current-voltage relationship to be found. If we assume that the most rapidly changing part of the system will be the phase of the electron between the barriers (i.e. $2ka$) and all other variables change slowly with respect to this then T has peaks (Fig. 8.2) when the sine in the denominator vanishes i.e. $\phi = 2n\pi$. Therefore $\tilde{T}$ has resonance peaks when

$$\tilde{T}_{res} = \frac{\tilde{T}_L \tilde{T}_R}{\left(1 - \sqrt{\tilde{R}_L \tilde{R}_R}\right)^2} \approx \frac{4\tilde{T}_L \tilde{T}_R}{\left(\tilde{T}_L + \tilde{T}_R\right)^2} \tag{8.16}$$

It may be seen that if $\tilde{T}_L = \tilde{T}_R$ i.e. both barriers have identical transmission coefficients then $\tilde{T}_{res} = 1$ i.e. perfect transmission. It can also be shown that if the individual transmission coefficients are small then $\tilde{T}$ may be approximated by

$$\tilde{T} \approx \frac{\tilde{T}_L \tilde{T}_R}{\frac{1}{4}\left(\tilde{T}_L + \tilde{T}_R\right)^2 + 4\sin^2\left(\frac{1}{2}\phi\right)}$$

$$= \tilde{T}_{res} \frac{1}{1 + \frac{16}{\left(\tilde{T}_L + \tilde{T}_R\right)^2}\sin^2\left(\frac{1}{2}\phi\right)} \tag{8.17}$$

If we use the approximation $\sin^2\left(\frac{1}{2}\phi\right) \approx \frac{1}{2}$ then

$$\tilde{T} \approx \frac{1}{2}\tilde{T}_L \tilde{T}_R \tag{8.18}$$

Thus the overall transmission when off-resonance is typically the product of the individual transmissions for the two barriers.Strong deviations may be expect from this when the sin approaches zero. Therefore if we set $\phi = 2n\pi + \delta\phi$ and expand the sin we find

$$\tilde{T} \approx \tilde{T}_{res} \frac{1}{1 + \frac{4(\delta\phi)^2}{\left(\tilde{T}_L + \tilde{T}_R\right)^2}} = \frac{\tilde{T}_{res}}{1 + \left(\frac{2\delta\phi}{\phi_0}\right)^2} \tag{8.19}$$

where $\phi_0 = \tilde{T}_L + \tilde{T}_R$. This shows a resonance peak with a Lorentzian shape which falls to half its peak value at $\delta\phi = \phi_0$. Thus ϕ_0 is the full width at half-maximum (FWHM) value. To translate this into energy

$$\Gamma = \frac{dE}{dk}\frac{dk}{d\phi}\phi_0 = \frac{\hbar v}{2a}\left(\tilde{T}_L + \tilde{T}_R\right) \tag{8.20}$$

where v is the velocity of the electron between the barriers and it is assumed that the phase is again dominated by $2ka$.

The width Γ can also be derived from basic physical considerations. We assume that an electron has managed to tunnel into a subband and is caught between the two barriers for an amount of time, τ. The velocity of an electron in the resonant subband state is v and the distance of a round trip is $2a$. Therefore the electron will hit the left barrier $v/2a$ times per second (and the right barrier a similar number of times) if it does not escape. The probability of escaping is $\tilde{T}_L$ on each occassion which gives a mean escape rate through the left barrier of $v\tilde{T}_L/2a$. To convert this into an energy uncertainty, the Heisenburg's uncertainty principle tells us that this value must be multiplied by $\hbar$. The total rate when the similar term for the right barrier is added gives the result $\Gamma = \frac{\hbar v}{2a}\left(\tilde{T}_L + \tilde{T}_R\right)$ as before. The lifetime of the state (the time the electron resides in the subband) is given by $\tau = {}^{\hbar}/_{\Gamma}$.

The transmission as a function of energy therefore becomes the Lorentzian with a FWHM of Γ

$$\tilde{T}(E) \approx \tilde{T}_{res}\frac{1}{1+\left(\frac{E-E_{res}}{\frac{1}{2}\Gamma}\right)^2}$$

$$= \tilde{T}_{res}\frac{\left(\frac{1}{2}\Gamma\right)^2}{\left(\frac{1}{2}\Gamma\right)^2 + (E-E_{res})^2} \tag{8.21}$$

where the resonance is centred on the energy E_{res}. E_{res} are the resonance energies in the quantum well and therefore correspond to the subband energies from (3.27) or (3.28). (8.21) is known as the Breit-Wigner formula in nuclear physics and a similar form is also used in the Fabry-Pérot etalon in optics.

Figure 8.5 shows schematically how the current voltage properties of the RTD change as a voltage is applied across the diode to vary the energy of the electrons in the left and right n-type regions which act as contacts to the RTD diode. These are degenerately doped n-type semiconductor and are therefore highly metallic with the Fermi level (chemical potential) above E_c. At small biases, if the subband in the quantum well is above the Fermi level of the electrons in the left or right contacts then no current flows (Fig. 8.5 far left). As the voltage across the diode is increased, the energy of electrons on the right of the structure decrease and the energy of the central subbands is also pulled down relative to the energy of the electrons on the left of the structure. It is therefore more energetically favourable to excite electrons into the first subband in the well (before tunnelling into the right contact at lower energy) and hence the current in the system increases (2nd diagram from left in Fig. 8.5).

Eventually the subband energy is aligned with the bottom of the conduction band which has the largest density of states. This maximises the tunnelling through the system and the RTD is in the resonant state (3rd diagram from left in Fig. 8.5). The current is therefore at a maximum or peak. As the voltage is further increased, the subband moves beneath the conduction band on the left and so only non-coherent electrons can be transported through the barriers (i.e. electrons must move to an empty subband state in the quantum well by tunnelling into the well and then losing energy through scattering). Finally after the resonance state, the current is reduced creating a valley before rising due to current increases from non-resonant components e.g. thermal activation over the barriers, scattering or tunnelling through surface or defect states (far right diagram Fig. 8.5).

Fig. 8.5 clearly shows that Ohm's law (current is proportional to voltage) does not hold for the RTD and that a region of negative differential resistance (NDR) occurs where the current is decreasing as the applied voltage is increased. This is due to the wave nature of electrons in the system and

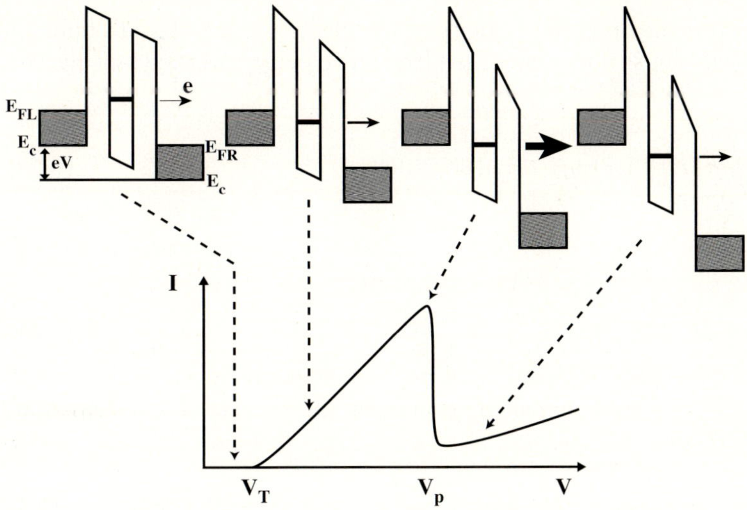

Fig. 8.5. A schematic diagram showing the current-voltage properties of a RTD along with the appropriate energy band diagrams for the electron transport. V_T is the threshold voltage for current transport and V_p is the voltage at which the current peaks

the interaction with the quantised states produced from a low dimensional system. It is this NDR which may be utilised in an amplifier or oscillator for circuit applications.

The voltage at which the peak current is observed is V_p (Fig. 8.5). An approximate value for this may be estimated by considering a voltage V across the device. Since the voltage is dropped from the left contact to the right contact corresponding to energy difference between the left and right chemical potentials of qV, half this voltage will be dropped between each contact and the quantum well. Therefore

$$\frac{qV_p}{2} \approx E_{res} \tag{8.22}$$

where $E_{res} = E_0$ in (3.27) or (3.28).

It is worth noting that this voltage is that which is dropped across the tunnel barriers and the measured value may be substantially different due to the finite resistance of the doped contact regions on either side of the wells and any contact resistances.

Up to this point we have only calculated the transmission coefficient while in experiments, the current is normally measured as a function of applied voltage. The general method used to calculate the current considers electrons impinging from the left and then adds the contribution due to electrons arriving from the right. The basic current equation is then related to the well known expression for current density of $J = nqv$ where n is the density, q is the electron charge and v is the velocity of the electrons or holes. To find the tunnelling current density we again use electron velocity and the transmission coefficient

$$J_L = q \int_0^\infty g(E) f(E) v_L \tilde{T}(E) \, \mathrm{d}E \tag{8.23}$$

with a similar expression for the right-hand side with the electrons moving in the opposite direction. The total current density is given by

$$J = J_L - J_R \tag{8.24}$$

For the full result the left and right current densities must be integrated over $x-$, $y-$ and $z-$directions but a number of approximations and simplifications can be made.

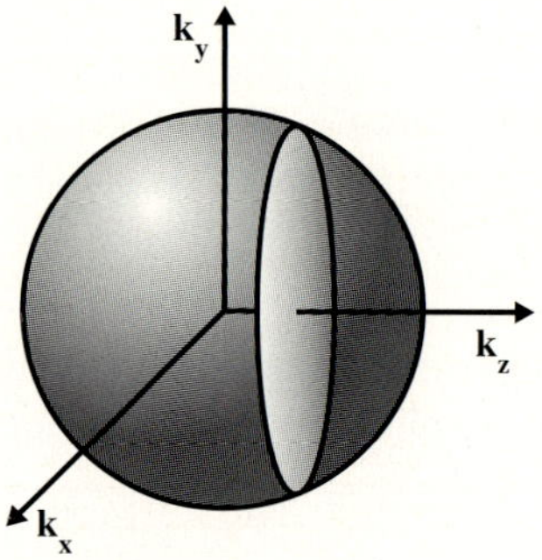

Fig. 8.6. The 3D Fermi sphere separating the filled from the unfilled states and a 2D slice along k_z which represents the quantised states in the quantum well

For the full three dimensional device it is worthwhile noting that while the leads on each side of the barriers are 3D, the quantum well has quantised states in the z-direction and hence the electrons are in a 2D layer. Therefore electrons tunnelling through the system start from 3D and tunnel through a 2D system back to a 3D system. One can think of this as each lead having a Fermi sphere which is sliced into a Fermi circle when the electrons tunnel into a resonant state (see Fig. 8.6). The density of states in the integral is therefore 2D. Again if we assume that a large bias is applied to the device so we may ignore the electrons in the right-hand contact then it may be shown that

$$\begin{aligned} J &= \frac{qm^*}{2\pi^2\hbar^3} \int_{E_{cL}}^{\mu_L} (\mu_L - E) \tilde{T}(E) \, \mathrm{d}E \\ &= \frac{qm^*}{2\pi^2\hbar^3} (E_{FL} - E_{res}) \frac{\pi}{2} \Gamma \tilde{T}_{res} \end{aligned} \tag{8.25}$$

8.2.4 The Resonant Tunnelling Diode (RTD)

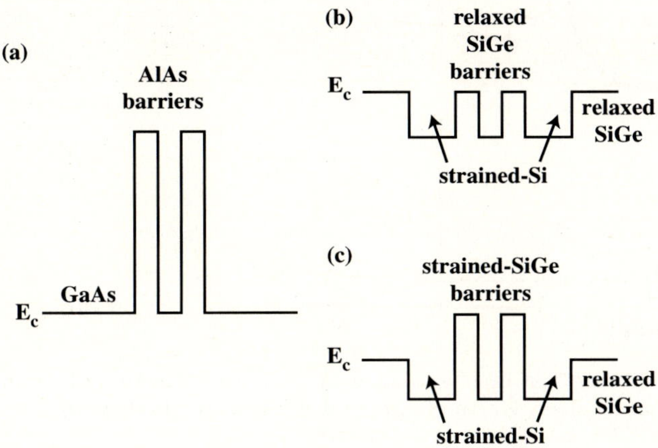

Fig. 8.7. A schematic diagram for the different band structures for RTDs with different materials showing the band alignments without any charges in the devices. (a) The typical and ideal band structure in the III-V system where the barriers have a height of over 1 eV. (b) The early Si/SiGe RTD structure where quantum wells are used to produce two barriers. (c) The use of strained barriers allows a much higher barrier than relaxed barriers greatly improving performance

The problems of producing a Si/SiGe RTD may be easily demonstrated when one considers the band structure typically used for high performance III-V RTDs. Figure 8.7(a) shows the typical GaAs/AlAs system where the AlAs layers provide a barrier to electrons which is over 1 eV high. The problem with Si/SiGe is that it is very difficult to produce a significant barrier for either electrons or holes. The typical method employs quantum wells (Fig. 8.7(b)) where the barriers use the same material as the contacts. This produces substantial problems since electrons may easily be excited over the barriers especially at higher temperatures of operation, a problem which is known as thermionic emission. The best performance of these types of devices has only just shown NDR at room temperature with a peak-to-valley current (PVCR) ratio of only 1.2. For use in circuits to be discussed in Sect. 8.2.6, a PVCR of at least 2 and preferably 3 or greater is required.

As an example of a typical III-V RTD, Fig. 8.8 shows the I-V for a n-GaAs RTD with 2 nm thick AlAs barriers and a 6 nm GaAs quantum well between the barriers. Two almost symmetrical current peaks are shown close to 1 V in both the positive and negative bias directions. The important figures of merit for an RTD are the peak current density (J_p), the peak to valley current ratio (PVCR) and the peak voltage (V_p).

Figure 8.7(c) shows the technique demonstrated by Douglas Paul at Cambridge where the barriers in the RTD are compressively strained. From the

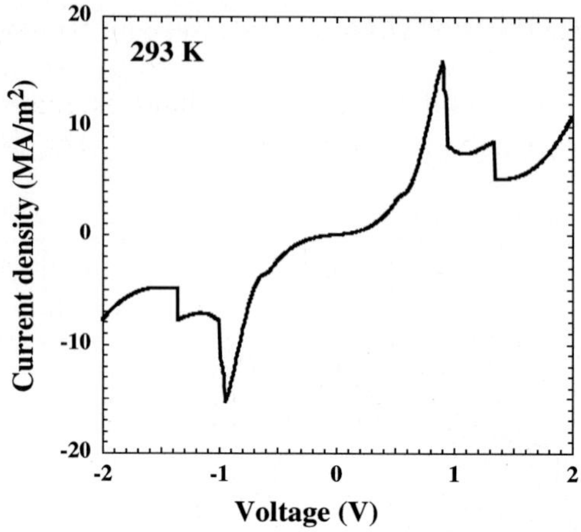

Fig. 8.8. The current density against bias voltage at 293 K for a n-GaAs RTD with 2 nm thick AlAs barriers and a 6 nm GaAs quantum well between the barriers

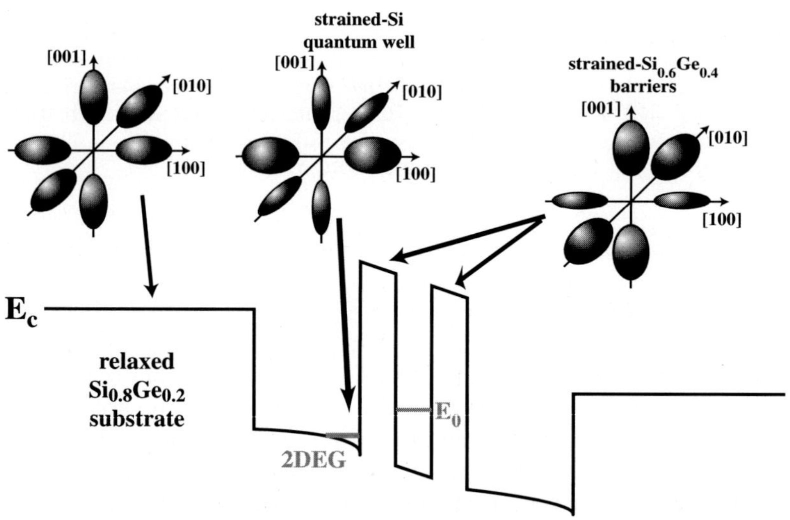

Fig. 8.9. A schematic diagram of the conduction band of a Si/SiGe RTD with strained wells and barriers with the different splitting of the valleys shown

band structure diagrams in Sect. 4.5, by compressively straining the barriers, the conduction band mimima, ΔE_c, is raised producing a larger barrier to electron tunnelling. This greatly reduces the thermionic emission over the tops of the barriers and produces significantly better performance at higher temperatures. Figure 8.9 shows the band structure of a typical RTD. Strained-Si quantum wells are placed at either side of the barriers which will help the barriers height through band bending when a bias is placed across the device. These spacers also reduce the capacitance of the diode which will become important for the implementation into circuits. The strain in the barriers split the valleys so that the Δ_4 valleys are the lowest in energy about 110 meV above the substrate conduction band while the Δ_2 valleys are raised further and may be ignored. The strained-Si quantum wells only have the Δ_2 valleys occupied.

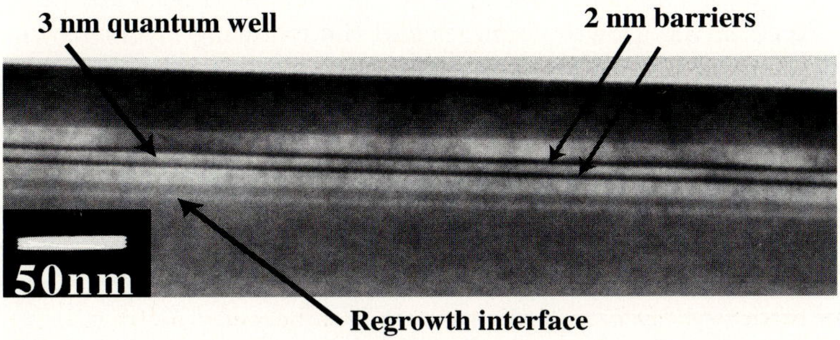

Fig. 8.10. A transmission electron micrograph (TEM) of a Si/SiGe RTD with 2 nm thick $Si_{0.4}Ge_{0.6}$ barriers on a relaxed $Si_{0.8}Ge_{0.2}$ substrate

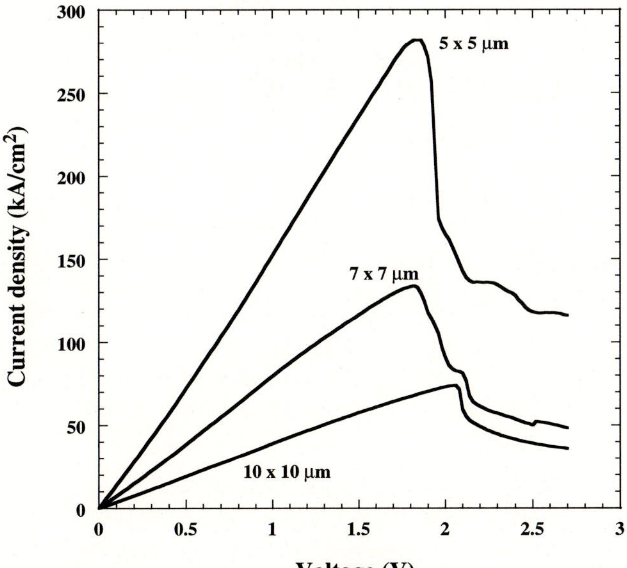

Fig. 8.11. The current voltage characterisitics from three different sizes of Si/SiGe RTD with 2 nm thick $Si_{0.4}Ge_{0.6}$ barriers on a relaxed $Si_{0.8}Ge_{0.2}$ substrate. The devices have sizes of $5 \times 5\,\mu m$, $7 \times 7\,\mu m$ and $10 \times 10\,\mu m$

The wafers for this work were grown by CVD. First a n-type thick virtual substrate with a Ge content up to $Si_{0.8}Ge_{0.2}$ was grown. The wafer was then removed from the growth chamber and given a modified RCA clean to remove excess As from the surface in an attempt to circumvent the As dopant segregation problems in CVD material. This can be seen as a white layer just below the lower strained-Si spacer as the RCA clean leaves a Si-rich layer at the interface in TEM pictures of the sample (Fig. 8.10). The wafers were replaced in the growth chamber and the following layers grown: 10 nm i-$Si_{0.8}Ge_{0.2}$ buffer, 10 nm i-Si spacer, 2 nm i-$Si_{0.4}Ge_{0.6}$ barrier, 3 nm i-Si well, 2 nm i-$Si_{0.6}Ge_{0.4}$ barrier, 10 nm i-Si spacer, 50 nm n-$Si_{0.7}Ge_{0.3}$ doped at $N_D \sim 3 \times 10^{18}\,cm^{-3}$ and a 4 nm n-Si cap. Figure 8.10 shows a TEM picture of the active layers of one such RTD grown by CVD at DERA, Malvern.

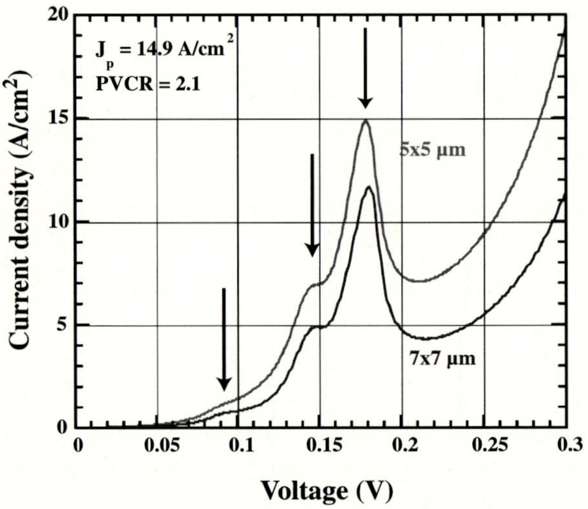

Fig. 8.12. A RTD grown by MBE with the structure modelled in Fig. 8.13. Three peaks are clearly shown which correspond to the peaks in Fig. 8.13. The series resistances along with Ge segregation and diffusion result in the peaks from the experiment not agreeing completely with the theory

The current voltage characteristics at 298 K are shown in Fig. 8.11. The peak that is observed in the positive bias direction has a peak current density of 282 kA/cm^2 for a PVCR of 2.4. A similar wafer with barriers of $Si_{0.8}Ge_{0.2}$ did not demonstrate any NDR at room temperature showing the additional benefits of the larger barrier heights with increase Ge content. The disadvantage of this system is that with the present design, the peak voltage is very high at about 2 V. For applications this needs to be less than 1 V. This is mainly due to the relatively small amount of n-type doping which CVD

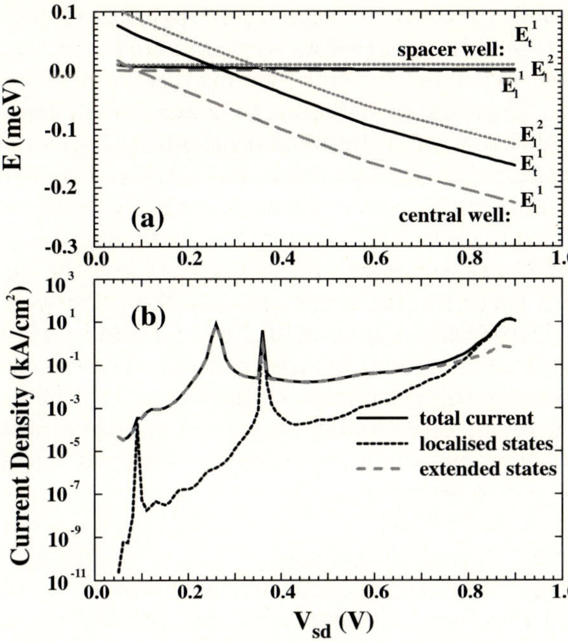

Fig. 8.13. The subband energy levels in the main quantum well and the emitter-spacer quantum well along with the current-voltage characterisitics calculated from theory for the RTD in Fig. 8.11. A subscript of l corresponds to the longitudinal mass and t to a transverse mass. The calculations are by Igor Zozoulenko

allows to be incorporated into the Ohmic contacts of less than $5 \times 10^{18}\,\text{cm}^{-3}$ which results in large contact resistances and therefore large peak voltages. Figure 8.12 shows the same design of RTD grown by solid source MBE. Here the doping has been increased to $5 \times 10^{19}\,\text{cm}^{-3}$ and the peak voltage has been significantly reduced.

When the RTD is modelled (see Fig. 8.13), the tunnelling of electrons through the system is actually dominated by electrons from the bulk virtual substrate. The splitting of the valleys in the strained-Si spacer results in only the Δ_2 valleys being populated. These have a low transverse effective mass of $0.198m_e$ along the quantum well but have a high longintudinal effective mass of about $0.98m_e$ in the vertical tunnelling direction. As was shown in section 3.1, the transmission coefficient for quantum mechanical tunnelling depends exponentially on the square root of the effective mass,

$$T \approx \frac{16E}{V_0} \exp\left(-2\sqrt{\frac{2m^*\left(V_0 - E\right)}{\hbar^2}}b\right) \tag{8.26}$$

Therefore the heavy mass in the strained-Si spacer is unlikely to tunnel compared to the light mass from the bulk substrate with 6 valleys all degenerate.

In addition, the selection rules in the system dictate that longitudinal effective mass electrons can only tunnel to longitudinal electron states and that transverse mass electrons can only tunnel to transverse mass electron states unless the electron is scattered (typically by a phonon to change the momentum of the electron). When all this is put together and modelled (Fig. 8.13) the localised states in the strained-Si spacer quantum well play little part in the observed I-V characteristics while the main NDR peaks are dominated by the extended states from the bulk virtual substrate. The three main peaks in Fig. 8.12 can all be identified in the theory plot of Fig. 8.13 but the absolute values of current are not correct in the theory. The difference between the experimental and theoretical voltages observed in Figs. 8.11 and 8.13 is because the theory does not take any account of series resistances in the circuit such as contact resistances or the resistances of the doped regions on either side of the barriers. Therefore a substantial amount of voltage can be dropped across these regions before any significant voltage is dropped across the active barrier regions.

The high peak voltage can be understood from the charging in the system. Figure 8.14 shows self-consistent Poisson-Schrödinger solutions for the RTD with voltages of 0.2 to 0.5 V applied across the device region shown. It can be observed that at low biases (0.2 V) there is a substantial charge build up on the right hand side of the second spacer. Basically this layer is the result of the ΔE_c discontinuity between the strained-Si spacer and the relaxed $Si_{0.8}Ge_{0.2}$ virtual substrate at the collector end of the device. This charge layer screens the applied potential from the barriers and therefore a larger potential must be applied to the totalsystem before the bands are brought sufficiently down that subbands align and electrons can tunnel through the system. As the applied voltage is increased to 0.2 and then 0.3 V, the charge at the collector is slowly reduced until by 0.5 V, it has almost completely disappeared. Figure 8.14 also shows the band bending in the strained-Si spacers which give the barriers a larger height compared to a RTD without spacers on either side of the barriers. One problem with the strained-Si spacers is that they allow charge to be built up also at the emitter side which again provides problems for the fast operation of such diodes.

A method to stop the charge build up was suggested by Neil Griffin and Barry Coonan at NMRC in Ireland. Basically the spacers are graded so that there is no ΔE_c discontinuity across the emitter or collector contact ends of the device. Figure 8.15 shows self-consistent Poisson-Schrödinger solutions for the structure both at a bias of 0 V (Fig. 8.15(a)) and 0.2 V (Fig. 8.15(b)). The band structure looks very similar to that of a III-V RTD (see Fig. 8.7(a)) when a bias is applied. More important is the charge density in the structure when a bias is applied as shown in Fig. 8.15(c). No charge is built up at the emitter or collector contact and so the device should operate with much smaller applied voltages.

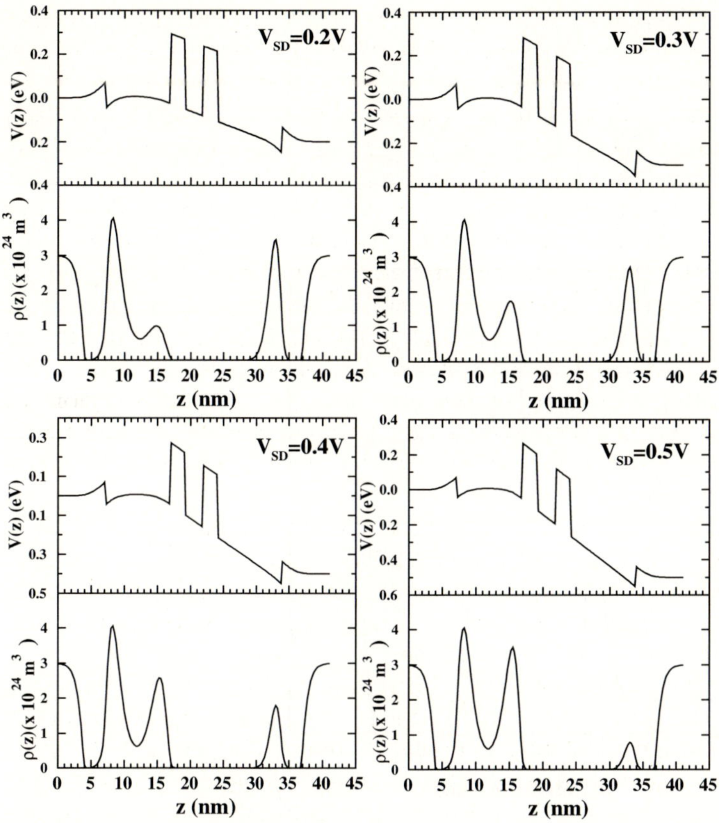

Fig. 8.14. Self-consistent Poisson-Schrödinger solutions to the RTD showing the band structure and charge density in the structure as a function of applied voltages across the diode. The calculations are by Igor Zozoulenko

To test this idea, a wafer was grown with graded spacers and the low temperature results are shown in Fig. 8.16. Higher doping was also grown into the cap so that the contact resistances could be reduced in the device. It is possible to get the peak with an applied bias of only 40 mV in this particular structure demonstrating the improvement to the peak voltage with graded spacers. More work is therefore still required to optimise the properties of SiGe RTDs.

8.2.5 Inter-band Esaki Tunnel Diodes

The Esaki tunnel diode dates back before the RTD to 1958 when Leo Esaki first observed negative differential resistance while studying the current-voltage characteristics of a degenerately doped p-n germanium diode. He explained the effect in terms of quantum mechanical tunnelling of carriers

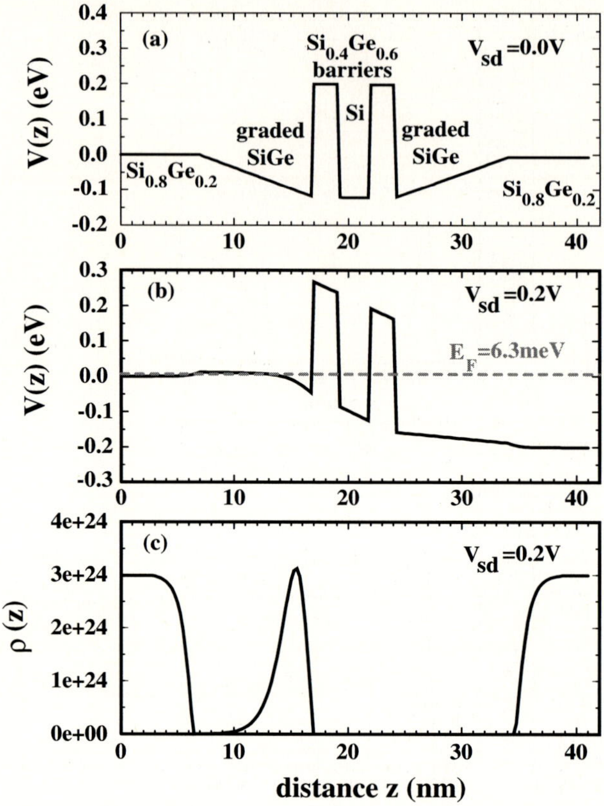

Fig. 8.15. Self-consistent Poisson-Schrödinger solutions to a RTD with graded spacer wells showing the band structure and charge density in the structure as a fucntion of applied voltages across the diode. The calculations are by Igor Zozoulenko

between the p and n-doped regions of the device.The tunnelling time for interband tunnelling is proportional to $\exp[2k(0)\,b]$ where b is the barrier thickness and $k(0)$ is the mean value of the momentum encountered in the tunnelling path corresponding to an incident carrier with zero transverse momentum and energy equal to the Fermi energy. This tunnelling time is very short and allows operation of tunnel devices in the 30 to 300 GHz regime.

The basic form of an Esaki diode is shown in Fig. 8.17. Two highly doped regions, one n-type and the other p-type are placed together and a depletion regions forms between the two. The doping must be high enough for both doped regions to be degenerate. The amount of degeneracy for electrons, V_e and holes V_h requires to be a few k_BT to produces a large density of states on both sides of the junctions. The depletion width is typically 10 nm or less, much narrower than many conventional p-n junctions. Therefore from Fig. 5.2 the doping must be greater than 10^{19} cm^{-3} in the doped regions.

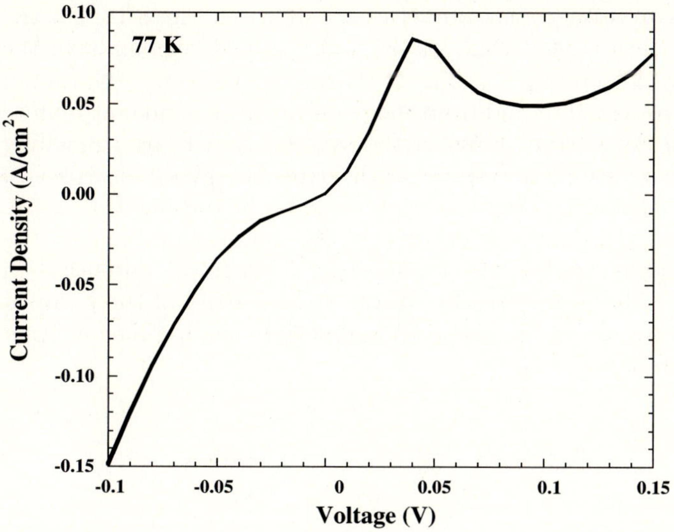

Fig. 8.16. The current voltage characteristics of a RTD with graded spacers at 77 K. The peak current has now shifted to 40 mV showing the effect of grading the spacers in the device

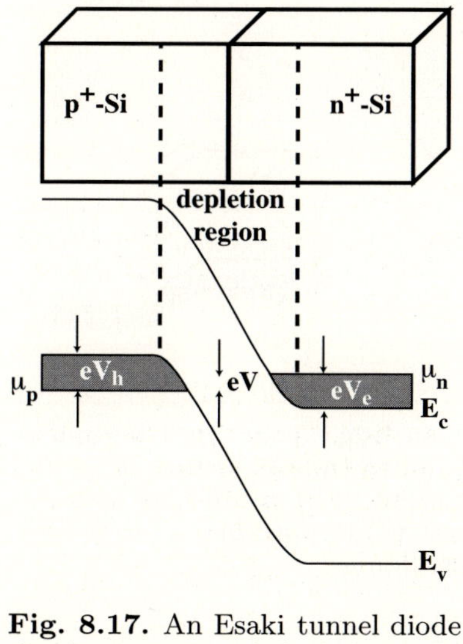

Fig. 8.17. An Esaki tunnel diode with the energy band structure. V_h and V_e are the degeneracies of the holes and electrons respectively

The typical current-voltage characteristics are shown in Fig. 8.18. For the explanation, it will be assumed that the n- and p-doped regions have the same doping density. At zero applied bias at thermal equilibrium, the rate of tunnelling from the n- to the p- and from the p- to the n- are equal at a finite temperature and no net current flows in the system. At 0 K no tunnelling would occur since there are no free states at the same energy for electrons to tunnel to since the chemical potentials on both sides of the junction align. When a negative bias is applied to the device, the conduction band in the n-doped contact is pulled below the Fermi energy (chemical potential) in the p-doped contact. Therefore electrons in the valence band of the p-doped contact have free states above the chemical potential in the n-doped contact and can tunnel creating a current.

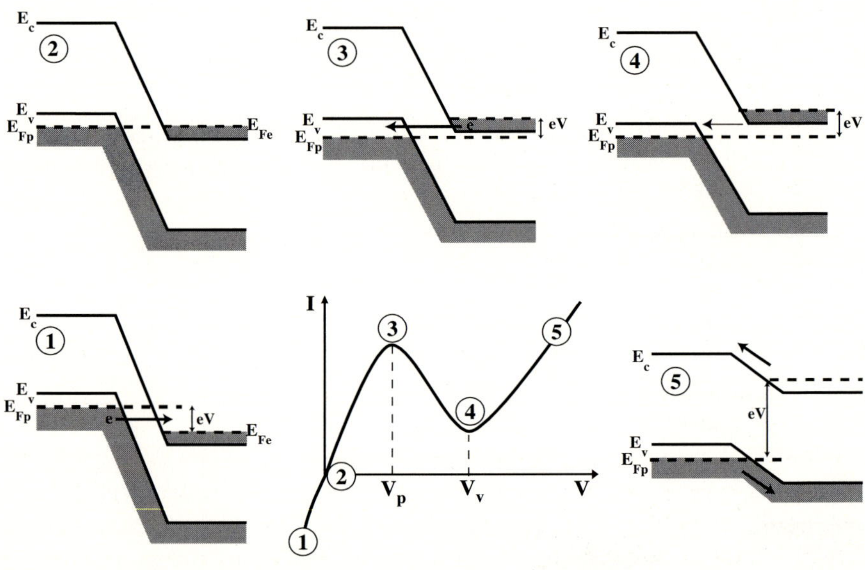

Fig. 8.18. The current-voltage characteristics of an Esaki diode. (1) reverse-bias with the electrons tunnelling from the valence band to the conduction band, (2) thermal equilibrium at zero bias with zero net current, (3) forward-bias with the maximum overlap of states giving the peak current, (4) forward-bias at the valley current and (5) forward-bias with thermal currents flowing

When a positive bias is applied to the device, the electrons below the Fermi level in the n-contact may tunnel to fill holes above the Fermi level in the p-doped contact. Therefore a net current flows which will increase until there is the maximum overlap between the filled states in the n-contact and the empty states in the p-contact. The current is therefore a maximum ((3) in Fig. 8.18). As the bias is increased, the overlap between the filled and empty states is reduced and the tunnelling current is reduced until the bottom of the conduction band in the n-doped contact aligns with the top of the valence band in the p-doped region. Therefore the current drops creating a negative differential region of conductivity. If further bias is applied, there are no states for the electrons in the n-doped contact to tunnel to and the band-to-band tunnelling can only take place through impurity or defect states in the bandgap or through scattering and thermal processes. Therefore the current flowing in an Esaki is made up of three components, (a) the band-to-band tunnelling, (b) the thermal current from phonon scattering and (c) the excess current from tunnelling through states in the bandgap (see Fig. 8.19).

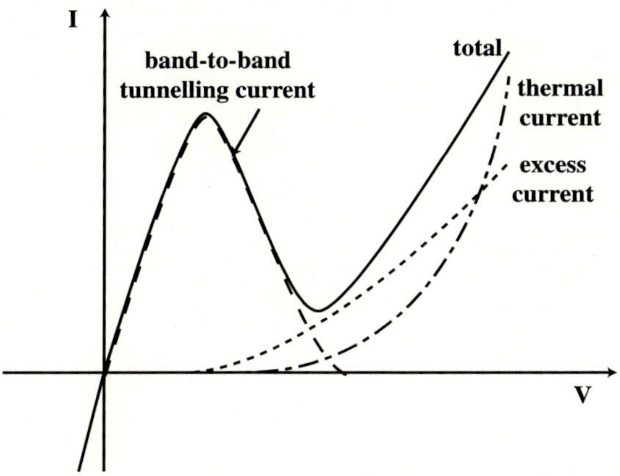

Fig. 8.19. The current-voltage characteristics of an Esaki diode broken into components

For a tunnelling process to proceed, there must be filled and empty states at the same energy, the potential barrier between these states must be small and/or thin enough to allow a reasonable probability of quantum mechanical tunnelling and momentum must be conserved in the tunnelling process. For direct bandgap semiconductors, direct tunnelling occurs between the Γ point in the conduction band and the Γ point in the valence band. For indirect semiconductor materials such as Si or Ge, the conduction band electrons have different momenta from the valence band holes and therefore tunnelling can only proceed through a scattering process such as phonon or electron-electron scattering. The probability for indirect tunnelling is generally much lower than the probability for direct tunnelling.

The band-to-band tunnelling will now be calculated using the WKB (Wentzel-Kramers-Brillouin) approximation. When a large electric field, F is applied to the device, the potential barrier $U(x)$ becomes triangular in form (Fig. 8.20). The tunnelling probablility from the WKB method is

$$T \approx \exp\left[-2\int_{-x_1}^{x_2} |k(x)| \mathrm{d}x\right] \tag{8.27}$$

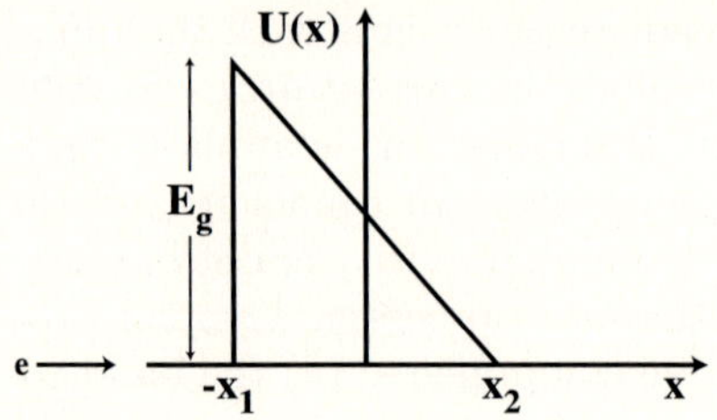

Fig. 8.20. The potential energy diagram of an Esaki diode with a triangular potential barrier

where $|k(x)|$ is the absolute value of the wave vector for carriers in the barrier and $-x_1$ and x_2 are the classical turning points which physically represent the edges of the barrier in Fig. 8.20.The tunnelling of an electron through the bandgap of forbidden states is the same as quantum mechanical tunnelling through a potential barrier. Therefore for the triangular barrier the wave vector is

$$k(x) = \sqrt{\frac{2m^*}{\hbar^2}(U(x) - E)} = \sqrt{\frac{2m^*}{\hbar^2}\left(\frac{E_g}{2} - qFx\right)} \tag{8.28}$$

for an electron with energy, E impinging on the barrier of a semiconductor with bandgap, E_g. Substituting (8.28) into (8.27) gives

$$\begin{aligned} T &\approx \exp\left[-2\int_{-x_1}^{x_2}\sqrt{\frac{2m^*}{\hbar^2}\left(\frac{E_g}{2} - qFx\right)}\mathrm{d}x\right] \\ &= \exp\left[\frac{4}{3}\sqrt{\frac{2m^*}{\hbar^2}}\left(\frac{E_g}{2} - qFx\right)^{3/2}\right]\Big|_{-x_1}^{x_2} \end{aligned} \tag{8.29}$$

The boundary conditions to the problem are

$$x = x_2 \qquad \left(\frac{E_g}{2} - qFx\right) = 0 \tag{8.30}$$

$$x = -x_1 \qquad \left(\frac{E_g}{2} - qFx\right) = E_g \tag{8.31}$$

and so solving the equations gives the tunnelling probability as

$$T \approx \exp\left[-\frac{4\sqrt{2m^*}E_g^{3/2}}{3q\hbar F}\right] \tag{8.32}$$

If a phonon of energy $\hbar\omega$ with frequency ω_q is involved in an indirect tunnelling event between a conduction band valley and the top of the valence band then (8.32) is modified to

$$T \approx \exp\left[-\frac{4\sqrt{2m^*}\left(E_g - \hbar\omega_q\right)^{3/2}}{3q\hbar F}\right] \tag{8.33}$$

The tunnelling current density will now be derived from the appropriate transmission coefficients above. The current density for tunelling from the conduction band to the valence band and from the valence band to the conduction band are

$$J_{c\to v} = \int_{E_c}^{E_v} f_c(E) n_c(E)\, T\left[1 - f_v(E)\right] n_v(E)\, \mathrm{d}E \tag{8.34}$$

$$J_{v\to c} = \int_{E_c}^{E_v} f_v(E) n_v(E)\, T\left[1 - f_c(E)\right] n_c(E)\, \mathrm{d}E \tag{8.35}$$

where $f(E)$ is the Fermi-Dirac distribution function and $n(E)$ is the density of states with the subscripts c for the conduction band and v for the valence band. When a bias is applied the total current density in the system is

$$J_{Esaki} = J_{c\to v} - J_{v\to c} \tag{8.36}$$

$$= \int_{E_c}^{E_v} T\left[f_c(E) - f_v(E)\right] n_c(E)\, n_v(E)\, \mathrm{d}E \tag{8.37}$$

A closed form of this equation is given by

$$J_{Esaki} = J_P \frac{V}{V_P} exp\left[\left(1 - {}^{V}\!/_{V_P}\right)\right] \tag{8.38}$$

where J_P is the peak current density and V_P is the peak voltage. The peak voltage is obtained by differentiating $n_c(E)$ and $n_v(E)$ with respect to energy when the number of electron states on the n-side and the number of empty states on the p-side has been maximised. This results in

$$V_P \approx \frac{V_n + V_p}{3} \tag{8.39}$$

The degeneracies can be evaluated from the Fermi-Dirac integral to be

$$V_n \approx \frac{k_B T}{q}\left[\ln\left(\frac{N_D}{N_c}\right) + \frac{7}{20}\left(\frac{N_D}{N_c}\right)\right] \tag{8.40}$$

$$V_p \approx \frac{k_B T}{q}\left[\ln\left(\frac{N_A}{N_v}\right) + \frac{7}{20}\left(\frac{N_A}{N_v}\right)\right] \tag{8.41}$$

The excess current of tunnelling through a defect or impurity states in the bandgap will now be calculated. If the defect state is at an energy, E_D in the bandgap then the tunnelling of an electron to this state is the same problem as calculating the tunnelling of electrons between the n- and p-doped regions of the device. If the electron is scattered to near the top of the valence band then

$$E_D \approx E_g - qV + q\left(V_n + V_p\right) \tag{8.42}$$

The potential barrier is triangular in form and the transmission coefficient is also given by (8.32) except the bandgap is replaced by the defect energy to give

$$T_D \approx \exp\left[-\frac{4\sqrt{2m^*}E_D^{3/2}}{3q\hbar F}\right] \tag{8.43}$$

The electric field in the system is given by

$$F = \frac{2\left(V_n + V_p - V\right)}{W_D} \tag{8.44}$$

where W_D is the depletion layer width given by (5.19) which for the present case is

$$W = \sqrt{\frac{2\varepsilon_0\varepsilon_r}{q}\left(\frac{N_A + N_D}{N_A N_D}\right)\left(V_n + V_p - V\right)} \tag{8.45}$$

If the volume density of the occupied defect levels at energy E_D is given by D_D then the excess current is

$$J_D = \alpha D_D T_D \tag{8.46}$$

α is a constant. Substituting (8.42) to (8.45) into (8.46) produces

$$J_D = \alpha D_D \exp\left[-\gamma\left(E_g - qV + \frac{3}{5}q\left(V_n + V_p\right)\right)\right] \tag{8.47}$$

with γ as a constant. This can be rewritten in the form

$$J_D = J_V \exp\left[\frac{4}{3}\frac{m^*\varepsilon_0\varepsilon_r}{N_A N_D}\left(N_A + N_D\right)\left(V - V_V\right)\right] \tag{8.48}$$

where J_V is the valley current density at a voltage V_V

The thermal current in the system is just the current from the Schockley equation for a p-n junction (5.40) which is in the form

$$J_{th} = J_0\left(\exp\left[\frac{qV}{k_B T} - 1\right]\right) \tag{8.49}$$

with J_0 a constant representing all the prefactor terms in (5.40).

The total current for the Esaki is now the combination of the three currents

$$\begin{aligned} J &= J_{Esaki} + J_D + J_{th} \\ &= J_P \frac{V}{V_P} \exp\left[1 - \frac{V}{V_P}\right] + J_V \exp\left[\kappa\left(V - V_V\right)\right] \\ &+ J_0 \left(\exp\frac{qV}{k_B T} - 1\right) \end{aligned} \quad (8.50)$$

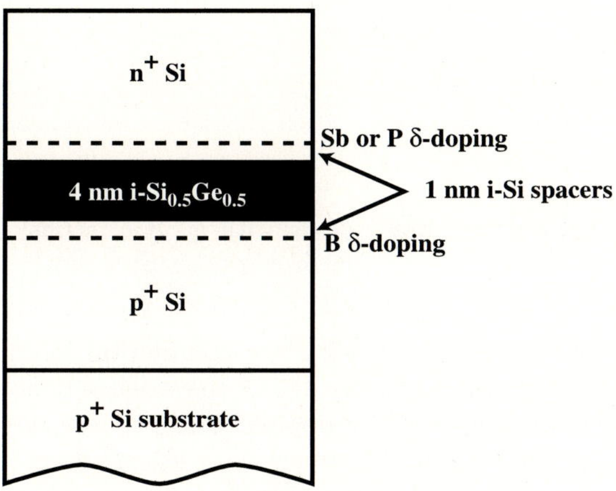

Fig. 8.21. The typical advanced Si/SiGe Esaki diode grown using MBE techniques to provide abrupt doping profiles

Most of the original Esaki diodes were fabricated by diffusion of dopants into the semiconductor material. PVCRs up to 4.0 in Si and 8.3 in Ge Esaki diodes have been reported. The Si diodes only show a peak current density of $1\,\text{kA/cm}^2$ and for most logic applications PVCRs of between 3 and 5 are required along with J_P of $10\,\text{kA/cm}^2$. Modern fabrication techniques and in particular epitaxial growth through MBE has allowed the performance figures of Si Esaki diodes to be improved considerably. The typical structure which is now used is shown in Fig. 8.21. Abrupt doping profiles are required and δ-doping in MBE provides an ideal solution. These layers are typically deposited at below $400\,^o\text{C}$ to prevent diffusion of the dopants. Unfortunately this produces a high point defect concentration which acts to trap electrons. Therefore as grown, these diodes show no NDR but once annealled at between $600\,^o\text{C}$ and $700\,^o\text{C}$ to anneal out point defects, NDR becomes observable.

The band structure of one of these devices is shown in Fig. 8.22. The δ-doping layers produce small quantum wells which quantise the energy levels at either side of the intrinsic region. The strained-$\text{Si}_{0.5}\text{Ge}_{0.5}$ layer also

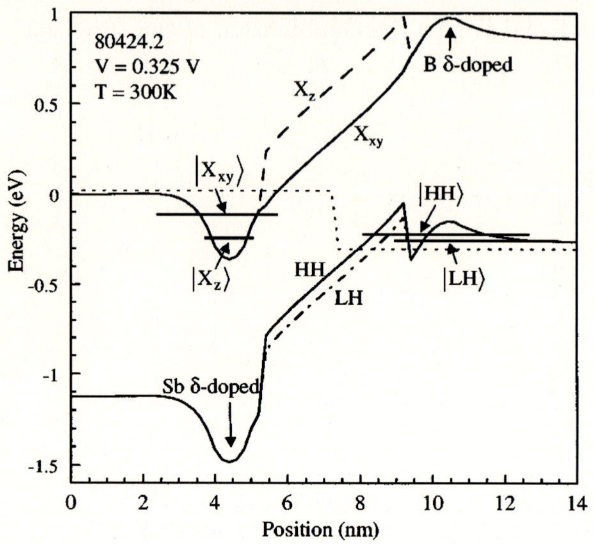

Fig. 8.22. The band structure of a δ-doped Esaki diode with a SiGe intrinsic region.©AIP 1998

splits the light hole and heavy hole levels along with the conduction band valleys. The strained-$Si_{0.5}Ge_{0.5}$ results in a quantum well in the valence band which will be modulation-doped because of the B δ-doping. This reduces the effective thickness of the doping layers. Since the reservoir of electrons and holes are both two-dimensional, the density of states of both are larger than a 3D system and increase the peak current density. Typical results from such devices are shown in Fig. 8.23 from a device fabricated by Günter Reitemann and Erich Kasper at Stuttgart University. The highest PVCRs are now over 8 with a peak current density of $8\,kA/cm^2$. Peak current densities up to $151\,kA/cm^2$ have been observed by Phil Thompson and colleagues working at NRL.

Another possibility for this advanced type of Esaki diode is to try and provide further quantisation in the intrinsic region. One method of achieving this is to produce quantum dots or 0D structures. The Stranski-Krastanov growth mode allows self-assembled quantum dots to be grown by MBE inside the intrinsic region as shown in Fig. 8.24. Devices have been fabricated by Günter Reitemann and Erich Kasper at Stuttgart but these have so far not produced better results than those shown by the 2D SiGe intrinsic regions (Fig. 8.23).

8.2.6 Tunnel Diode High Frequency Performance

The performance of tunnel diodes in a circuit does not just depend of the abruptness of the doping profiles or the thickness of tunnel barriers. It also

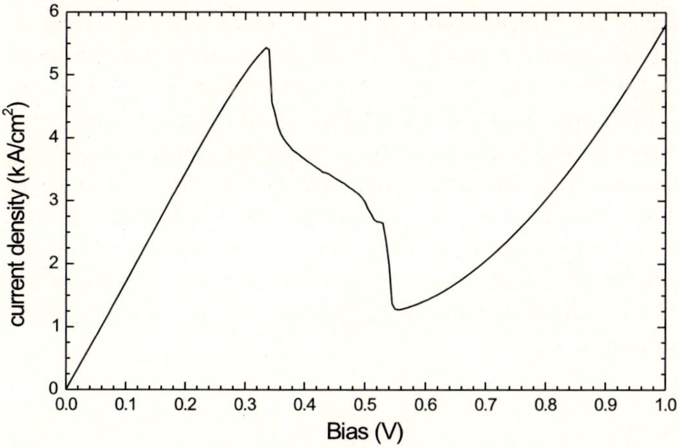

Fig. 8.23. The current-voltage measurements from an Esaki tunnel diode with a thin intrinsic SiGe between the doped regions. The device has a peak current density of 5.4 kA/cm^2 and a PVCR of 4.25 at 300 K. The device was fabricated and measured by Günter Reitemann at Stuttgart

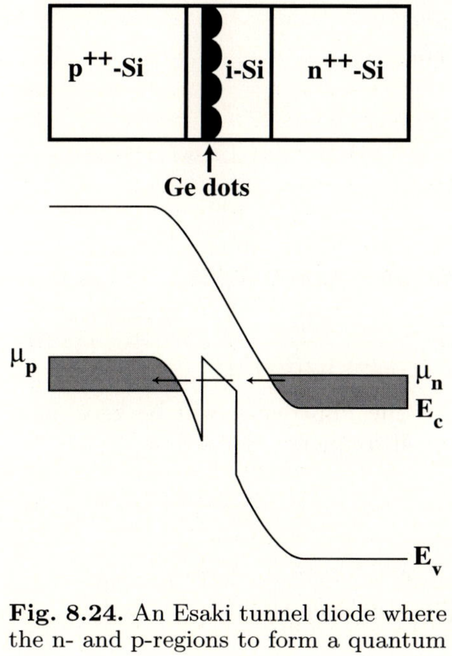

Fig. 8.24. An Esaki tunnel diode where Ge dots have been self-assembled between the n- and p-regions to form a quantum well through which electrons can tunnel

depends on the quality of the Ohmic contacts to the device and any series resistances in the circuit. Fabricating shallow Ohmic contacts to tunnel diodes is a major problem as if the contact diffuses or spikes too far then it can easily short through the Esaki diode or RTD. Therefore relatively thick cap layers are normally used for such diodes. These, however, form series resistances which decrease the high frequency performance of such devices. In addition, the contact resistance of any metal contact to the doped semicondutor also reduces the high frequency performance.The equivalent circuit for tunnel diodes is shown in Fig. 8.25. The contact and series resistances are lumped together with the wires in the circuit under R_s and any inductive components are combined in L_s.

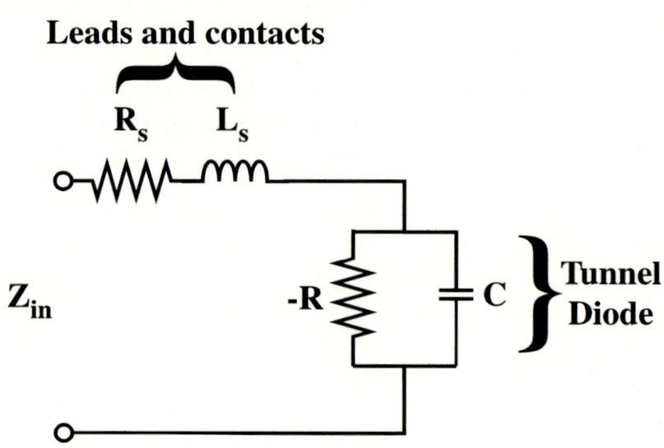

Fig. 8.25. Equivalent circuit of a tunnel diode

The input impedence of the equivalent circuit is given by

$$Z_{in} = \left[R_s + \frac{-R}{1 + (\omega RC)^2}\right] + \mathrm{i}\left[\omega L_s + \frac{-\omega RC^2}{1 + (\omega RC)^2}\right] \tag{8.51}$$

From this equation, the resistive part of the impedence will be zero at a certain frequency termed the resistive cutoff frequency defined as

$$f_r = \frac{1}{2\pi RC}\sqrt{\frac{R}{R_s} - 1} \tag{8.52}$$

It is also possible to define a reactive cutoff frequency where the reactive or imaginary part of the impedance becomes zero such that

$$f_x = \frac{1}{2\pi}\sqrt{\frac{1}{L_s C} - \frac{1}{(RC)^2}} \tag{8.53}$$

A figure of merit which is typically used for tunnel diodes is the speed index which is defined as the peak current density divided by the capacitance at the valley voltage. Another important parameter is the noise figure (NF) defined as

$$NF = 1 + \frac{q}{2k_B T} |RI|_{\min} \tag{8.54}$$

where $|RI|_{\min}$ is the minimum value of the negative resistance-current product on the current-voltage characteristic.

8.2.7 Comparison of Tunnel Diode Results

To finish this section on tunnel diodes, a brief comparison of experimentally demonstrated performance from RTDs and Esaki diodes will be reviewed with some of the values from the III-V system (Table 8.1). Recent results from Si/SiGe RTD and SiGe interband diodes are now competing with many of the best results from the III-V materials. Table 8.2 shows the main parameters which must be optimised for applications. In addition, values calculated by Christian Pascha and Karl Goser at Dortmund University for high speed logic and low power memory are also shown for comparison. These suggest that for high speed logic peak current densities of 10 kA/cm^2 and PVCR of 3 are ideal while for low power memory, peak current densities of 0.1 kA/cm^2 with a PVCR of 3 are required. It is clear from the table that III-V devices easily meet the requirements for applications.

Table 8.1. A comparison of the highest J_P values reported for RTDs in different material systems

Material	InGaAs	InAs	Si/SiGe	GaAs	Si Esaki
$J_p(kA/cm^2)$	460	370	282	250	151
PVCR	4	3.2	2.4	1.8	2.0
$\Delta I \Delta V$	5.4	9.4	43.0	4.0	1.1
R_D (Ω)	1.5	14.0	12.5	31.8	79.5
Area (μm^2)	16	1	25	5	2.2
Group	MIT	Caltech	Cambridge	Stanford	NRL

To allow tunnel diodes to be fabricated on CMOS fabrication lines or integrated with Si-based transistors for circuits will require a Si-based technology. The SiGe RTDs need substantial reduction of the peak voltage before they can be used in circuits while the Si Esaki diodes grown by MBE already meet the appropriate criteria for making circuits. The Si Esaki diodes, however, may suffer from high capacitance since the δ-doped layers are so close together. This may limit the ultimate high speed performance of such devices. There is also the issue of whether either of these technologies can survive processing with transistors. This will be reviewed in the next chapter where a number of tunnel diode circuits will also be investigated.

Table 8.2. A comparison of present III-V RTD results for both logic and memory circuits compared to the ideal RTD devices for those applications. Also included are the best Si and SiGe tunnel devices for comparison

	High Speed RTD Logic	Theory High Speed Logic	Low Power Memory	Theory Low Power Memory	nm Scaled RTDs	SiGe RTDs	SiGe Esaki Tunnel Diodes
PVCR	4	3	2	3	3	2.4	2.0
J_p (kA/cm^2)	40	10	0.0002	0.0001	10	282	151
Min. Feature Size	2 μm	0.2 μm	0.5 μm	0.2 μm	50 nm	5 μm	2.2 μm
Peak Voltage	0.35	0.16	0.20	0.20	0.20	1.8	0.28
Max. Clocking Freq.	12.5 GHz	6.25 GHz	592 kHz	56.8 MHz	6.25 GHz	-	0.5 GHz
RTD Time Constant	0.02 ns	0.04 ns	422 ns	4.4 ns	0.04 ns	-	0.5 ns
Ref.	Pacha	Pacha	Raytheon	Pacha	Pacha	Paul	Jim

8.3 Real Space Transfer (RST) Devices

As the gate length of field effect transistors is scaled down, the temperature of the electrons between the source and drain is increasing since the electric field in the device is increasing. While hot electron effects are detrimental to the operation of FETs, real space transfer (RST) devices use the hot electrons to perform logic functions. The concept is to apply a large source-drain bias, V_{sd} along a quantum well device so that at some applied bias, hot electrons are excited out of the well and detected at a collector (I_c). This is the principle behind the operation of a three terminal heterojunction device named the charge injection transistor (CHINT).

A SiGe CHINT fabricated and measured by Bell Laboratories is shown in Fig. 8.26 along with the band diagram of the device. Pseudomorphic layers have been used in this particular design so that holes are the majority carrier in the device. Hot electron CHINTs would require virtual substrates and strained-Si quantum wells. The most important part of fabricating a CHINT is to achieve shallow Ohmic contacts which contact the upper quantum well but do not allow holes to be excited into the barrier. As a bias is applied

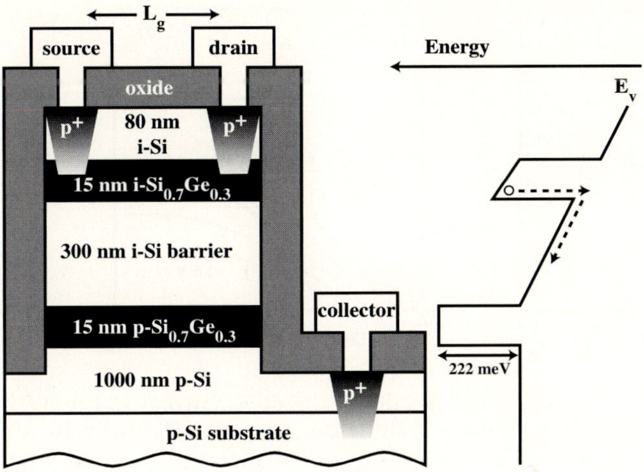

Fig. 8.26. A charge injection transistor (CHINT) designed and fabricated at Bell Laboratories along with the valence band structure. By application of a bias between the source and drain, V_{sd}, electrons are excited out of the upper quantum well and with an appropriate collector bias (V_c), are swept into the collector by the electric field to be measured as a current, I_c

between the source and drain, at some point the holes become hot enough to be excited out of the quantum well and are swept down to the collector by an appropriate bias. Typical current-voltage characteristics are shown in Fig. 8.27. Negative differential resistance is observed in the I_{ds} versus V_{ds} plot as holes are excited out of the quantum well at approximately -1 V. As the collector voltage is increased to more negative values, the collector is pulled up in energy thereby increasing the electric field to sweep holes into the collector (Fig. 8.27(b)).

Monte Carlo modelling of the CHINT device has shown that the intrinsic short circuit current gain cut-off frequency, f_T is higher than that of a FET fabricated from the same material with an equivalent channel length. High frequency measurements demonstrated a f_T of 6 GHz for a channel length of 0.5 μm for the simple device structure shown in Fig. 8.26. The results from Bell Laboratories demonstarted on/off current ratios of 3.2 at room temperature and 1.3×10^5 at 77 K. The value at room temperature requires substantial improvement before such transistors can be efficiently used in circuits. Figure 8.28 shows simulations on SiGe CHINTs. For comparision the experimental results are also shown. Doping in the upper quantum well clearly reduces the performance of the device which is demonstrated by the increase in I_{ds} and a reduction in I_c. This is the result of additional scattering in the quantum well from the impurities which reduces the energy of the holes and therefore the hole temperature. Reduction of the silicon cap layer is also observed to improve the performance as this potentially offers an alternative

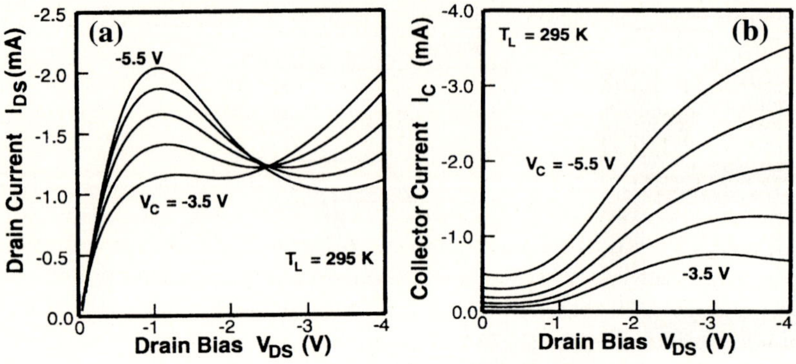

Fig. 8.27. Room temperature characteristics of (a) the drain current, I_{ds} and (b) the collector current, I_c for different source-drain biases, V_{ds} and collector voltages, V_c

(parasitic) path for increasing, I_{ds}. The silicon cap is beneficial from the fabrication point of view as it relaxes the requirements on the Ohmic contacts to the upper quantum well from being abrupt and shallow to only requiring abrupt.

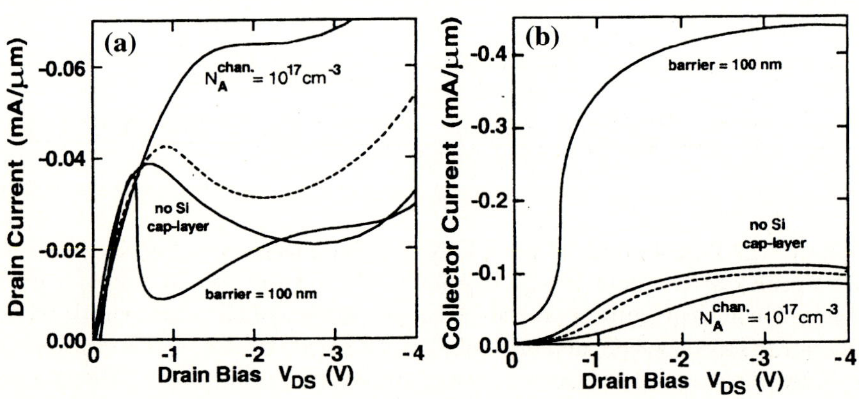

Fig. 8.28. (a) The drain current, I_{ds} and (b) collector current, I_c simulated using Poisson, drift-diffusion and energy balance equations for $L_g = 0.5\,\mu$m. The device from Fig. 8.27 is also shown for comparison. Reduction of the barrier layer and silicon cap will increase the device performance. V_c= -4 V has been used in all the simulations

The largest increase in performance is observed for a decrease in the barrier thickness. One must be careful with the simulations of the barrier thickness because the important parameter for collecting holes is the electric field in the device. Therefore a constant, V_c with a reduced barrier thickness has a substantially larger electric field and will give better performance. Leakage currents are reduced as V_c is reduced and therefore the thinner barriers do aid the overall performance of the device as they may be operated at lower voltages and with lower leakage currents.The CHINT in Fig. 8.26 has an in built symmetry in that the source and drain can be exchanged without any change to the operation of the device. While a FET can also have the source and drain exchanged, the difference with the CHINT is that the output terminal is the collector and therefore the ouput current is independent of an exchange of input voltages to source and drain. Thus the device operates as an exclusive-OR (XOR) gate with the I_c as the output for the binary logic signals applied as the input voltages to the source and drain.

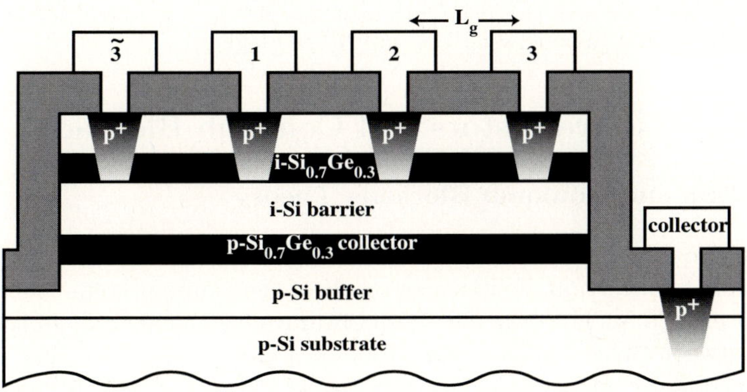

Fig. 8.29. A schematic diagram of an expanded CHINT device with 3 source contacts which define an OR-NAND gate. By tying contacts 3 and $\tilde{3}$, cyclic symmetry results from the periodic boundary conditions

The CHINT design may be expanded to having many more source contacts (Fig. 8.29). As an example of the flexibility of this concept, a logic gate with three contacts will be considered (Fig. 8.30). This device has a cyclic three-fold symmetry and may operate either as an OR gate or as a NAND gate. One electrode (let us choose number 3) is used as a control gate and will switch the device between an OR gate and a NAND gate. The truth table (Fig. 8.30(b)) shows the output at I_c as the input voltages V_1 and V_2 are varied. The output is low (logic 0) in two states when $V_1 = V_2 = V_3$ and high (logic 1) for all other input values.

The CHINT concept of real space transfer of charge offers a substantially different approach to logic architectures and potentially may be a superior approach at small gate lengths where hot electrons effects may dominate. At present the research has been very limited and leakage and on/off currents are quite poor. These need to be substantially improved before CHINTs could ever be used in real systems.

(a)

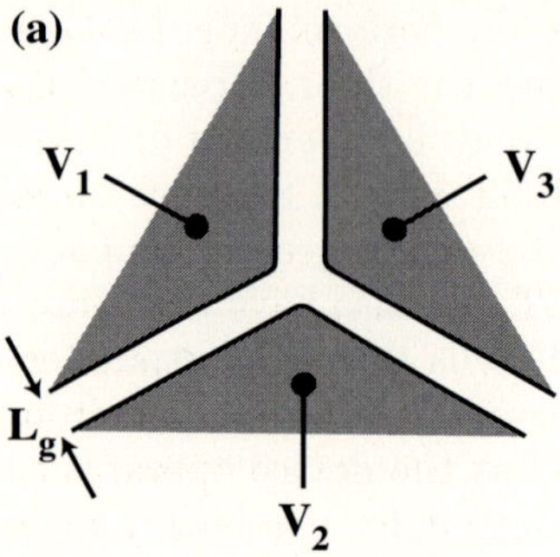

(b)

function		OR	NAND
control	V_3	0	1
input	V_2	1 0 0 1	1 0 0 1
	V_1	0 0 1 1	0 0 1 1
output	I_c	1 0 1 1	1 1 1 0

Fig. 8.30. The principle behind a multiterminal OR-NAND logic gate fabricated using real space transfer of charge. (a) A schematic of a possible layout of the device with the 3 voltages defined. (b) The truth table for the device. V_3 is used as a control which switches the gate between OR and NAND logic

8.4 Single Electron Transistors and Coulomb Blockade

8.4.1 Introduction and Coulomb Blockade Theory

The discreteness or quantisation of the electron charge manifests itself in the electrical conductance of certain devices as a result of the Coulomb repulsion of the individual electrons. The transfer by quantum mechanical tunnelling of one electron between two initially neutral regions of mutual capacitance, C increases the electrostatic energy of the system by an amount, $q^2/2C$. When this electrostatic energy is larger than the thermal fluctuations in the system then conduction is suppressed for small applied voltages. This phenomenon is known as the Coulomb blockade of single electron tunnelling. While single electron tunnelling might be considered a classical effect due to the capacitance, it is a true quantum device because it is the quantisation of the system that produces the Coulomb blockade effect.

The conventional treatment of single electron tunnelling in a tunnel junction starts from the current biased tunnel junction of Fig. 8.31. In this system, the charge on the junction, Q is a continuous variable. A small shift in the electron distribution relative to the positive ionic background created by the current, I flowing in the system can change the junction charge by a fraction of the discrete charge of an electron, q. The charge tunnelling across the junction, however, must be discrete and equal to q or multiples of q. The charging or Coulomb energy of the junction is

$$E_{ch} = \frac{Q^2}{2C} \tag{8.55}$$

Hence, for a normal junction, the reduction of the charging energy when one electron tunnels through the barrier is

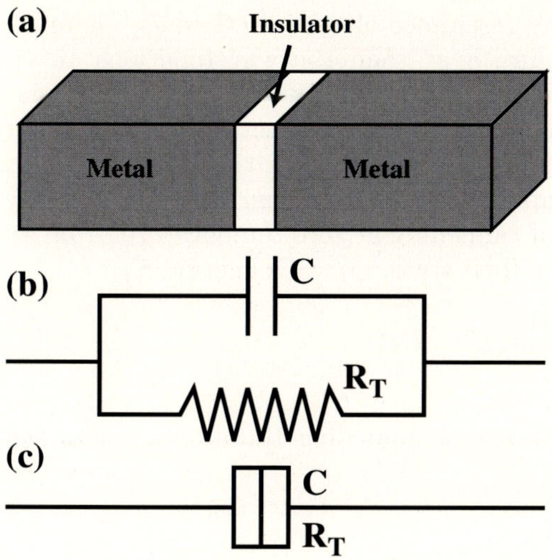

Fig. 8.31. The equivalent representations of the single tunnel junction with charge, Q, an effective capacitance, C and tunnel resistance, R_T

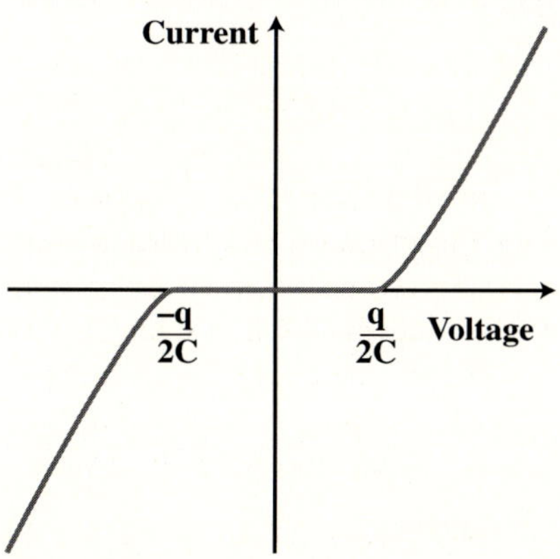

Fig. 8.32. The current-voltage characteristics showing Coulomb blockade

$$\Delta E_{ch} = \frac{Q^2}{2C} - \frac{(Q-q)^2}{2C} = \frac{q\,(Q-q/2)}{C} \tag{8.56}$$

ΔE_{ch} must be positive for tunnelling to occur and so for $Q < \frac{q}{2}$ no current flows, hence the name Coulomb blockade. A Coulomb gap to single electron tunnelling appears in the current voltage characteristics of the junction as shown in Fig. 8.32 so that

$$I = 0 \qquad \text{for} \qquad -\frac{q}{2C} < V < \frac{q}{2C} \tag{8.57}$$

If we assume that the tunnelling resistance of the junction, R_T, is much greater than the resistance quantum, h/e^2 then the wavefunctions on either side of the tunnel junction only mix very weakly. We also assume that charge equilibrium is established before a tunnelling event occurs and that the tunnelling rate, Γ may be calculated by considering the Hamiltonian as a perturbation. Under these assumptions, Fermi's Golden Rule (3.223) may then be used to calculate the rate of tunnelling at zero temperature, from an initial state, ψ_i with energy, E_i to a final state, ψ_f with energy, E_f

$$\Gamma_{i\to f} = \sum_{i,f} \frac{2\pi}{\hbar} \left|\left\langle \psi_f|\hat{H}_T|\psi_i\right\rangle\right|^2 \delta\left(E_i - E_f\right) \tag{8.58}$$

The solution to this using the appropriate tunnelling Hamiltonian, $\hat{H}_T$ can be shown to be

$$\Gamma = \frac{\Delta E}{q^2 R_T}\left[\frac{1}{\exp\left(\frac{\Delta E}{k_B T}\right) - 1}\right] \qquad \text{for} \Delta E > 0 \tag{8.59}$$

$$= 0 \qquad \text{for} \Delta E < 0 \tag{8.60}$$

where

$$\Delta E = -qV + \frac{q^2}{2C} \tag{8.61}$$

for the tunnelling of positive charge q in the direction of the voltage drop across the junction.

It should be noted that this particular tunnelling model has a number of shortcomings. In particular:-

- the dimensions and geometry of the tunnel junctions are ignored (it is a 0D model).
- the finite tunnelling time of the electron is ignored.
- the electric charge is assumed to redistribute instantaneously on each electrode after a tunnelling event.
- the energy in the electrodes and the source are assumed to be continuous without any quantisation.
- the source is considered as a fixed classical function of time.

Despite these shortcomings, the model has given an adequate description of the experimental observations of Coulomb blockade.

When a current source is connected to the junction, the source will charge the capacitor until the charge threshold of $q/2$ is reached when an electron will tunnel across the insulator. The junction charge then becomes $-q/2$ and a new charging cycle can begin. All the above analysis assumes that the capacitance of the junction is small enough that the charging energy is substantially larger than the thermal fluctuations in the system, $k_B T$.

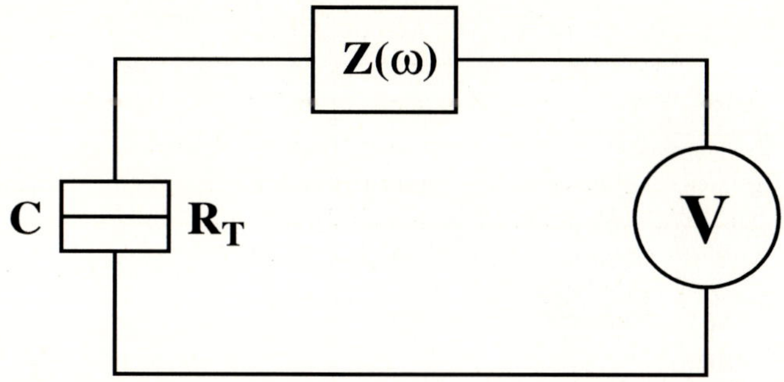

Fig. 8.33. The equivalent circuit of a single tunnel junction attached to a voltage source, V with an impedence, $Z(\omega)$ produced by the leads

Coulomb blockade is never observed in a single tunnel junction, however. This is because the single tunnel junction is strongly effected by the leads which are attached to the junction. In particular the capacitance of the leads is always several orders of magnitude larger than the junction capacitance and hence all junction effects are dwarfed by the circuit. The system can be modelled by treating the leads as an impedence, $Z(\omega)$ in series with the junction (Fig. 8.33). For experimental single tunnel junctions, the capacitance of the leads and the tunnel junctions act like a system with a large capacitance and hence no Coulomb blockade is observable in any single tunnel junction. It is this impedence, $Z(\omega)$ which takes account of the environment, i.e. the classical coupling of the tunnel junction to the external circuit.

Two different types of analysis for explaining SET effects have been developed:- the **local** view and the **global** view. If only the tunnel junction through which an electron is tunnelling is considered and the rest of the world is ignored, then we are considering the **local** view of Coulomb blockade.

After a tunnelling process, a nonequilibrium situation occurs since the charge on the junction, $Q-q$ and the charge imposed by the voltage source, $Q=CV$ are different. The voltage source must therefore do work to re-establish the equilibrium in the system by transferring an electron to recharge the junction capacitor to the charge, Q. Overall there is no change in the charging energy but an amount of work, qV has appeared between the Fermi levels of the two electrodes straddling the junction. Here the circuit has been viewed **globally**.

8.4.2 The Quantum Dot, Double Tunnel Junction System

Let us consider a system with a small island of charge between two tunnel junction (Fig. 8.34) where the size of the island is small enough to be zero dimensional in nature. Hence the electron energy levels on the island are quantised. This is the case in a number of semiconductor samples which have

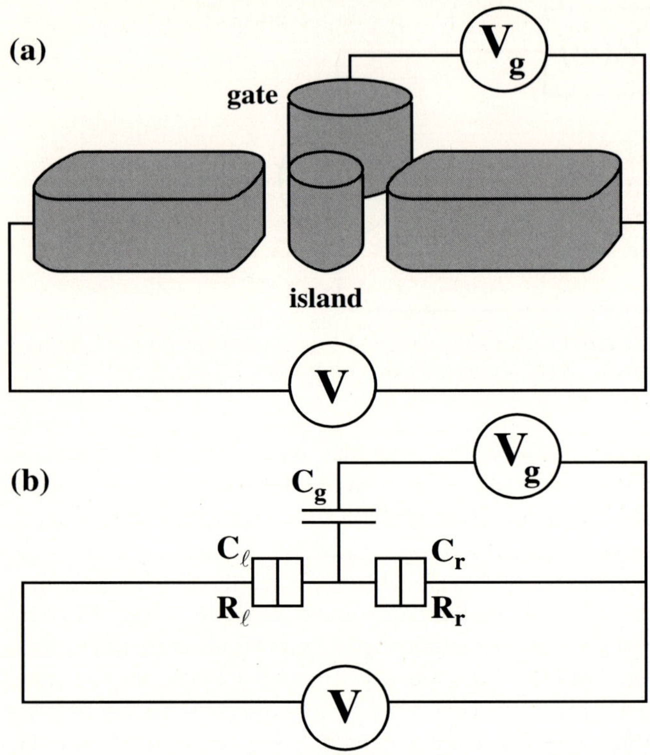

Fig. 8.34. (a) A schematic diagram showing an island of charge with a gate and (b) the equivalent circuit diagram

islands of comparable size to the Fermi wavelength, λ_F and the number of electrons may be suppressed to only a few by use of a gate electrode.

Let the island have N electrons, N_0 electrons at zero gate voltage, V_g and zero bias voltage, V (i.e. $V_g = V = 0\,\mathrm{V}$ and $N_0 > N$) which compensates for the positive background charge originating from the donors and n excess electrons. C_g is the capacitance of the gate and also any other external capacitances while C_ℓ and C_r are the capacitances of the left and right tunnel junctions respectively. Let the electrochemical potential or Fermi energy of the island be $E_{Fd}\,(N)$ so that the continuous part of the excess charge on the island, Q_0 induces a voltage difference of $qV_\ell = E_{F\ell} - E_{Fd}\,(N)$ and $qV_r = E_{Fd}\,(N) - E_{Fr}$. The electrostatic energy of the island is

$$E_{es} = \frac{(-qn + Q_0)^2}{2C_\Sigma} \tag{8.62}$$

where $n = N - N_0$, $Q_0 = C_\ell V_\ell + C_r V_r + C_g V_g$ and $C_\Sigma = C_\ell + C_r + C_g$. Setting $n = 1$ and Q_0 in (8.62) gives $q^2/2C_\Sigma$ which is the charging energy of a single electron.

In experiments a small bias voltage is applied across the device such that

$$V = \frac{E_{F\ell} - E_r}{q} \tag{8.63}$$

is the voltage dropped across the two tunnel junctions while the gate voltage is varied. By simplifying, $Q_0 = C_g V_g$, the ground state for N electrons in the island at zero temperature is the sum of the single particle energies, E_p relative to the bottom of the conduction band, and the electrostatic energy (Fig. 8.35),

$$U(N) = \sum_{p=1}^{N} E_p + \frac{(-qn + C_g V_g)^2}{2C_\Sigma} \tag{8.64}$$

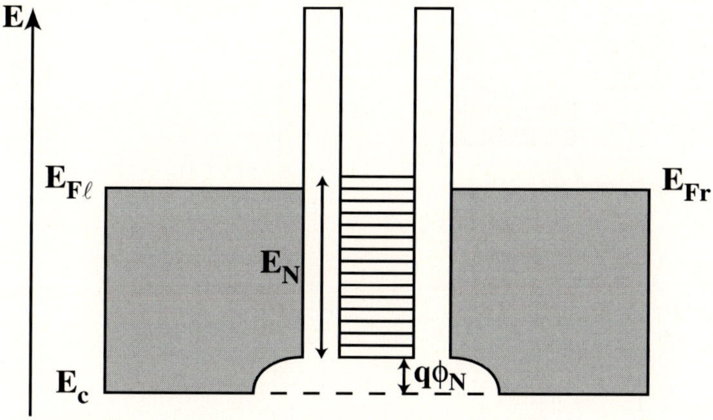

Fig. 8.35. The potential energy diagram of a semiconductor double barrier system for Coulomb blockade where the island is 0D. ϕ_N is the electrostatic potential with N electrons on the island relative to the bottom of the conduction band, E_c. E_N is the sum of the single particle energy levels on the quantum dot island

The minimum energy to add the Nth electron to the island is the electrochemical potential

$$E_{Fd}(N) = U(N) - U(N-1) \tag{8.65}$$

$$= E_N + \frac{\left(n - \frac{1}{2}\right) q^2}{C_\Sigma} - q\frac{C_g}{C_\Sigma} V_{gn} \tag{8.66}$$

Hence the change in the electrochemical potential when one electron is added to the island for a fixed gate voltage is

$$E_{Fd}(N+1) - E_{Fd}(N) = E_{N+1} - E_N + \frac{q^2}{C_\Sigma} \tag{8.67}$$

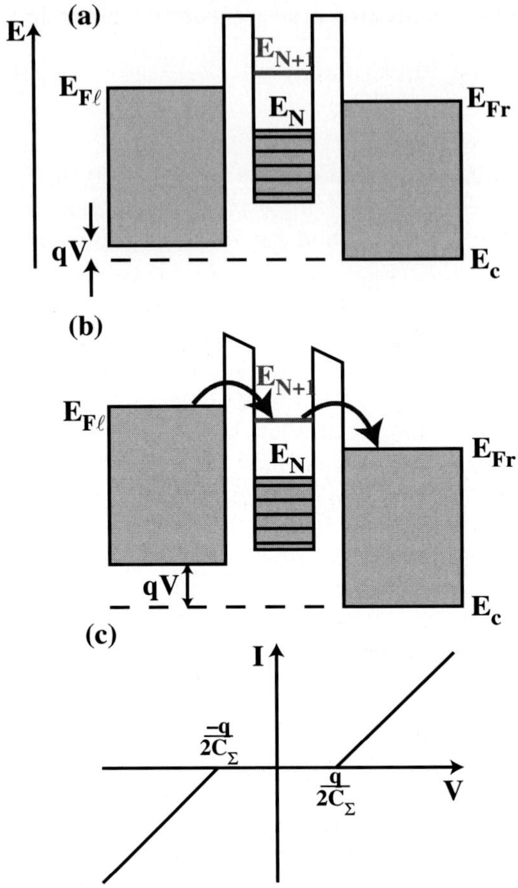

Fig. 8.36. (a) Coulomb blockade of electrons tunnelling from the left to the right electron reservoir occurs when $E_{Fd}(N) < E_{Fr} < E_{F\ell} < E_{Fd}(N+1)$. (b) Single electrons can tunnel onto the island when $E_{Fr} < E_{Fd}(N+1) < E_{F\ell}$. (c) The current-voltage characteristics with Coulomb blockade for $|V| < \frac{q}{2C_{Sigma}}$

The change in the electrochemical potential indicates an energy gap to add an extra electron to the island with charging energy of q^2/C. This leads to the Coulomb blockade of the the tunnelling of electrons into and out of the island (Fig. 8.36). By changing the gate voltage, V_g or increasing the total circuit voltage, V, the Coulomb blockade may be lifted when $E_{Fr} < E_{Fd}(N+1) < E_{F\ell}$. An electron can then quantum mechanically tunnel from the left reservoir into the island because $E_{F\ell} > E_{Fd}(N+1)$ and the electrochemical potential of the island increases as in (8.67) ($q\phi_{N+1} - q\phi_N = \frac{q^2}{C}$). The electron on the island can now tunnel out of island to the electron reservoir on the right because $e_{Fd}(N) < E_{F\ell}$ and the electrochemical potential of the island falls to $E_{Fd}(N)$. The process can now be repeated with the island being charged and then discharged through single electron tunnelling or single charge tunnelling.

The conditions to observe Coulomb blockade are

$$\frac{q^2}{C} \gg k_B T \qquad \text{charging energy} \gg \text{thermal fluctuations} \tag{8.68}$$

$$R_T \gg \frac{h}{q^2} \qquad \text{tunnel resistance} \gg \text{quantum fluctuations} \tag{8.69}$$

The first condition above prevents electrons from being thermally excited into the $N+1^{\text{th}}$ state and then tunnelling out of the island even when the Coulomb blockade condition of $E_{Fd}(N) < E_{Fr} < E_{F\ell} < E_{Fd}(N+1)$ exists. The second condition ensures that the electron wavefunction on the quantum dot island is localised on the island. If the tunnel resistance, R_T is much lower than $\frac{h}{q^2}$ =25.8 kΩ then delocalised states exist with lower Coulomb energy and the electrons can be transported through the quantum dot without having to be raised to the charging energy.

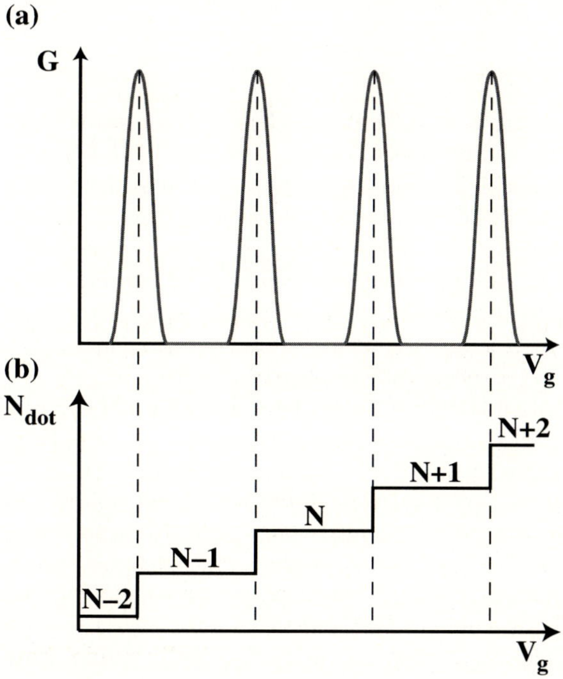

Fig. 8.37. The variation of (a) the conductance, G between the source and drain as a function of the gate voltage, V_g and (b) the number of electrons on the quantum dot island

Coulomb blockade can be used to produce a transistor effect in the quantum dot structure of Fig. 8.34. As the gate voltage, V_g is changed, the conduction of the quantum dot island will oscillate between the Coulomb blockade regime and the single electron tunnelling regime (Fig. 8.37) as the number of electrons on the island is changed. Hence the device characteristics as a function of gate voltage switch between conduction or on and Coulomb blockade of current or off. Hence the device operates like a transistor with on and off. The device is extremely sensitive to charge close to the quantum dot island. It therefore is not a robust transistor for circuits but it is one of the best charge sensors. The main applications for such single electron transistors of this design may be for sensing charge.

It should be clear from the above analysis that Coulomb blockade is the quantum mechanical tunnelling of single electrons where the electron wavefunctions and the probablity for tunnelling are controlled by the electrostatics in the system. While at first it may appear to be a classical effect due to electromagnetism, the discreteness or quantisation of the electron charge along with a small enough island or quantum dot to produce a Coulomb gap are essential for the effect to exist. Coulomb blockade is therefore clearly a quantum effect.

8.4.3 Single Electron Transistors

While CMOS transistors now have sub-100 nm gate-lengths, the number of electrons which are used in a switching operation are still tens of thousands. If this could be reduced to a situation where only one electron is used (or a few) then the energy required to switch the device between on and off should be much lower. This is the basic philosophy of single electron transistors (SETs). There are a number of different types of SETs. The original type rely on Coulomb blockade while a second type are literally miniature flash memory where the addition of a single electron to the gate-memory node results in a large change to the current in the measuring transistor.

When a gate is used to control the single electron tunnelling through an island of charge the device is called a single electron transistor or SET (Fig. 8.38). A large number of demonstrators of such transitors have been fabricated in a number of different material systems. The best performance has been produced in the silicon system due to the strong Coulomb interactions compared to a number of other standard semiconductor materials. For room temperature operation of such devices, the island must be below about 10 nm in diameter. Numerous SETs using silicon and silicon dioxide which operate at room temperature have been demonstrated. SETs are more likely to be used for memory applications because they have no gain (i.e. amplification). It is therefore difficult to create logic circuits where the gain in a transistor or logic device overcomes the losses in the circuit and interconnects (i.e. resistance in the interconnect wires). While a number of logic architectures have been proposed and a few have been demonstrated, it is questionable about

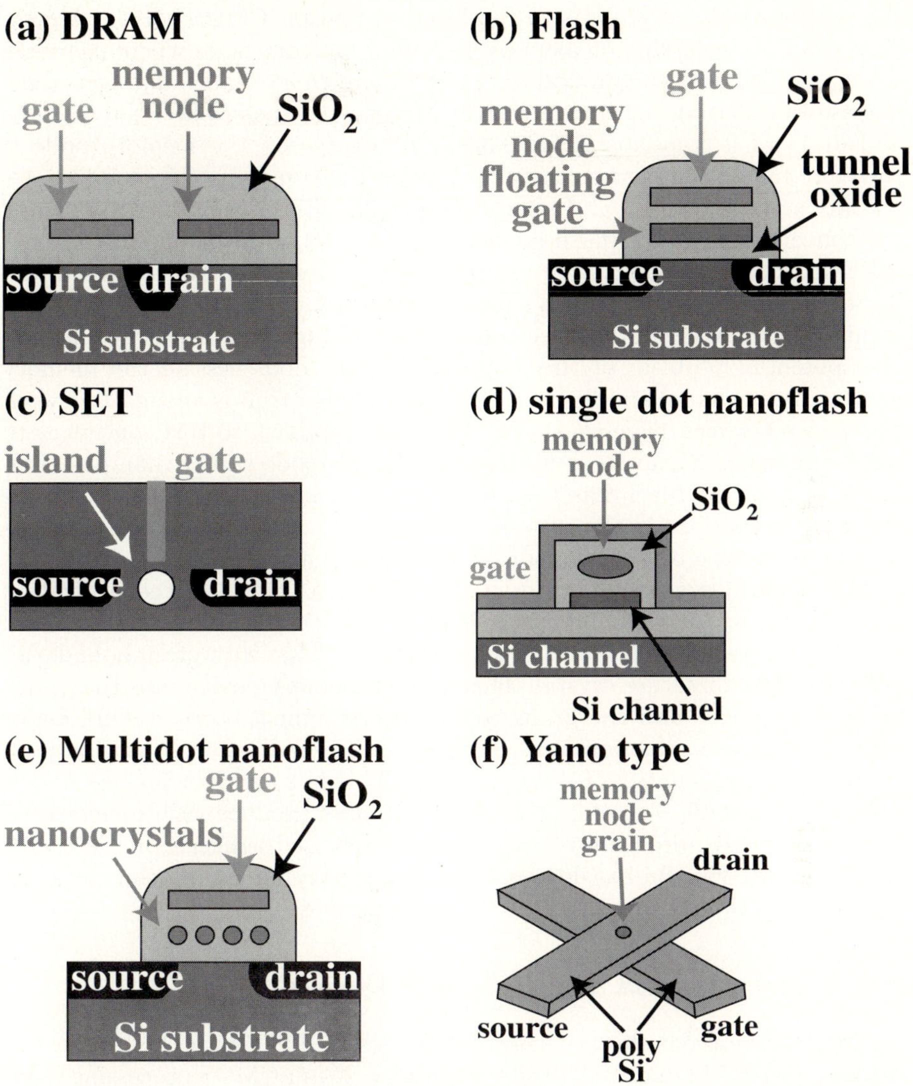

Fig. 8.38. Different types of conventional and single electron memories. (a) A cross section of a DRAM with lateral capacitor in the oxide. (b) A cross section of a CMOS flash memory. (c) The lateral pattern of a SET on top of a Si substrate. (d) A cross section of a multidot nanoflash. (e) A cross section of a single dot nanoflash. (f) A lateral view of a Yano type memory made from two crossed poly-Si stripes of material

the scalability of the circuits to the levels of present CMOS MPUs. Almost all types of semiconductor memory require a memory node where charge is stored to represent the on and off (1 and 0) memory states and then some means of measuring the charge with a current which passes between a source and drain of a transistor (Fig. 8.38). The charge on the memory node is therefore frequently used as the gate to switch a transistor channel.

The second type of SET memory is really just a miniature version of the conventional CMOS flash memory which is found in mobile phones and MPEG music stick players, for example. The structure is that shown in Fig. 8.38(d) where the addition of a single electron to the memory node results in a substantial change to the electron current through the transistor channel. One potential problem of this approach is the robustness of the memory node to stray charge and fluctuations since one electron is enough to switch the device between memory states. A second approach to this concept is to use a number of Si nanocrystals as nodes in the oxide rather than one (Fig. 8.38(e)). This approach has the advantage that it is more robust to single electron fluctuations in the system.

The next type of SET memory is that demonstrated by Yano at the Hitachi Central Research Laboratories (Fig. 8.38(f)). It involves the fabrication using standard CMOS fabrication lines of two crossed poly-Si wires. The poly-Si consists of small grains of single-crystal silicon with grain boundaries between each miniature crystal. This type of memory device uses the grains as memory nodes and the grain boundaries as tunnel barriers. One major problem is that conduction in the channel is the result of a percolation path through a large number of poly-Si grains. This is a random process and is difficult to control. Therefore the ability to mass produce such memory devices with the required control of properties may be the major problem with this apporach. Hitachi has demonstrated a 128 Mbyte memory chip with the technology although only half the devices operated.

8.4.4 Comparisons of Single Electron Devices

Table 8.3 summaries the experimental results in SETs with the production memory of DRAM and flash produced using 0.25 μm CMOS processing lines. Some of the performance is comparable to present CMOS DRAM and flash although noise because of the single electron nature is still a major barrier to manufacturable devices.

There are a number of major semiconductor manufacturers researching different types of singles electron transistors as shown in Fig. 8.38. As there are real problems in scaling DRAM to smaller dimensions due to the requirement of increasing the capacitance of the capacitor used as the memory node, SETs appear a natural progression of conventional silicon based memory. Many of the proposed devices are very close to conventional flash memory and indeed if flash is scaled down to the dimensions predicted in the ITRS

Table 8.3. A comparison of the performance of different types of SET and conventional memory technology. The DRAM and flash memory results are taken from the 0.25 μm technology node devices

	Conventional memory		Single electron based memories			
				Nanoflash		
	DRAM	Flash	SET	Multidot	Single Dot	Yano type
Read time	~6 ns	~6 ns	1 ns	~10 ns	~10 ns	~20 μs
Write time	~6 ns	1 ms	1 ns	~100 ns	<1 μs	~10 μs
Erase time	<1 ns	~1 ms	<1 ns	~1 ms	<1 ms	~10 μs
Retention time	250 ms	~10 years	~1 s	~1 week	~5 s	~1 day
Endurance cycles	∞	10^6	∞	10^9	10^9	10^7
Operating V	1.5 V	10 V	1 V	5 V	10 V	15 V
V for state inversion	0.2 V	3.3 V	<0.1 V	0.65 V	0.1 V	0.5 V
Electron no. to write bit	10^4	250	1 (exclude gate V)	10^3	1 (exclude gate V)	2 (exclude gate V)
Cell size (F^2/bit)	8.5	~9	9-12	9	9	2

roadmap, single electron effects will eventually dominate the devices when they reach dimensions of 10 nm or so.

8.5 Further Reading

1. M.J. Kelly, Low Dimensional Semiconductors: Materials, Physics, Technology, Devices" OUP, Oxford (1995)
2. J.H. Davies, The Physics of Low Dimensional Semiconductors, CUP, Cambridge (1998)
3. S.M. Sze, Semiconductor Devices: Physics and Technology, 2nd Edition, John Wiley and Sons, New York (2002)
4. D.J. Paul, Semiconductor Science and Technology **19**, R59 (2004)

9. Optoelectronics

9.1 Photonic Devices

Optoelectronics and optical components are used by almost everyone in the western world in everyday life. While compact discs (CD) and digital versatile disc (DVD) players are perhaps the most obvious examples of systems using lasers in the home, most telecommunication systems rely on optical fibres with laser transmitters at one end and semiconductor detectors at the other. Switches and routers for telecommunication applications are also heavily relianton the optical properties of semiconductor materials. In this section, a number of basic optical properties of semiconductor materials will be reviewed before the main silicon based optical devices are covered. At the end of the chapter, a number of contemporary device concepts will be discussed.

9.1.1 Basic Photonic Properties

There are a number of different ways in which photons may be absorbed by a semiconductor. Carriers may be generated either by intrinsic processes such as band-to-band transitions Fig. 9.1(a) and intersubband transitions in quantum wells (Fig. 9.1(d)) or by extrinsic processes involving states inside the forbidden bandgap (Fig. 9.1(b) and (c)). A more complete list of processes is discussed in Sect. 3.7.

For an intrinsic photoconductor the conductivity can be written as

$$\sigma = ne\mu_e + pe\mu_h \tag{9.1}$$

and the increase in conductivity when the semiconductor is illuminated is predominantly due to the increase in the number of carriers. The bandgap defines a long wavelength cut-off which for this case is given as

$$\lambda_c = \frac{hc}{E_g} = \frac{1.24}{E_g\,(eV)} \qquad (\mu\text{m}) \tag{9.2}$$

All radiation with wavelength below λ_c is absorbed by the semiconductor and electron-hole pairs are generated. Photoexcitation may occur for the extrinsic absorption of a photon with energy greater than or equal to the forbidden state in the bandgap and the conduction or valence band edge. A similar

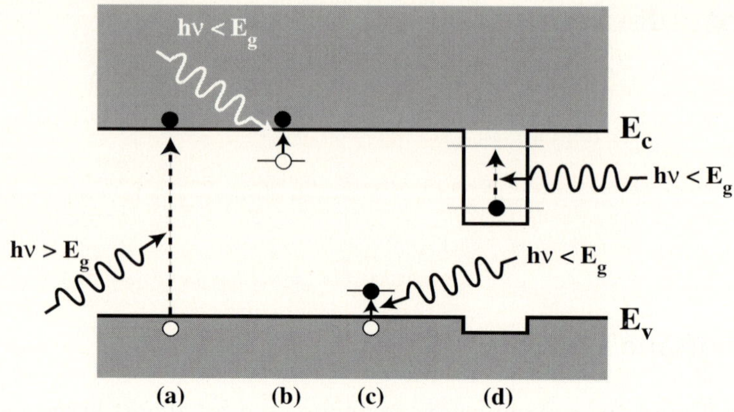

Fig. 9.1. The different absorption mechanisms in a semiconductor. The grey areas and lines represent allowed states. (a) band-to-band absorption, (b) and (c) are extrinsic absorption by dopants or defects and (d) is intersubband absorption

long wavelength cut-off also occurs but it is determined by the energy of the forbidden state in the bandgap.

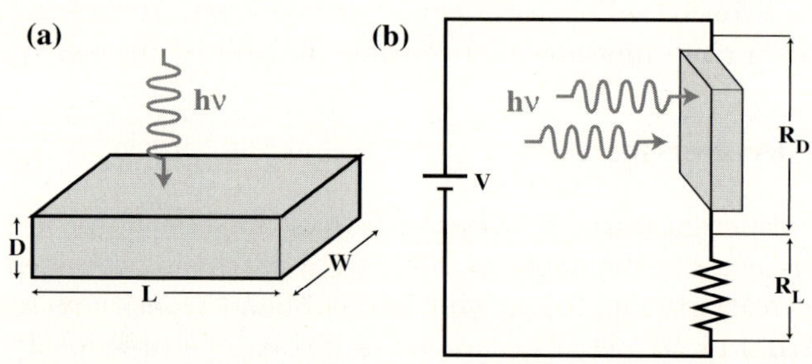

Fig. 9.2. (a) A piece of semiconductor material of dimensions length L, width W and depth D absorbing photons of energy $h\nu$. (b) The typical measurement circuit where the semiconductor photodetector has resistance R_D including any series resistances and the load resistance across which a voltage is measured for output is R_L

Photodetector performance in general is typically determined by three main criteria: the quantum efficiency or gain, the response time and the sensitivity (or detectivity). Fig. 9.2 demonstrates the basic absorption of a semiconductor where the two ends connected to a circuit have Ohmic contacts. At zero time the number of carriers generated in a unit volume by a flux of photons is n_0. The number of carriers, $n(t)$ at a later time t in the volume decays with the recombination of carriers as

$$n = n_0 exp\left(\frac{-t}{\tau}\right) \tag{9.3}$$

where τ is the carrier lifetime. Therefore the recombination rate is the inverse of this, i.e. $1/\tau$. If the area of the semiconductor absorbing a steady flow of photons is WL (Fig. 9.2) the total number of photons with energy $h\nu$ arriving at the surface is $(P_{opt}/h\nu)$ per unit time where P_{opt} is the incident optical power.

In equilibrium the generation rate of carriers must equal the recombination rate. If the semiconductor has a thickness, D which is much greater than the penetration depth of the light (i.e. 1/absorption $(1/\alpha)$), the total steady-state generation rate of carriers per unit volume is

$$G = \frac{n}{\tau} = \frac{\eta\,(P_{opt}/h\nu)}{WLD} \tag{9.4}$$

where η is defined as the quantum efficiency which is the number of carriers generated per photon. In an experimental setup, it is the photocurrent, I_p which is measured. This is given by

$$I_p = (\sigma F)\,WD = (q\mu_n nF)\,WD = (qnv_d)\,WD \tag{9.5}$$

where F is the electric field inside the photoconductor and v_d is the drift velocity of the carriers. Subsituting for the carrier density from (9.4) into the photocurrent gives

$$I_p = q\left(\eta\frac{P_{opt}}{h\nu}\right)\left(\frac{\mu_n\tau F}{L}\right) \tag{9.6}$$

It should be clear from rearranging this equation that the quantum efficiency is given by

$$\eta = \left(\frac{I_p}{q}\right)\left(\frac{h\nu}{P_{opt}}\right) \tag{9.7}$$

Examples of quantum efficiency for the main photoconductive materials are shown in Fig. 9.3 as a function of wavelength. Si has the highest quantum efficiency but the cut-off wavelength is too low for operation at the communications wavelengths of 1.3 μm and 1.55 μm. Related to the quantum efficiency is the responsivity which is the ratio of the photocurrent generated relative to the optical power incident on the photodetector as is given by

$$\Re = \frac{I_p}{P_{opt}} = \frac{\eta q}{h\nu} = \frac{\eta\lambda\,(\mu m)}{1.24} \qquad \text{A/W} \tag{9.8}$$

Sensisitivity is dependent on noise in the photodetector. It becomes impossible to get a signal when it equals the noise. Therefore the noise equivalent power (NEP) is defined as the incident rms optical power to produce a signal-to-noise ratio of one in a 1 Hz bandwidth. More generally this is defined as the power of a sinusoidally modulated monochromatic radiation which would result in the same rms output signal as an ideal noise-free detector. The bandwidth of the detector, Δf, is inversely proportional to the response time of

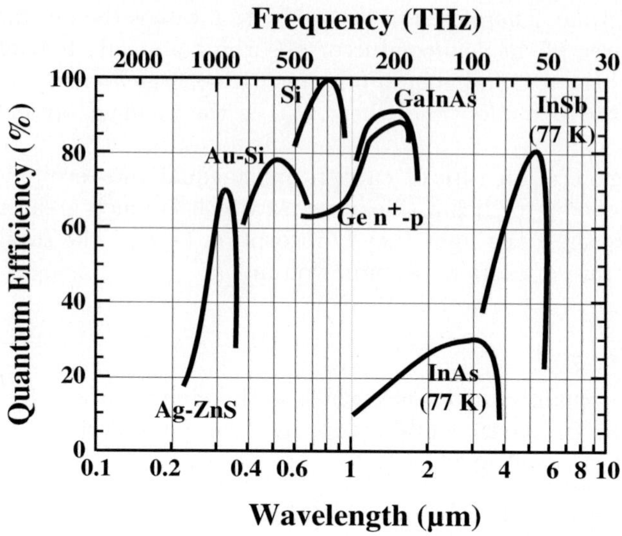

Fig. 9.3. The quantum efficiency for a number of the major photodetectors as a function of wavelength

the detector. If we assume that the noise power generated in a photodetector is proportional to the cross sectional area A then the noise current varies as $A^{1/2}$. Thus the NEP* is defined as a function of the area and the bandwidth as

$$NEP^* = \frac{NEP}{\sqrt{A \Delta f}} \tag{9.9}$$

More commonly the quantity that is quoted is the specific detectivity, D^* defined as

$$D^* = \frac{\sqrt{A \Delta f}}{NEP} \qquad \mathrm{m(Hz)^{1/2}/W} \tag{9.10}$$

In quoting D^*, one must state whether the radiation is from a black-body source or a monochromatic source and also the modulation frequency. It is recommended that D^* be expressed as $D^*\,(\lambda, f, 1)$ or $D^*\,(T, f, 1)$ where λ is in μm, f is the frequency of modulation in Hz, T is the black-body temperature in K and the reference bandwidth is always 1 Hz.

For a background limited photoconductor, the ideal D^* for unit quantum efficiency can be shown to be (see Sze in Further Reading)

$$D^*\,(\lambda, f, 1) = \frac{c \exp\left(\zeta/2\right)}{2\sqrt{\pi h k_B T \nu^2}\sqrt{(1 + 2/\zeta + 2/\zeta^2)}} \qquad \mathrm{m(Hz)^{1/2}/W} \tag{9.11}$$

where c is the velocity of light and $\zeta = h\nu/k_B T$. The ideal D^* curves are plotted in Fig. 9.4 along with the main photodetectors, bolometers and quantum well infrared photodetectors which will be discussed later in the chapter.

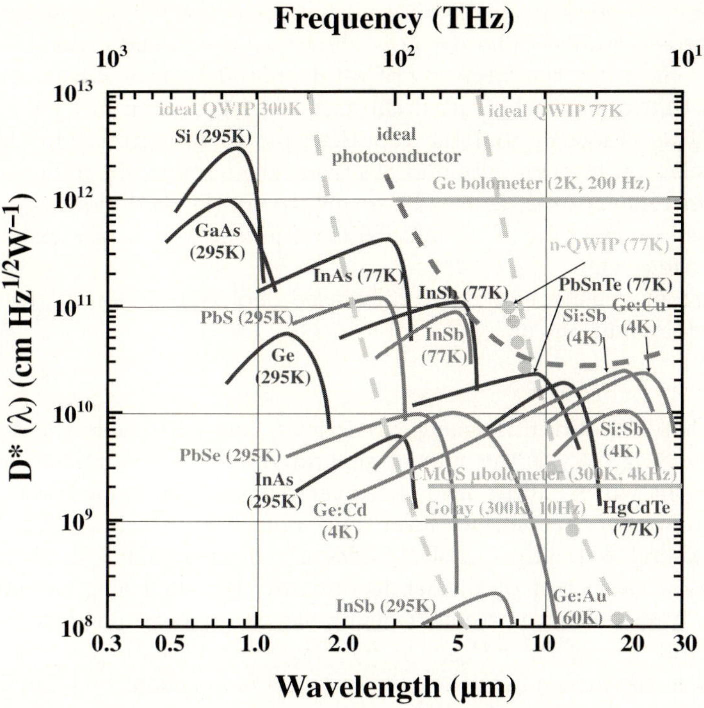

Fig. 9.4. The detectivities of a number of different photodetectors, bolometers and quantum well infrared photodetectors (QWIPs) at room temperature and 77 K

9.1.2 p-i-n Photodiodes

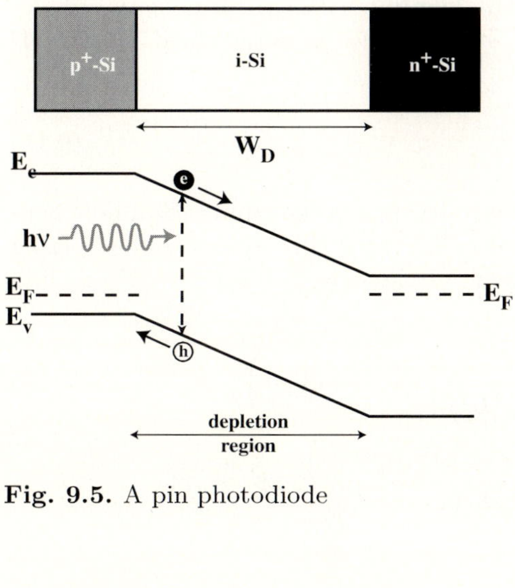

Fig. 9.5. A pin photodiode

The p-i-n photodiode is probably the most common of all photodetectors used in microelectronics. The diode is shown as a schematic diagram in Fig. 9.5. The reason for the large use of p-i-n photodetectors is that the depletion region thickness can be varied through the fabrication process allowing the quantum efficiency and the reponse time to be optimised. The basic principle relies on the generation of electron-hole pairs in the undoped "i" semiconductor region of the device. Those created in the depletion region are then separated by the built in or applied electric field which produces a photocurrent in the circuit.

The total current density in the reverse-biased depletion region is given under steady-state conditions as

$$J_{total} = J_{drift} + J_{diffusion} \tag{9.12}$$

where J_{drift} is the drift current formed by carriers generated inside the depletion region and $J_{diffusion}$ is the current generated by carriers outside the depletion region but which diffuse into the reverse-biased junction. For the device to be useful the p-type surface layer must not be too thick to absorb all the light and therefore requires to be significantly thinner than $1/\alpha$ where α is the absorption coefficient of the semiconductor. We shall also assume that the thermally generated current is negligible. If we have an incident photon flux per unit area of $\Phi_0 = P_{opt}(1-R)/Ah\nu$ where R is the reflection coefficient and A is the device area then the electron-hole generation rate is

$$G(x) = \Phi_0 \alpha e^{-\alpha x} \tag{9.13}$$

The drift current density can be found by integrating $G(x)$ to produce

$$J_{drift} = -q \int_0^{W_D} G(x)\,\mathrm{d}x = q\Phi_0 \left(1 - \exp\left[-\alpha W_D\right]\right) \tag{9.14}$$

where W_D is the depletion layer width. The minority carrier density of holes in the bulk undoped semiconductor for $x > W_D$ is determined by the 1D diffusion equation

$$D_p \frac{\partial^2 p_n}{\partial x^2} - \frac{p_n - p_{n0}}{\tau_p} + G(x) = 0 \tag{9.15}$$

where D_p is the diffusion coefficient for holes, p_{n0} is the equilibrium hole density and τ_p is the lifetime of excess holes. We solve this equation by applying the boundary conditions $p_n = p_{n0}$ for $x = \infty$ and $p_n = 0$ for $x = W_D$ to find

$$p_n = p_{n0} - \left(p_{n0} + \left[\frac{\alpha \Phi_0 L_p^2}{D_p\left(1-\alpha^2 L_p^2\right)}\right] \exp\left[-\alpha W_D\right]\right) \exp\left[\frac{W_D - x}{L_p}\right] + \left[\frac{\alpha \Phi_0 L_p^2}{D_p\left(1-\alpha^2 L_p^2\right)}\right] \exp\left(-\alpha x\right) \tag{9.16}$$

where $L_p = \sqrt{D_p \tau_p}$.

The diffusion current density is $J_{diffusion} = -qD_p \left(\partial p_n / \partial x\right)|_{x=W_D}$ which gives

$$J_{diffusion} = q\Phi_0 \frac{\alpha L_p}{1+\alpha L_p} e^{-\alpha W_D} + qp_{n0}\frac{D_p}{L_p} \tag{9.17}$$

and hence the total current density of a p-i-n diode is given by

$$J_{total} = q\Phi_0 \left(1 - \frac{e^{-\alpha W_D}}{1+\alpha L_p}\right) + qp_{n0}\frac{D_p}{L_p} \tag{9.18}$$

The term involving P_{n0} is typically significantly smaller than the first term and can therefore be ignored in many cases. Hence the total photocurrent is proportional to the photon flux. The quantum efficiency for a p-i-n diode can be obtained from (9.7) and (9.18) which gives

$$\eta = \frac{J_{total} A h\nu}{qP_{opt}} = (1-R)\left(1 - \frac{e^{-\alpha W_D}}{1+\alpha L_p}\right) \tag{9.19}$$

Hence for a high quantum efficiency a low reflection coefficient and $\alpha W_D \gg 1$ is desirable. If $W_D \gg 1/\alpha$ then the transit time of carriers across the drift region can be extremely slow.

The dark current is dominated by the diffusion currents from the n- and p-doped Si regions of the device and can be calculated as

$$J_{dark} = q\left(\frac{D_p}{L_p} p_{n0} + \frac{D_n}{L_n} n_{p0}\right) \tag{9.20}$$

where L_n is the equivalent n-type variable to L_p.

The complete way to estimate the response time of such p-i-n photodiodes is to first consider the capacitance of the diode and then obtain the frequency response assuming a RC time delay. Analysis of such techniques demonstrates that the optimum trade-off between absorption depth, $1/\alpha$ and speed occurs when the width of the undoped i-Si is $W_D \sim 1.5/\alpha$. The maximum speed at which the photodetector can respond to a changing optical signal is determined by the transit time of the generated carriers across the device. If the saturation velocity is v_{sat} then the modulation bandwidth (the maximum response to a changing optical signal) is given by

$$f_{3dB} \sim \frac{0.4 v_{sat}}{W_D} \tag{9.21}$$

There are many examples of SiGe p-i-n photodiodes (Fig. 9.6). The main use of SiGe is to use the smaller bandgap to move the aborption edge to longer wavelengths, in particular to the communication wavelengths of 1.3 μm and 1.5 μm where optical fibers have minima in absorption coefficients. Pure Ge photodiodes have an absorption edge at 1.88 μm and SiGe technology should allow p-i-n interband photodiodes to operate at least up to that wavelength.

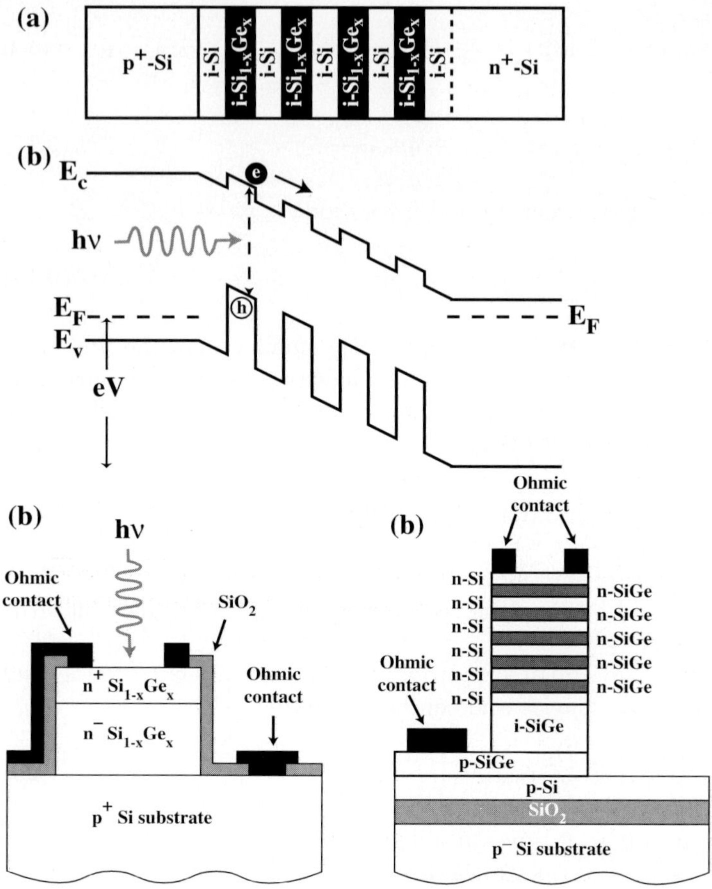

Fig. 9.6. Different SiGe pin photodiodes. (a) is a multiple hole quantum well structure to reduce the bandgap using pseudomorphic SiGe strained layers. This increases the operating wavelength of the detector as shown in (b) by reducing the bandgap. (c) demonstrates a simple pseudopmorphic SiGe p-n diode which is limited by the critical thickness to low Ge contents. (d) demonstrates the use of a buried reflector using silicon-on-insulator wafers with a buried SiO_2 layer in the substrate. Such devices can be grown with or without virtual substrates although the wavelength is still very short if virtual substrates are not grown

The major problem is that the thickness of a pseudomorphic SiGe layer is quite small especially for the significant Ge contents required for the communication wavelength. The first method to get around this is to grow a finite number of quantum wells with the total thickness below the critical thickness. This allows absorption in the quantum wells at the longer wavelength Fig. 9.6(a). A second method just grows a thick SiGe layer (Fig. 9.6(c)) but the Ge content is very limited by the critical thickness constraints. A third method is to produce the p-i-n diode on a virtual substrate, however, the dislocations are detrimental to the diodes operation and performance. Superlattice structures have also been used and silicon-on-insulator layers where a buried SiO_2 layer is used as a reflector in the substrate to increase the absorption of carriers in the drift region of the device (Fig. 9.6(d)). The superlattice structures also allow pronounced wavelength selectivity which for communications where a number of closely spaced frequencies can be used for broadband, parallel communication down an optical fiber is extremely useful. Since III-V photodiodes perform extremely well, SiGe p-i-n diodes require high performance and low prices to compete with the presently available technology.

9.1.3 Avalanche Photodetectors

The avalanche photodetector is similar in design to a p-i-n photodiode but are operated at high reverse-bias voltages where avalanche multiplication can take place. This process provides multiplication or amplification of the photocurrent but the dark current and noise are also multiplied by the device in addition to the measured signal. Ideally an avalanche photodiode should be a unipolar device with the carrier concentration increasing if impact ionisation occurs in with a large electric field. In semiconductors, however, both electrons and holes can impact ionise more electron-hole pairs as shown in the schematic diagram of Fig. 9.7.

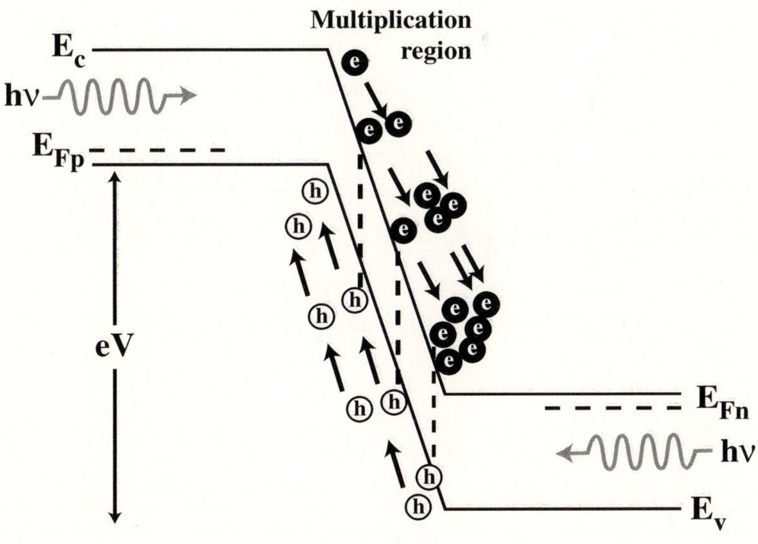

Fig. 9.7. A schematic diagram of the band structure in an avalanche diode showing the multiplication region with some recombination of electron-hole pairs across the bandgap reducing the multiplication factor

A feedback process occurs since electrons and holes travel in opposite directions. Let us define
α_n = the electron ionisation coefficient (m^{-1})
= the number of electron-hole pairs generated by one incident electron per unit distance
α_p = the hole ionisation coefficient (m^{-1})
= the number of electron-hole pairs generated by one incident hole per unit distance

Unfortunately $\alpha_n \neq \alpha_p$ for most semiconductors. Both are functions of the electric field, F with exponential dependencies of the form

$$\alpha_n = \alpha_{n0} \exp\left(\frac{-C_n z}{F}\right) \quad \text{and} \quad \alpha_p = \alpha_{p0} \exp\left(\frac{-C_p z}{F}\right) \tag{9.22}$$

with the constants α_0, C_{n0}, α_p and C_{p0} dependent on the semiconductor material being used.

The multiplication factor for a p-i-n avalanche photodiode can be shown to be

$$M = \frac{\left(1 - \frac{\alpha_p}{\alpha_n}\right) \exp\left[\alpha_n W_D \left(1 - \frac{\alpha_p}{\alpha_n}\right)\right]}{1 - \left(\frac{\alpha_p}{\alpha_n}\right) \exp\left[\alpha_n W_D \left(1 - \frac{\alpha_p}{\alpha_n}\right)\right]} \tag{9.23}$$

which reduces to the simpler form for equal ionisation coefficients ($\alpha_p = \alpha_n$) of

$$M = \frac{1}{1 - \alpha_n W} \tag{9.24}$$

The breakdown voltage, V_B corresponds to the situation where $\alpha_n W_D = 1$. The noise is also multiplied in the structure and it can be shown that the noise factor, $NF(M)$ for a multiplication factor, M for electron injection in the photodiode along is given by

$$NF(M) = \frac{\alpha_p}{\alpha_n} M + \left(2 - \frac{1}{M}\right)\left(1 - \frac{\alpha_p}{\alpha_n}\right) \tag{9.25}$$

It can also be shown that the noise equivalent power for avalanche photodiodes is

$$NEP = \sqrt{2} \frac{h\nu}{\eta} \sqrt{\frac{I_{av}}{q\left(NF(M)\right)^2}} \tag{9.26}$$

where

$$J_{av} = \left(I_B + I_{dark}\right) NF(M) + 2 \frac{k_B T}{q R_{equivalent} M^2} \tag{9.27}$$

with I_B the current from background radiation, I_{dark} is the dark current and $R_{equivalent}$ is the equivalent resistance of the photodiode in the operating circuit (Fig. 9.2 $1/R_{equivalent} = 1/R_D + 1/R_L$ where R_D is the resistance of the diode including the resistance of the pin diode and any series resistances).

The best Si avalanche photodiodes have quantum efficiencies near 100% at a wavelength of about 750 nm in the visible part of the spectrum. They use a SiO_2 - Si_3N_4 antireflection coating to improve number of photons being absorbed in the diode. The ratio of α_p/α_n is about 0.04 and so the noise factor from (9.25) is 2.3 for $M = 10$.

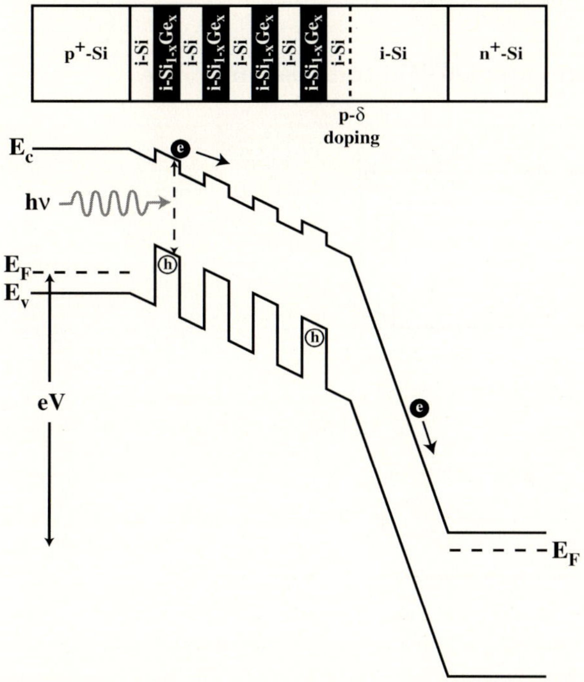

Fig. 9.8. A SiGe avalanche photodiode designed for use at 1.3 μm using pseudomorphic SiGe quantum wells to move the absorption edge to longer wavelengths than Si photodiodes

Figure 9.8 shows a schematic diagram of a SiGe quantum well avalanche photodiode with separate absorption and multiplication (SAM). This has been designed for operation at around 1.3 μm. The SiGe quantum wells provide the main absorption region for the photons and the undoped i-Si region beneath the wells is the avalanche region. In this case a p-type δ-doped layer has been added to provide a small electric field across the quantum wells and allow a significantly larger electric field across the avalanche region. The large valence band offset reduces the hole current while the small conduction band discontinuity does not provide any significant impediment to the flow of electrons. This allows the diode to get closer to the ideal case of being a unipolar device at least in the absorption region.

Most SiGe p-i-n diodes which have been demonstrated in the literature operating at 1.3 or 1.55 μm have been designed with speed as the main design parameter and the quantum efficiency has been traded off for high speed operation. Typical examples have quantum efficiencies ranging between 10 and 40% operating at frequencies of 2 GHz or above at room temperature. Significant improvements are required if these devices are to compete with the present III-V diodes on the market place.

9.1.4 The Heterojunction Internal Photoemission Diode

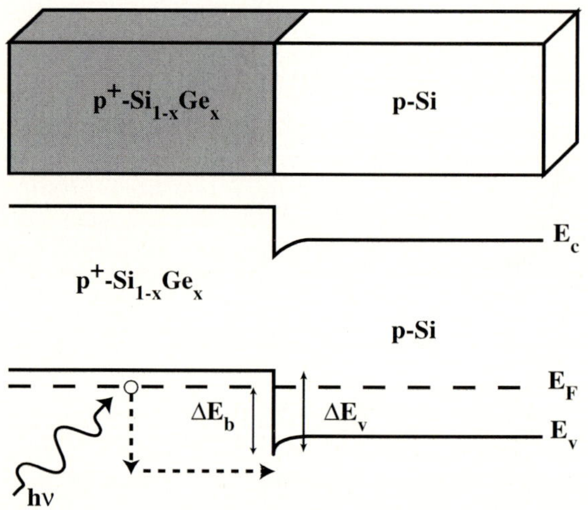

Fig. 9.9. A SiGe heterojunction internal photoemission diode using a p-type pseudomorphic SiGe quantum well where the valence band discontinuity sets the wavelength response of the device

The heterojunction internal photoemission diode is based on excitation of holes in a quantum well out of the well to form a photocurrent which can be collected. As such it is designed as a below-bandgap device and can operate at wavelengths greater than 1.67 μ in the SiGe system when the valence band is used. The basic structure, band structure and operation is shown in Fig. 9.9. Results in the literature demonstrate the operation of such devices at 77 K between 3 and 5 μm with quantum efficiencies of around 1%, dark currents down to 10^{-8} A/cm^2 and detectivity values of 10^9 cm(Hz)$^{1/2}$/W which is better than Schottky barrier Pt:Si photodiodes at 4 μm. The performance is not high enough to compete with the best III-V devices at such frequencies and recent microbolometers fabricated using CMOS technology are substantially cheaper and perform with significantly better performance at 300 K and frequencies of 4 kHz in focal plane arrays.

9.1.5 Quantum Well Infrared Photodetectors (QWIPs)

For transitions above the bandgap energy of Si, SiGe or Ge then interband p-i-n photodiodes can be used. Below the bandgap energy (0.66 eV for Ge) then intersubband transitions in quantum wells allow much longer wavelength photodetectors to be created (Fig. 9.1(d)). In particular in the mid- and far-infrared there are few practical semiconductors with appropriate bandgaps for absorption or emission and intersubband transitions are then used. For III-V materials typically n-type QWIPs are used but for SiGe the valence band is more appropriate with the large band offsets available in easily grown pseudomorphic quantum wells. As was demonstrated in Chap. 3, the subband energy levels can be approximated in the infinite well case to be

$$E_n = \frac{(n+1)^2 \pi^2 \hbar^2}{2m^* w^2} \qquad \text{for } n = 0, 1, 2, 3, \ldots \tag{9.28}$$

demonstrating that the emission between 2 different subbands is controlled completely by the effective mass, m^* of the carriers and the quantum well width, w. This is only strictly true for unstrained materials. The strain in SiGe quantum wells also splits the valence band states and both heavy-hole (HH) and light-hole (LH) have different subband spacings in the quantum well due to the strained and differing effective masses.

For intersubband transitions of HH to HH or of electron to electron, it can be shown that the radiation only couples to the TM mode. Basically there is only a dipole matrix element (the integral in Fermi's Golden rule (3.223) which determines the strength of the transition) for edge absorption into the quantum well as the electrons or holes have to oscillate in the vertical direction and through Maxwell's equations the Pointing vector must be perpendicular to the dipole, hence only edge absorption is possible. For a LH to HH intersubband absorption, the dipole has a finite component in the plane of the quantum well and surface normal absorption is possible. A surface-normal QWIP is much easier to fabricate into arrays of detectors, hence in SiGe most demonstrations have been on the LH to HH intersubband detectors.

There are also the possibility of having bound-to-bound transitions and bound-to-continuum transitions (Fig. 9.10). QWIPs are typically set up in a similar manner to the p-i-n diode with vertical current transport but are unipolar. For the bound-to-bound transition shown in Fig. 9.10(a) the photoexcited electron in the upper, excited $n = 1$ level requires to either tunnel out of the quantum well or get excited out by thermionic emission if it is to form part of the photocurrent. The escape probability can be optimised if

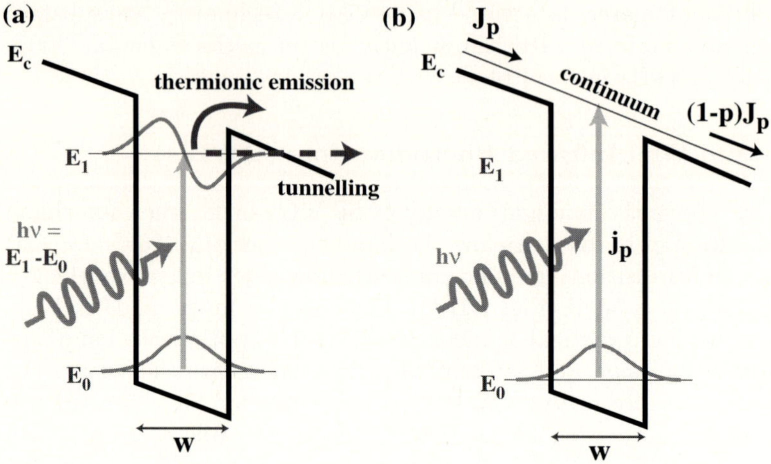

Fig. 9.10. (a) A bound-to-bound transition where any photoexcited hole in E_1 requires to either tunnel into the continuum or be thermally excited to the continuum to be extracted to the collector. (b) A bound-to-continuum transition in which the photoexcited current density, j_p has an effective capture probability of p_c to remain in the quantum well. Therefore $p_c J_p$ is the fraction of current captured by the well and $(1 - p_c) J_p$ as the remain current density transmitted to the next well in the QWIP

the excited subband state is designed to be close to the top of the quantum well so that there is a small barrier to either tunnel or be thermally excited out of the quantum well.

For a bound-to-continuum transition (Fig. 9.10(b)) there is a higher probability of the holes being transported to the collector in the device but the dipole matrix element which determines the probability of a photon being absorbed and exciting a hole to the continuum state is less than in the bound-to-bound case. A simple analysis can be used to obtain the current in the photodiode for both the bound-to-bound and bound-to-continuum case by using the diagram in Fig. 9.10(b) where p_c is defined as the effective capture probability for an incident current density of J_p. We obtain $p_c J_p$ as the fraction of the incident current which remains captured by the quantum well while $(1 - p_c) J_p$ is the proportional of remaining current density which is transmitted to the next period. The incident infrared radiation of photon flux density, Φ_0 creates a photocurrent,

$$j_p = q\eta_w \Phi_0 \tag{9.29}$$

where η_w is the quantum efficiency of a single well. Using current continuity we have

$$J_p = (1 - p_c) J_p + j_p \tag{9.30}$$

Hence

$$j_p = p_c J_p \tag{9.31}$$

and the total net photocurrent is

$$J_p = q\eta\Phi_0 G \tag{9.32}$$

where G is defined as the overall photoconductive gain. For a structure with N quantum wells the total quantum efficiency is $\eta = \eta_w N$ provided $\eta \ll 1$ since it is proportional to the absorption of the structure. The gain in the QWIP is therefore given by

$$G = \frac{1}{p_c}\frac{\eta_w}{\eta} \simeq \frac{1}{p_c N} \tag{9.33}$$

For the dark current density for a bound-to-bound transition, let us assume that only tunnelling allows carriers to escape from the quantum well. Then we have the transmission coefficient for tunnelling if the barrier from the excited subband, E_1 to the top of the quantum well is E_B of

$$\begin{aligned} \tilde{T}(E) &= 1 \qquad \text{for } E > E_B \\ &= 0 \qquad \text{for } E < E_B \end{aligned} \tag{9.34}$$

From Sect. 8.2 the dark current is obtained from the integral

$$J_{dark} = q \int f(E)\tilde{T}(E)\, v \mathrm{d}E \tag{9.35}$$

which can be shown to be

$$J_{dark} = qv\left(\frac{m^* k_B T}{\pi\hbar^2 w}\right)\exp\left(-\frac{E_v - E_F}{k_B T}\right) \tag{9.36}$$

The important point from (9.36) is that the dark current decreases exponentially as the temperature of the QWIP is reduced. Many QWIPs are operated at 77 K or using Peltier coolers due to this exponential dark current dependence.

A number of surface-normal incidence SiGe QWIPs have been demonstrating operating in the important 8 to 12 μm infrared atmospheric window. In particular using a buried silicide layer, higher responsivities than GaAs QWIPs have been demonstrated at significantly lower voltages potentially allowing very low power focal plane arrays. The buried silicide was demonstrated to increase the responsivity by a factor eight by acting as a substrate reflector. The use of such silicide layers also reduced the dark current in such devices by around an order of magnitude. Again while the responsivities are better than some III-V technologies, better technologies do exist especially the new CMOS microbolometers which are already being fabricated into large focal plane arrays and offer video rate (> 30 frames per second) imaging at room temperature. These are broadband detectors (see Fig. 9.4) which are good for a number of applications including firefighting and night vision. There are a few niche applications where the ability of the QWIP to distinguish between different frequencies may allow them to obtain some market share.

9.2 The Quantum Cascade Laser

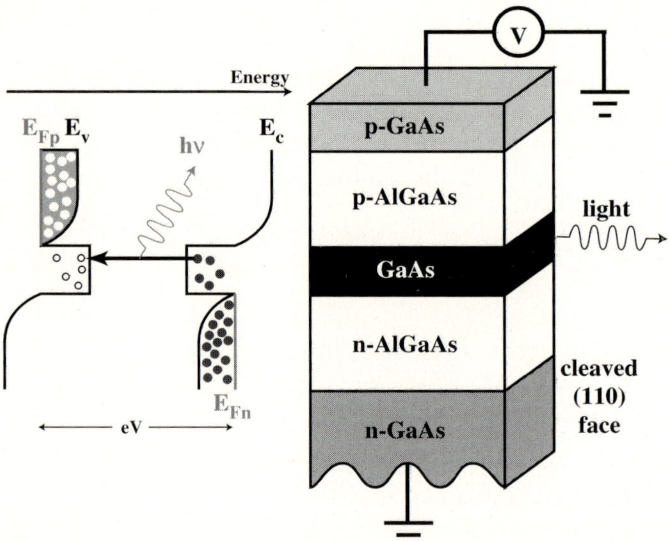

Fig. 9.11. A schematic diagram of a GaAs / AlGaAs p-n diode laser along with the appropriate bandstructure for the system. Electrons and holes recombine across the bandgap and emit photons of frequency, $\upsilon = \frac{E_g}{h}$

The final section in this chapter will be looking at a relatively new concept in the field of lasers and one which is potentially realisable in the Si system with heterostructure band engineering. This section is on unipolar lasers which are known as quantum cascade lasers since an electron cascades down a number of periods of the structure emitting photons. The first cascade laser was demonstrated by Fredric Capasso and coworkers at Bell Laboratories in 1992. The cascade idea is substantially different from the standard interband p-n diode lasers made from GaAs and other direct bandgap III-V materials for applications such as CD-ROMS and DVD players. These diodes basically pull electrons and holes into the one quantum well layer (in a type I system) between the p- and n- regions of the device where they recombine emitting a photon of energy corresponding to the bandgap energy divided by Planck's constant (Fig. 9.11). The indirect bandgap of silicon and SiGe precludes the efficient recombination of electrons and holes for the emission of photons as an appropriate phonon is also required.

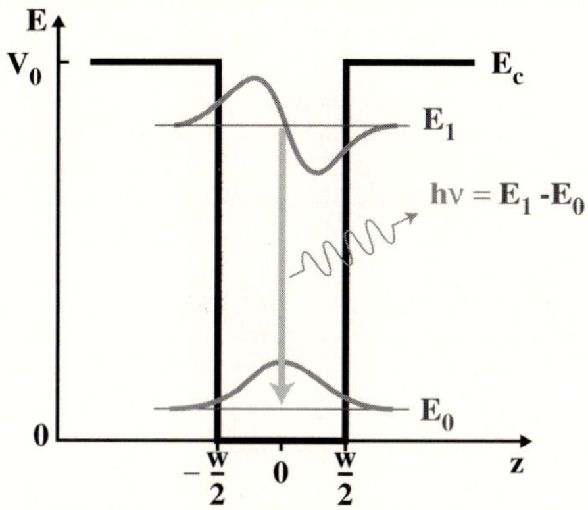

Fig. 9.12. Intersubband emission between the upper state, E_1 and the lower energy state, E_0. A photon of frequency, $\nu = (E_1 - E_0)/h$ is emitted by this intersubband transition

The cascade laser uses intersubband transitions to emit photons (Fig. 9.12). This allows much smaller energies to be realised, particularly in the mid- and far-infrared regions of the electromagnetic spectrum. In addition, the emission frequency or wavelength can be tuned by the thickness of the quantum well and so one semiconductor material can be tuned to emit photons over a wide range of the spectrum. This also allows lasers to be realised in large parts of the electromagnetic spectrum where no materials exist with appropriate bandgaps for p-n diode lasers.

9.2.1 Basic Laser Physics

There are three main interactions light can have with semiconductors. In the previous sections (9) the absorption process of a photon has been discussed with regard to photodetectors. Fig. 9.13(a) shows the absorption process in a schematic diagram where a carrier is excited by the photon from a group state, E_0 to an excited state E_1 if the photon has frequency, $\nu > E/h$. For a carrier sitting in an upper energy level (Fig. 9.13(b)) this state is unstable. At some time without any external influences it will make a transition to the ground state emitting a photon in the process of frequency given by $h\nu_{10} = E_1 - E_0$. This is termed spontaneous emission and is the emission process in light emitting diodes (LEDs). If a photon of energy $h\nu_{10}$ is incident on the carrier in the upper energy state E_1 then this photon can stimulate the emission of another photon of the same frequency and phase (i.e. it is coherent with the first photon) with the carrier making a transition to the ground state (E_0). This is the process in a laser. The radiation is monochromatic because all the photons have energy $h\nu_{10}$ and is coherent because all the photons are in phase.

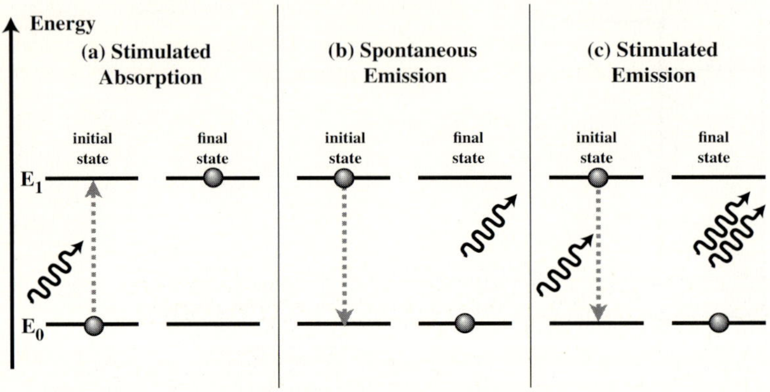

Fig. 9.13. A schematic diagram of the photonic transitions between two energy (subband) states. (a) If a carrier in the lower level absorbs a photon of energy corresponding to the difference in the energy levels then the carrier can be excited to the upper energy level. (b) If a carrier is in the upper level then it can fall to the lower energy level emitting a photon with frequency $\nu = E/h$ where E is the energy difference between the two levels. (c) If a carrier is in the upper level it can be stimulated to fall to the lower level by another photon of frequency $\nu = E/h$ emitting a second photon of the same frequency which is coherent (in phase) with the first photon

Let us assume that the instantaneous populations of the states, E_0 and E_1 are n_0 and n_1 respectively. Under thermal equilibrium and for $(E_1 - E_0) > 3k_BT$ the population of the states is given by the Boltzmann distribution

$$\frac{n_1}{n_0} = \exp\left(-\frac{E_1 - E_0}{k_BT}\right) = \exp\left(-\frac{\nu_{10}}{k_BT}\right) \tag{9.37}$$

as plotted using a dashed line in Fig. 9.15. As is consistent with a Boltzmann distribution, there are more carriers in the lower energy states.

In the steady state, the stimulated emission rate and the spontaneous emission rate must be balanced by the rate of absorption to maintain the populations n_0 and n_1 constant. The transition rates between the states E_1 and E_0 can be calculated using Fermi's Golden rule (3.223). It was Einstein who first defined the transition rate per incident photon for the transition from E_0 to E_1 as

$$B_{01} = \frac{2\pi}{\hbar}\left|\langle\psi_1|\,\hat{H}\,|\psi_0\rangle\right|^2 \tag{9.38}$$

The stimulated emission rate is proportional to the photon-field energy density in the system defined as $\rho(h\nu_{01})$. The stimulated emission rate can therefore be written as $B_{10}n_1\rho(h\nu_{01})$. The spontaneous emission rate from E_1 to E_0 is defined as $A_{10}n_1$ with A_{10} a constant and is independent of the photon density. A and B are named the Einstein coefficients. The absorption rate is proportional to the carrier population in the lower level and the photon-field energy density. Hence it is given by $B_{01}n_1\rho(h\nu_{01})$. In the steady state we have

stimulated-emission rate + spontaneous-emission rate = absorption rate that is

$$B_{10}n_1\rho(h\nu_{01}) + A_{10}n_1 = B_{01}n_1\rho(h\nu_{01}) \qquad (9.39)$$

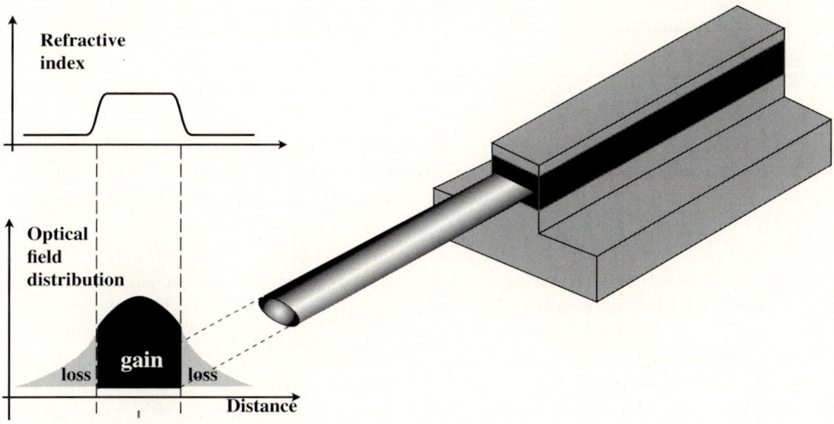

Fig. 9.14. Modal confinement into a ridge waveguide using a change in refractive index between the semiconductor and air

For a laser it is stimulated emission that we are interested in dominating over spontaneous emission. Therefore rewriting equation 9.39 as

$$\frac{\text{stimulated} - \text{emission} - \text{rate}}{\text{spontaneous} - \text{emission} - \text{rate}} = \frac{B_{10}}{A_{10}}\rho(h\nu_{01}) \qquad (9.40)$$

it is clear that to enhance the stimulated emission, the photon-field density requires to be very large. The easiest way to enhance the photon-field density is to produce a resonant optical cavity. The simplest type of cavity is a Fabry-Perot ridge waveguide (Fig. 9.14) where a ridge is etched out of the active semiconductor material. The refractive index change between the semiconductor and air is enough to reflect around 30% of the radiation at each facet. Hence along the length of the cavity standing wave modes are set up such that the frequency $\nu = c/2L$ where L is the length of the cavity and c is the speed of light in the medium.

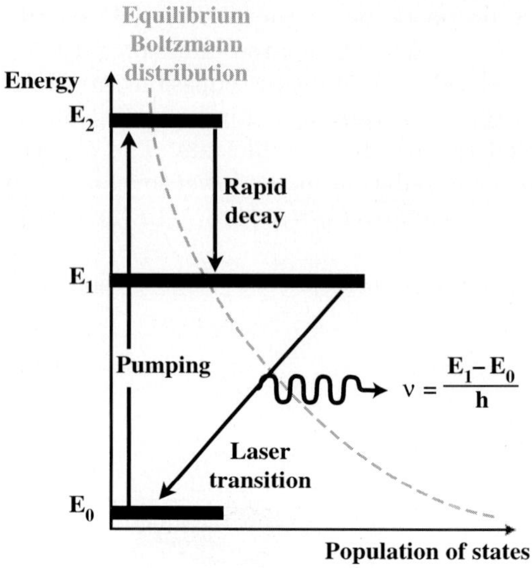

Fig. 9.15. The basic concept behind a three level laser population inversion scheme. In thermal equilibrium the population of states for a number of subbands in a quantum well is a Boltzmann distribution which results in significantly higher number of states in the lower energy subbands. For a laser, a higher number of states is required in the upper state of the laser transition (in this case between E_1 and E_0).To achieve this, some mechanism is required to pump electrons into a higher energy state (E_2) where the electrons may relax into E_1. If the lifetime of the upper laser transition state, E_1 is made longer than either E_2 or E_0 then population inversion occurs. This creates a plentiful supply of carriers for the laser transition. In most lasers, a fourth state which is lower than E_0 is used to create fast depopulation of E_0 to give the required lifetime difference between E_1 and E_0 for population inversion

Equation (9.39) can also be rearranged to give

$$\frac{\text{stimulated} - \text{emission} - \text{rate}}{\text{absorption} - \text{rate}} = \frac{B_{10}}{B_{01}}\frac{n_1}{n_0} \tag{9.41}$$

Therefore for the stimulated emission to dominate over the absorption of photons, the upper energy state, E_1 requires a larger population of states than the lower energy state, E_0 i.e. population inversion. This normally requires at least three energy levels rather than two to guarantee population inversion. Carriers are pumped either optically or electrically from the lowest energy state to a higher energy state, E_2 (Fig. 9.15). This pumping changes the population density of the states from the thermal equilibrium Boltzmann population. Provided the lifetime of the upper E_2 states is much shorter than the upper laser transition state, E_1 then the population of E_1 will be greater than E_2. In a similar fashion the lifetime of E_0 requires to be shorter than E_1 for population inversion. While pumping can achieve this, the typical

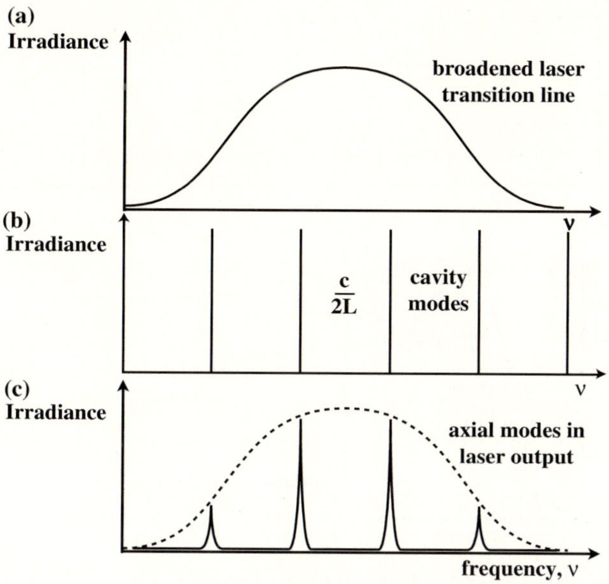

Fig. 9.16. The cavity modes as the laser achieves threshold. (a) is the output from the intersubband transition being used for the laser (i.e. the E_1 to E_0 transition in Fig. 9.15). This electroluminescence is extremely broad compared to a laser. (b) is the cavity modes from a simple ridge waveguide Fabry-Perot cavity. The ends of the ridge reflect the light creating cavity modes with frequency spacing of $c/2L$ where L is the length of the ridge. (c) When threshold is reaches, the laser emits a convolution of (a) the electroluminescence and (b) the cavity modes to produce the output demonstrated in (c)

way of achieving this in real systems is to have a fourth energy level below E_0 which is used to create fast depopulation of E_0 and then the pumping is from this fourth state. This decouples the pumping mechanism from the lower energy state of the laser transition.

Once a cavity has been fabricated and an active semiconductor heterostructure with population inversion has been placed in the cavity, the cavity modes (Fig. 9.16(b)) will convolve with the broadened electroluminescence spectrum of the transition E_1 to E_0 (Fig. 9.16(c)) forming the laser spectrum shown in Fig. 9.16(c). This is a Fabry-Perot cavity mode spectrum with 4 sharp peaks corresponding to the modes along the length of the cavity.

To achieve lasing, the current needs to be increased in the system so that the density of carriers and the population inversion can increase to increase the stimulated emission in the cavity. Below threshold the spectrum of the laser is a broad electroluminescence (Fig. 9.17(b)). As threshold approaches, the gain in the cavity starts to overcome the losses and absorption of photons and the broad spectrum starts to narrow with the cavity modes beginning to be observed convoluting the broad electroluminescence spectrum (Fig.

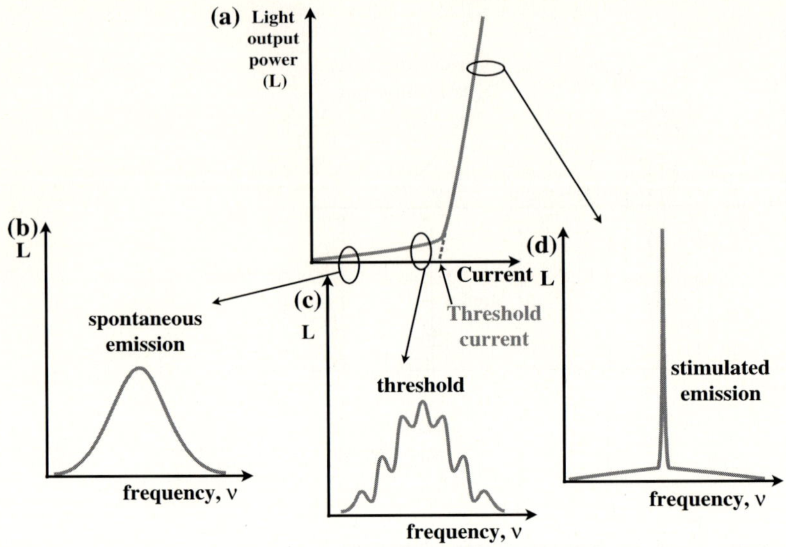

Fig. 9.17. A demonstration of the threshold of a laser

9.17(c)). Once above threshold the power of the laser increases substantially as the optical gain in the cavity amplifies the output power (Fig. 9.17(a)) and the emission spectrum involves a very sharp δ-function like peak (Fig. 9.17(d)) or peaks if multiple Fabry-Perot modes can be supported (Fig. 9.16(c)).

9.2.2 The Si/SiGe Quantum Cascade Laser

The original quantum cascade lasers were fabricated in a number of different III-V heteromaterials. The first laser operated in the mid-infrared but more recently the concept has been expanded and quantum cascade lasers emit radiation across the electromagnetic spectrum from around 2 THz (150 μm) to $\sim$ 2 μm wavelength (150 THz). Figure 9.18 shows the typical heterolayer structure of a III-V quantum cascade laser. Only three periods are shown but it should be clear from the diagram why it is called a cascade since the electrons cascade down the heterolayers. In this particular structure minibands and minigaps are used. These are the same as the bands and bandgaps in semiconductors but are created by the heterolayers. If quantum wells are placed close enough together with thin barriers between the wells that allows the wavefunctions of carriers in each well to overlap as in the tight binding model (Sect. 3.2.5) then minibands and minigaps are formed from the artificially grown heterostructure in the same manner to bandgaps and bands in semiconductors. Hence carriers have allowed and disallowed energies in the quantum wells in addition to the subband states. These can be used to great effect as an injector of electrons into a subband state which is designed to be the upper state of a laser transition.

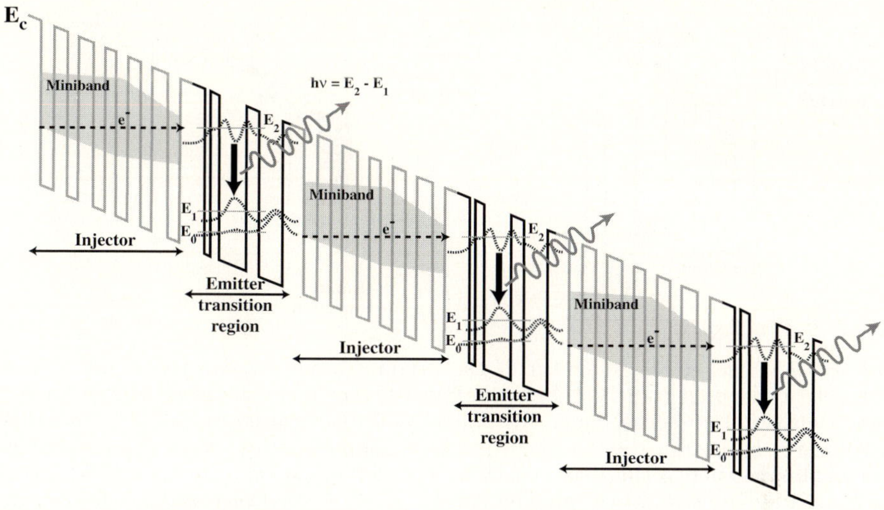

Fig. 9.18. A schematic diagram of the conduction band in a three period cascade emitter. The injector region is designed to pull the electrons from an emitted active region back to the upper level in the next active region. The whole structure is designed so that the bands align when an appropriate electric field is placed across the whole structure

For a single quantum well emitter, the power is proportional to the frequency of the emitted radiation. For visible frequencies this provides no problems with laser powers. As the frequency is reduced, however, the power is also reduced in a linear fashion. The advantage of the cascade is that a number of periods, N can be used (Fig. 9.18) and the power scales as the number of periods. Therefore the power is given by

$$P = N\eta \frac{I - I_{th}}{2e} h\upsilon \tag{9.42}$$

where I is the applied current, I_{th} is the threshold current for lasing and η is the power efficiency. This is extremely important for far-infrared or terahertz lasers where the frequency can be a thousand times lower than the visible part of the electromagnetic spectrum. Hence the power output is predicted to be very low. By having a large number of periods, that is a few hundred active periods, the power can be scaled back up much closer to the power output from visible p-i-n lasers.

The most important part of the heterolayer design in a quantum cascade laser is to design a structure which allows population inversion between two energy subbands in quantum wells. There are a number of different ways this can be achieved. Fig. 9.19 shows two standard ways of achieving population inversion. The first in Fig. 9.19(a) uses a diagonal or interwell laser transition followed by two wells to produce fast depopulation of the lower laser transition state, E_1. Above the optical phonon energy of 62 meV in Si, fast

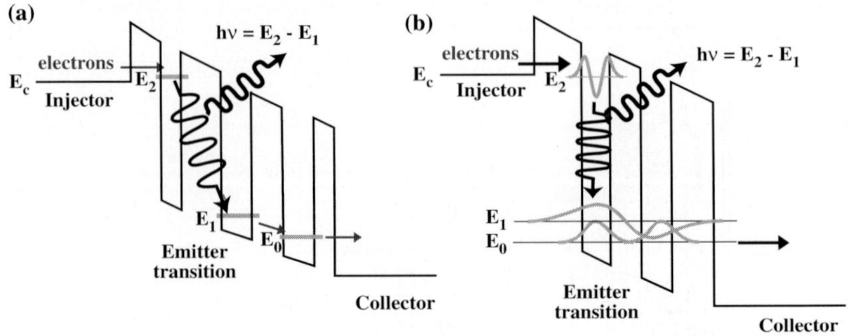

Fig. 9.19. A schematic diagram of the original quantum cascade lasers. (a) A interwell or diagonal transition between two different quantum wells is used for the laser transition between E_2 and E_1. A third well is then used for fast depopulation of the lower laser transition state to produce population inversion. (b) Above the optical phonon energy, this can be used for fast depopulation if the E_1 and E_0 levels are spaced by the LO optical phonon energy, E_{LO}. A vertical transition in the one quantum well between E_2 and E_0 can then be used for the laser transition which has a larger matrix element and hence stronger emission

depopulation of the lower laser transition state can be achieved by making the two lower energy subbands spaced by the optical phonon energy E_{LO}. This produces a very fast resonant depopulation (Fig. 9.19(b)) and hence population inversion can be achieved. The added advantage is that a vertical or intrawell laser transition can be produced which has a larger matrix element for the intersubband transition if calculated from Fermi's Golden rule (3.223) resulting in higher efficiency and therefore higher power operation from the laser.

There are a number of other possible mechanisms. In SiGe it is the valence band which is typically studies for quantum cascade laser because the valence band offsets are larger than the conduction band offsets for small Ge fractions and the high electron effective mass in the tunnelling direction forces extremely thin (<0.5 nm) quantum mechanical tunnel barriers if significant tunnelling is required. As this is not easy to achieve in the conduction band of Si/SiGe, hole cascades are being studied. Figure 9.20 shows four different active regions which can be achieved using Si/SiGe heterostructures. The first (Fig. 9.20(a)) is the resonant LO optical phonon depopulation previously studied by this time designed for the valence band. The second, Fig. 9.20(b), is a diagonal transition. The third is a miniband design, Fig. 9.20(c), which has the added advantage of laser transitions over a number of quantum wells. This improves the matrix element calculated from Fermi's Golden rule and hence improves the current and therefore the efficiency and power output of the device as more than one quantum well can emit photons. The final design shown here is a negative effective mass design, Fig. 9.20(d). The idea is to use the strain in a SiGe quantum well to split the LH and HH states. The

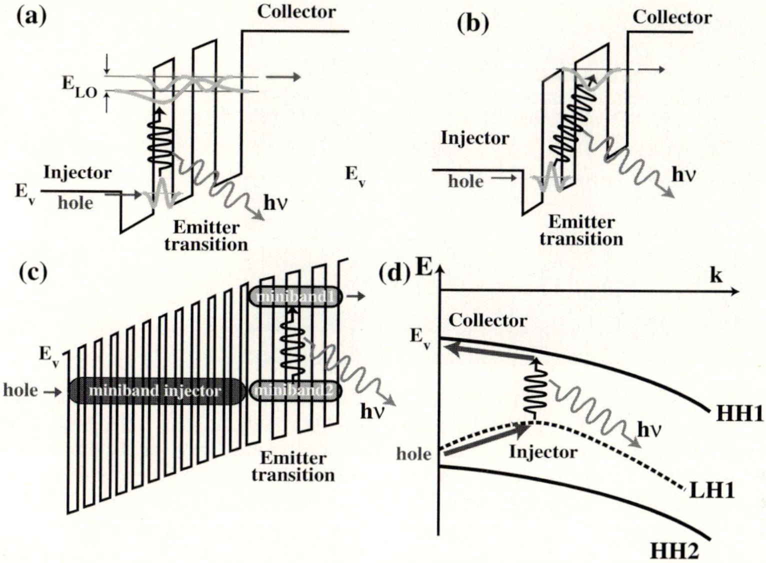

Fig. 9.20. (a) a resonant optical phonon depopulation active quantum cascade region. (b) an interwell or diagonal quantum cascade active region. (c) a miniband quantum cascade active region. (d) a negative effective mass quantum cascade active region

strain is then chosen through the Ge content to try and push the LH1 state higher in hole energy (i.e. lower in real or electron energy so down the page) above the HH2. As this is forbidden from the quantum mechanics selection rules (i.e. the matrix element in Fermi's Golden rule is zero) the LH1 band produces an anti-crossing and it bends up in real energy i.e. down to a lower energy state for holes. Therefore tunnelling into a quantum well must happen at $k = 0$. Scattering from the SiGe alloy can scatter the hole to the hole minimum in $k-$space where the laser transition can occur from the LH1 to HH1 state before the hole is alloy scattered back to $k = 0$ and it can tunnel to next quantum well.

Figure 9.21 shows an transmission electron micrograph of a negative effective mass quantum cascade emitter grown using SiGe heterostructures. The light areas are tensile strained-Si barriers and the darker regions between the barriers are $Si_{0.72}Ge_{0.28}$ quantum wells grown on top of a $Si_{0.8}Ge_{0.2}$ virtual substrate. The structure was grown using LPCVD with boron doped contacts on either side of the quantum wells and barriers. Fig. 9.22 shows the electroluminescence from the edge of the fabricated device. The spectrum has been taken using a Fourier transform infrared spectrometer using a cooled Si-bolometer as a detector for the far-infrared (terahertz) radiation. A larger voltage has been applied across the device than that required to line up the appropriate subband states in neighbouring quantum wells. This allows all

Fig. 9.21. A transmission electron micrograph of a strain symmetrised quantum cascade emitter designed for a negative effective mass transition. The light areas are 5 nm thick tensile strained-Si barriers and the darker regions between the barriers are 8 nm $Si_{0.72}Ge_{0.28}$ quantum wells grown on top of a $Si_{0.8}Ge_{0.2}$ virtual substrate. The structure has 30 active quantum wells

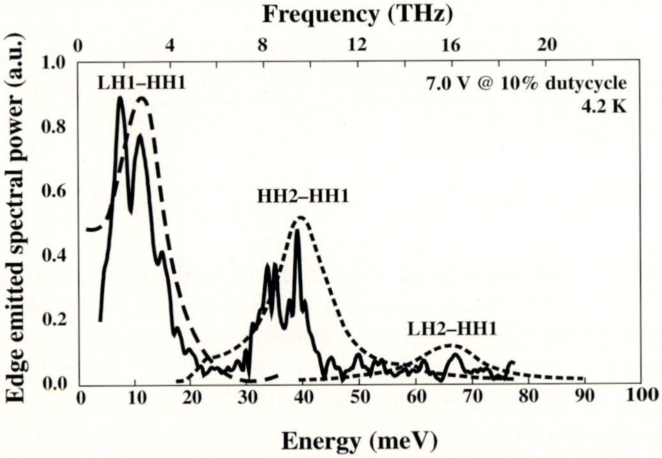

Fig. 9.22. The edge electroluminescence from a Si/SiGe negative effective mass quantum cascade emitter at 4.2 K with 7 V applied bias. The LH1 to HH1 and HH2 to HH1 transitions are clearly visible. The dotted lines are the theoretical calculated spectrum from **k.p** theory

the intersubband transitions including higher energy ones to be mapped out. In the edge emission LH to HH and HH to HH transitions can be observed. As was previously discussed with regard to polarisation, HH to HH transitions can only be observed in the edge or TM mode. Therefore by observing the surface-normal emission, only the LH to HH transition should be observed. This is indeed what is observed in experiments as shown in Fig. 9.23. The advantage of such surface-normal emission is that vertical cavity surface emitting lasers (VCSELs) may be fabricated into arrays of lasers which are ideal for imaging applications.

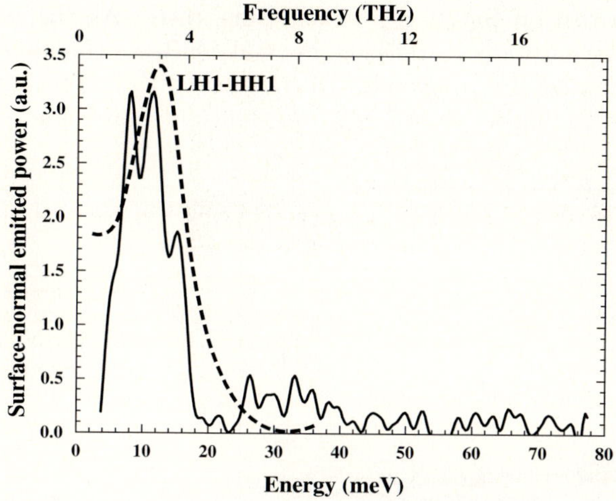

Fig. 9.23. The surface-normal electroluminescence from the negative effective mass quantum cascade emitter with 4 V applied across the structure. The dotted line is the theoretical spectrum calculated using **k.p** theory

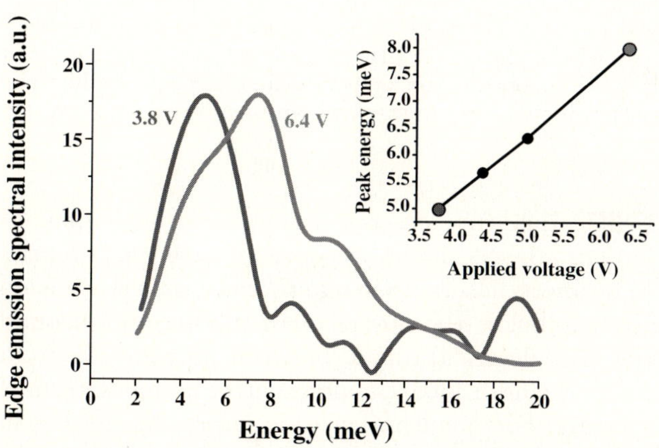

Fig. 9.24. A diagonal or interwell intersubband transition between a LH1 and a HH1 state in neighbouring quantum wells. On the left are two different spectrum for two different applied voltages across the 100 period quantum well structure. As expected for a diagonal transition, the peak shifts to higher energy as the bias across the wells is increased. The insert on the right demonstrates how the centre of the peak shifts as the bias is increased

In addition to negative effective mass transitions, diagonal transitions have also been demonstrated in the Si/SiGe system. Figure 9.24 shows a spectrum from interwell intersubband transitions between strain-symmetrised 4.4 nm $Si_{0.7}Ge_{0.3}$ compressively-strained quantum wells with 2.2 nm tensile-strained-Si barriers grown on $Si_{0.8}Ge_{0.2}$ virtual substrates. As the voltage across the 100 period structure is increased, the peak shifts as the alignment of the subbands is increased. The movement of the peak with increasing voltage is demonstrated in the insert to Fig. 9.24 on the right hand side.

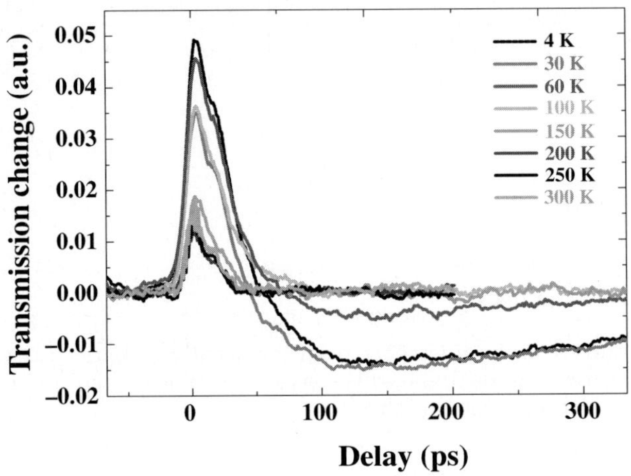

Fig. 9.25. Non-radiative lifetime measurements on a Si/SiGe quantum cascade emitter structure with $Si_{0.7}Ge_{0.3}$ compressively-strained quantum wells. A free electron laser is used to excite carriers to the upper LH1 state and then a time resolved absorption measurement is used to obtain the non-radiative lifetime. Unlike the GaAs results, the non-radiative lifetime is constant at around 30 ps between 4 K and 300 K

While to date no Si/SiGe quantum cascade lasers have been demonstrated, particularly for terahertz lasers the Si-based lasers have the potential to be far superior to III-V terahertz lasers. In group IV materials there is no polar optical phonon scattering below the optical phonon energy of around 62 meV. In III-V materials, the electrical dipole between a group III atom and a group V atom results in the additional phonon scattering mode termed polar optical phonon scattering. For quantum cascade lasers this results in the non-radiative lifetime between the laser transition states having a very strong temperature dependence. For lasing the non-radiative transition requires to be significantly longer than the radiative transition so that the radiative transition (i.e. the photon emission) dominates the device. Measurements have shown that above about 40 K in GaAs quantum wells, polar optical phonon scattering dominates and the non-radiative lifetime collapses. Hence III-V terahertz quantum cascade lasers can only operate at present at relatively low temperatures. Fig. 9.25 shows results from Carl Pidgeon at Heriot-Watt University where the non-radiative lifetime has been measured in compressively-strained $Si_{0.7}Ge_{0.3}$ quantum wells. Unlike the GaAs results, the non-radiative lifetime is longer than the III-V results but it is also approximately constant between 4 K and 300 K. This bodes well for room temperature operation from Si/SiGe terahertz quantum cascade lasers.

9.3 Further Reading

1. S.L. Chuang, Physics of Optoelectronic Devices, John Wiley and Sons, New York (1995)
2. J.H. Davies, The Physics of Low Dimensional Structures" Cambridge University Press (1998)
3. M.J. Kelly, Low Dimensional Semiconductors: Materials, Physics, Technology Devices, OUP, Oxford (1995)
4. L. Pavesi, J. Phys.: Condens. Matter **15**, R1160 (2003)
5. S.M. Sze, Physics of Semiconductor Devices 2nd Edition, John Wiley and Sons, New York (1981)
6. S.M. Sze Semiconductor Devices: Physics and Technology 2nd Edition, John Wiley and Sons, New York (2002)
7. D.J. Paul, Semiconductor Science and Technology **19**, R59 (2004)

10. Integration

This chapter is devoted to the fabrication processes involved in producing complete systems in addition to providing a brief overview of typical circuit architectures used for different applications. Much of the chapter will be devoted to conventional technology particularly with bipolar, HBT, CMOS and BiCMOS although more modern systems such as RTDs and fault tolerent architectures will also be briefly reviewed.

10.1 The CMOS Inverter and MOS Memory Circuits

Fig. 10.1. The representations of the n- and p-MOS transistors as drawn in circuit architecture diagrams

CMOS is the most common circuit architecture used in modern microelectronics. It is based on the combination of p-MOS and n-MOS which have the circuit symbols shown in Fig. 10.1. By combining the n- and p-MOS transistors in different ways, all standard logic gates can be fabricated with appropriate numbers of transistors. By combining the standard logic gates into simple or complex circuits, real products ranging from microcontrollers to complete single chip computers can be designed. The most important logic gate is the CMOS inverter (Fig. 10.2) which only consists of one n- and one p-MOS device (Fig. 10.2(a)). Any input voltage, V_{in} of '1' is translated into a V_{out} of '0' and vice versa. The circuit symbol for an inverter is shown in Fig. 10.2(b).

The importance of CMOS circuits and the CMOS inverter becomes clear when the power dissipation of the inverter is considered. All other CMOS logic gates dissipate power in a similar manner to the inverter, therefore understanding the inverter will provide a general basis for CMOS logic circuits.

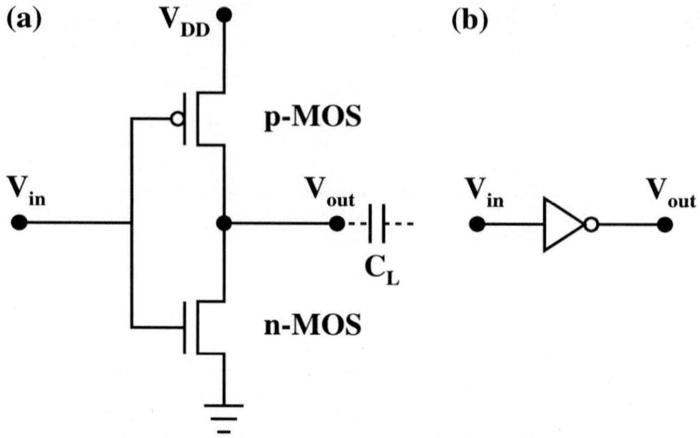

Fig. 10.2. (a) A schematic diagram of the CMOS inverter with one p-MOS and one n-MOS transistor. (b) The circuit symbol for an inverter. When an inverter is connected to the rest of the circuit by interconnects, a load capacitance results as shown by the dashed capacitor in (a)

There are three main parts to the power being dissipated in the inverter. The first is the static power dissipation which is the power dissipated when no switching is taking place. This is solely related to the sum of all leakage currents in the transistors or interconnects, $I_{leakage}$ and the applied voltage, V_{DD} as

$$P_{static} = \Sigma I_{leakage} V_{DD} \tag{10.1}$$

For an ideal device and circuit technology the leak current should be zero. In reality it is always finite and correspond to the I_{off} of each transistor which for MOSFETs depends on the subthreshold slope, S and the threshold voltage, V_T as

$$P_{staticMOS} = V_{DD} I_{on} W 10^{-V_T/S} \tag{10.2}$$

for transistor of gate width, W. MOS devices (I_{off}) have the lowest leakage currents of any technology while for MODFET or MESFET technology, the off currents are orders of magnitude higher due to the poorer isolation between the source and drain. Also Schottky gates have higher leakage currents than gates with insulator layers such as silicon dioxide. For a CMOS microprocessor with a million 0.25 μm transistors operating at 3 V, this static contribution corresponds to a few milli-Watts of power. The power consumed by switching or dynamic power dissipation is significantly higher at least for CMOS technologies of 90 nm or higher. As the gate oxide is reduced below 1.5 nm the static power dissipation grows significantly and for deep submicron CMOS technologies may dominate.

The dynamic power dissipation may be considered by assuming that the interconnect of the output of the inverter has a load capacitance of C_L (Fig.

10.2(a)). When the load capacitor is charged during the low-to-high transition through the p-MOS, the voltage of the load capacitor rises from 0 to V_{DD}. While part of the energy is stored in the load capacitor, most of the power is dissipated into the p-MOS transistor. For the high-to-low transition, the capacitor is discharged and the stored energy is dissipated through the n-MOS device. Therefore if the gate is switched on and off at a frequency, f_c per second, the dynamic power consumption is

$$P_{dynamic} = C_L V_{DD}^2 f_c \tag{10.3}$$

Frequently a factor $\frac{a}{2}$ is also added to the above equation where a represents the transition probability of the logic gates.

There is a third element to the power dissipation of a CMOS inverter. There is a finite time for the inverter to switch between the on and off states or vice versa. When the inverter is switching between the two states, both the n- and p-MOS transistors are dissipating dc power in the circuit. As the current is varying and different types of clock pulses are used for different applications, the easiest way of considering a general term for this power dissipation is to consider an average current during the switching of I_{mean} and then the short circuit power dissipation is given by

$$P_{short-circuit} = I_{mean} V_{DD} \tag{10.4}$$

The short circuit term can generate noise by leakage into the substrate. This can couple to power supply buses or the semiconductor substrate and is responsible for cross-talk, where the digital and analogue parts of a circuit couple to each other creating errors. By careful design of the threshold voltage and V_{DD}, these effect can be minimised.

The total power dissipation of the CMOS inverter is the sum of the three terms giving

$$P_{total} = \Sigma I_{leakage} V_{DD} + C_L V_{DD}^2 f + I_{mean} V_{DD} \tag{10.5}$$

The power is predominantly dissipated in the form of heat. Modern microprocessors with millions of transistors actually operate at up to 100 °C, significantly above room temperature. All this heat needs to be dissipated either through the substrate or the interconnects. Most of the dissipation is through the substrate and therefore the thermal conductivity of the substrate is an important parameter in microprocessors. The amount of heat which can be dissipated actually determines the integration density of the transistors as a higher density would result in the substrate being too hot for the transistors to operate with lower error rates.

The inverter is the simplest CMOS circuit but three or more terminal gates are required if any type of logic operation is to be performed. If a NAND gate can be fabricated then every other standard logic gate can be created by combining NAND gates. For instance, if the two input terminals are tied together then an inverter is created. Examples of CMOS NAND

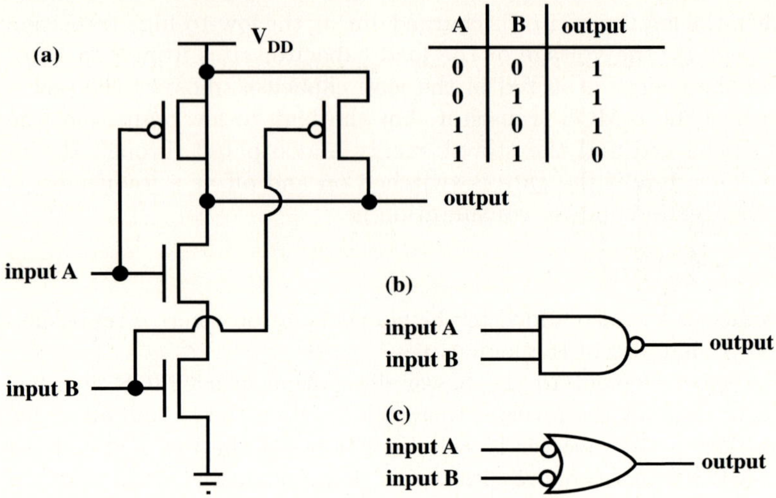

A	B	output
0	0	1
0	1	1
1	0	1
1	1	0

Fig. 10.3. (a) The circuit diagram of a NAND gate along with (b) the standard circuit symbol and (c) an alternative circuit symbol

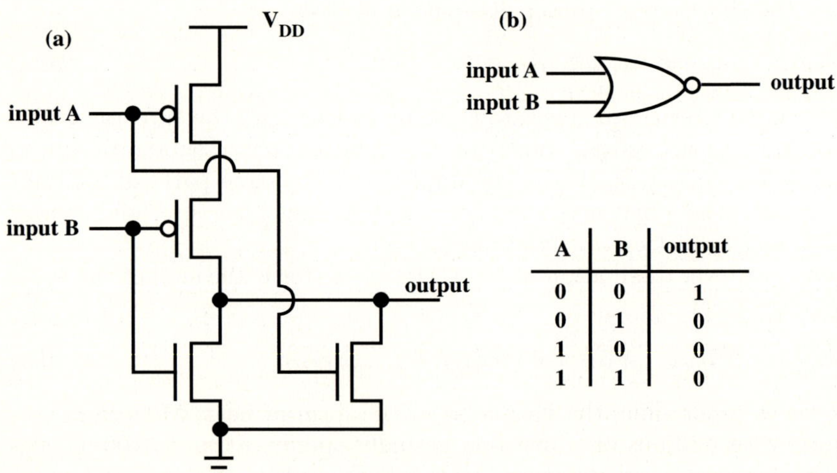

A	B	output
0	0	1
0	1	0
1	0	0
1	1	0

Fig. 10.4. (a) The circuit diagram of a NOR gate along with (b) the circuit diagram

and NOR gates are shown in Figs. 10.3 and 10.4 respectively. Both these gates can be expanded to have N input gates rather than two. Most modern microprocessors will not just use inverters, NANDs and NOR gates. The real power of CMOS circuit design is the ability to design complex functions in an efficient manner using n- and p-MOS transistors.

In addition to switching and logic, microprocessors also require memory. In modern computing, data is transfered from one memory location, a logic operation is performed and then the answer is written into another mem-

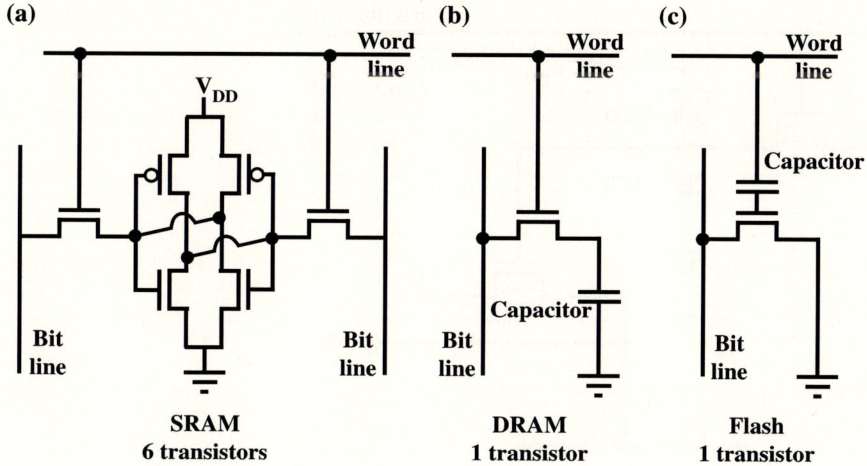

Fig. 10.5. Schematic diagrams of (a) static random access memory, (b) dynamic random access memory and (c) flash memory

ory location. By combining these operations with different logic operations, a general microprocessor can be programmed to perform most tasks by reconfiguring different standard operations. Therefore memory is also very important. Figure 10.5 shows schematic diagrams of the three most important types of memory, (a) SRAM, (b) DRAM and (c) flash.

Dynamic random access memory (DRAM) is the cheapest and has the lowest static power but needs to be refreshed very frequently. It is also very difficult to integrate with high speed CMOS transistors on a microprocessor as the capacitor structure requires significantly different processing from the transistor. Therefore static random access memory (SRAM) is typically used on most microprocessors. It is significantly faster than DRAM or flash but consumes much more power (both static and dynamic power). Flash memory is non-volatile but is slow and requires a large voltage to write operations. Flash memory is used in mobile phones and portable devices as a large amount of energy is required to write to the memory but being non-volatile, no energy is required to maintain the information. Only a small amount of energy is consumed in the reading process compared to DRAM or SRAM. The typical structure has a poly-Si floating island inserted into a MOSFET between the main gate and the channel. By applying a large voltage to the gate, the island can be charged and the threshold of the MOSFET changes. Therefore applying a source-drain bias will produce a readout dependent on the charge on the island.

We will finish this section by showing how tunnel diodes such as RTDs can be made into memories. Figure 10.6 shows a schematic of a tunnelling static random access memory (TSRAM). The two RTDs operate to create a charging node between them. The advantage of the TSRAM is that the

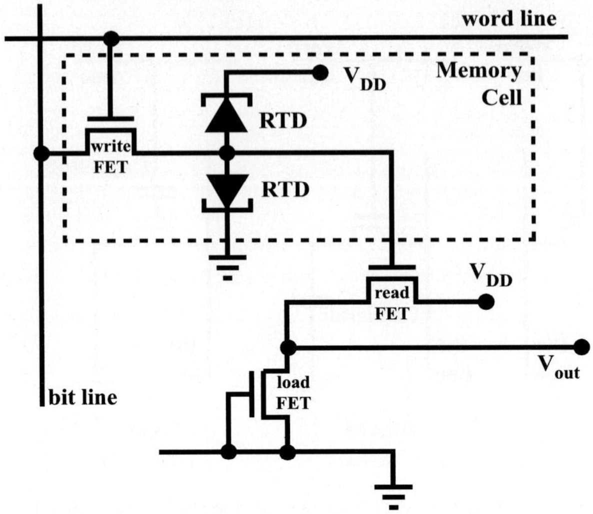

Fig. 10.6. The circuit diagram for a tunnelling static random access memory

static power dissipation is significantly lower than SRAM, DRAM or flash while operating speeds comparable to SRAM can be obtained. In particular, the number of transistors and device is significantly lower than a CMOS SRAM which partially accounts for the reduction in power. To some extent the RTDs are operated as capacitors with the back to back current-voltage characteristics creating two stable points for the '1' and '0' at the valley of each RTD. No demonstration of such a circuit has been achieved in silicon technology although III-V devices have shown very low static power dissipation and high switching speeds. Table 10.1 compares the main parameters in different types of memory. Standby power is becoming an important parameter especially as the minimum feature size is reduced and leakage currents are increasing. Speed is also important especially for the memory on a microprocessor which now accounts for over 80 % of the transistors.

10.2 Silicon Process Technology

Most of the thin film deposition or epitaxial growth methods required for fabricating different silicon-based devices were discussed in Chap. 2. In this section the remaining techniques required for fabricating different silicon-based devices will be briefly reviewed before a number of standard processes will be demonstrated.

Table 10.1. A comparison of a number of the main parameters affecting the performance of different types of memory

Technology	Access time	Retention time	Speed (GHz) (1-2 μm pitch)	Density (Mbit/cm^2)	Standby power (W/Mbit)
16 MB SRAM	1 ns	>10 years	0.03 to 0.1	5 to 10	0.01
256 MB DRAM	10 ns	<256 ms	0.01 to 0.02	60 to 150	2×10^{-4}
256 MB flash	10 ns	>10 years	0.01 to 0.02	60 to 150	
256 MB 1T Si TSRAM	10 ns	>10 years	0.01 to 0.04	50 to 100	10^{-9}
256 MB 1T strained-Si TSRAM	4 ns	>10 years	0.03 to 0.12	50 to 100	10^{-9}

10.2.1 Thermal Oxidation

Thermal oxidation is one of the key processes which may be used many times in the fabrication of a device. Not only is it used for the gate oxide in MOSFETs but is also used for insulating layers to electrically isolate devices. There are two main types of oxidation reaction termed dry and wet which are respectively

$$\mathrm{Si\,(solid)} + \mathrm{O_2\,(gas)} \longrightarrow \mathrm{SiO_2\,(solid)} \tag{10.6}$$

$$\mathrm{Si\,(solid)} + \mathrm{2H_2O\,(gas)} \longrightarrow \mathrm{SiO_2\,(solid)} + \mathrm{2H_2\,(gas)} \tag{10.7}$$

Wet oxidation is faster than dry and is used for thick oxide layers such as electrical isolation. Dry oxidation has a lower interface state density which is important as the interface states can trap electrons and degrade the electrical properties at the interface. Good oxides are amorphous in nature and have interface state densities of around $10^{14}\,\mathrm{m}^{-2}$ for thermal gate oxides. The Si/SiO_2 interface is constantly moving into the silicon as the oxide is grown with any contamination on the original surface ending up on the top of the oxide surface. Hence the technique can end up with high quality interfaces with few impurities or trapped charges. Fast diffusers such as sodium or gold can get to the interface at relatively low temperature and therefore such impurities must be kept well away from silicon processing.

Thermal oxidation of SiGe also creates problems as a number of detrimental effects occur. It is worthwhile noting that germanium oxide is water soluble which is not ideal for many of the silicon processing steps where water is the solvent. The second is that an electrical defect is created in the silicon dioxide for every Ge atom that is incorporated as germanium oxide. Even at quite low concentration of Ge in the layer being oxidised can end up creating a surface state density comparable to the electron charges at the interface in a MOSFET. A third effect is called snow-ploughing where as the Si/SiO_2 interface moves further into the SiGe layer, not all the germanium is incorporated in the oxide film and a higher concentration of Ge is driven into the substrate in front of the oxide interface. All these effects result in devices having to be designed where either a deposited oxide must be used or a silicon cap is required which can be consumed during oxidation.

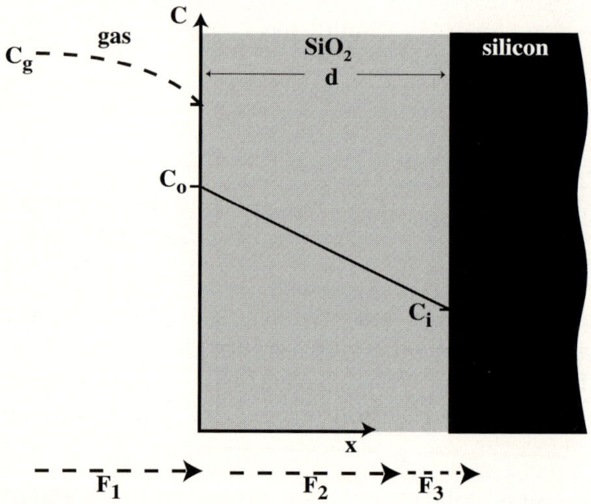

Fig. 10.7. The Deal and Grove three stage model of oxidation. F_1 is the flux from the flow of gas stream to the surface. F_2 is the flux of the diffusion of oxidising species through the formed oxide and F_3 is the flux for the reaction at the Si - SiO_2 interface. The concentration of the oxidising species varies from C_g in the gas to C_0 at the oxide surface and C_i at the Si - SiO_2 interface

The kinetics of thermal oxidation of silicon were first described by Deal and Grove in 1965 and form a three stage process (Fig. 10.7). For oxidation the oxidant must: (1) travel from the gas phase to the gas-oxide interface with flux F_1, (2) move across the SiO_2 film towards the silicon with flux, F_2 and (3) react with silicon at the Si / SiO_2 interface with flux F_3. The model assumes that the oxidation process is in the steady state and therefore all the fluxes must be equal (i.e. $F_1 = F_2 = F_3 = F$).

The gas-phase oxidant flux, F_1, is given by

$$F_1 = h\,(C^* - C_0) \tag{10.8}$$

where h is the gas-phase mass-transport coefficient, C_0 is the oxidation concentration in the oxide at the outer surface and C^* is the equilibrium oxidant concentration in the oxide. Henry's law is used to relate the partial pressure of the oxidant in the gas at pressure, p with the equilibrium oxidant concentration in the oxide as

$$C^* = kp \tag{10.9}$$

where k is the Henry's law constant. This law only holds when the oxidant does not associate or dissociate at the outer surface, implying the oxidant is molecular.

Transport of the oxidising species to the Si - SiO_2 interface occurs by diffusion in a similar fashion to electron and hole diffusion described in earlier chapters. The flux of the diffusing species can be written as the product of the concentration gradient across the oxide, $(C_0 - C_i)/x_{ox}$. Fick's law is then used to describe the flux of oxidant across the oxide layer as

$$F_2 = \frac{D_{ox}(C_0 - C_i)}{x_{ox}} \tag{10.10}$$

where D_{ox} is the diffusivity, C_i is the concentration of oxidising species at the Si/SiO_2 interface and x_{ox} is the oxide thickness. The flux corresponding to the reaction at the Si/SiO_2 interface is given by

$$F_3 = k_s C_i \tag{10.11}$$

where k_s is the chemical reaction rate constant.

By solving the above equations for the flux, the oxide growth rate, R_{ox} may be determined from

$$R_{ox} = \frac{dx_{ox}}{dt} = \frac{F}{N_{ox}} = \frac{k_s C^*/N_{ox}}{1 + k_s/h + k_s x_{ox}/D_{ox}} \tag{10.12}$$

where N_{ox} is the number of oxidation molecules incorporated per unit volume into the oxide. Solving this differential equation while assuming that an initial thickness of oxide (x_i) may be present ($x_{ox} = x_i$ at t=0), results in the following equations for the oxide thickness as a function of time, t

$$x_{ox} = \frac{A}{2}\left[\sqrt{\left(1 + \frac{t+\tau}{\frac{A^2}{4B}}\right)} - 1\right] \tag{10.13}$$

where the standard variables as used for oxidation have been defined as

$$A = 2D\left(\frac{1}{k_s} + \frac{1}{h}\right) \tag{10.14}$$

$$B = \frac{2D_{ox}C^*}{N_{ox}} \tag{10.15}$$

$$\tau = \frac{x_i + Ax_i}{B} \tag{10.16}$$

The parameter τ is used to include any initial oxide thickness that is present on the surface. If the oxidation time is short then F_3 is the rate limiting step and (10.13) can be approximated by a linear relationship of

$$x_{ox} = \frac{B}{A}(t+\tau) \tag{10.17}$$

The fraction B/A is termed the linear rate coefficient and is related to breaking the bonds at the Si/SiO_2 interface. The values will therefore depend on the crystal orientation of the surface. The linear rate coefficient is smaller for the Si(100) surface compared to the Si(111) surface since the later has fewer bonds between adjacent planes.

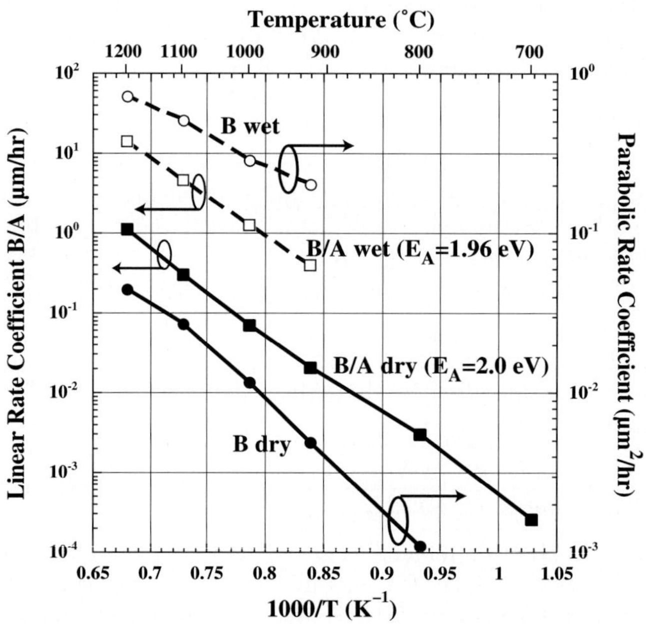

Fig. 10.8. The variation of the linear and parabolic rate coefficients with inverse temperature. Both wet and dry data are at pressures of 8.5×10^4 Pa

For long oxidation times (10.13) reduces to the parabolic form

$$x_{ox} = \sqrt{B(t+\tau)} \approx \sqrt{Bt} \tag{10.18}$$

The coefficient B is therefore termed the parabolic rate coefficient and depends on the diffusion across the already formed oxide layer. Since SiO_2 is amorphous it does not depend on crystal orientation. Both the linear and parabolic coefficients are shown in Fig. 10.8. Figure 10.9 shows plots of (10.13) for both wet and dry oxidation for a number of different oxidation times. The

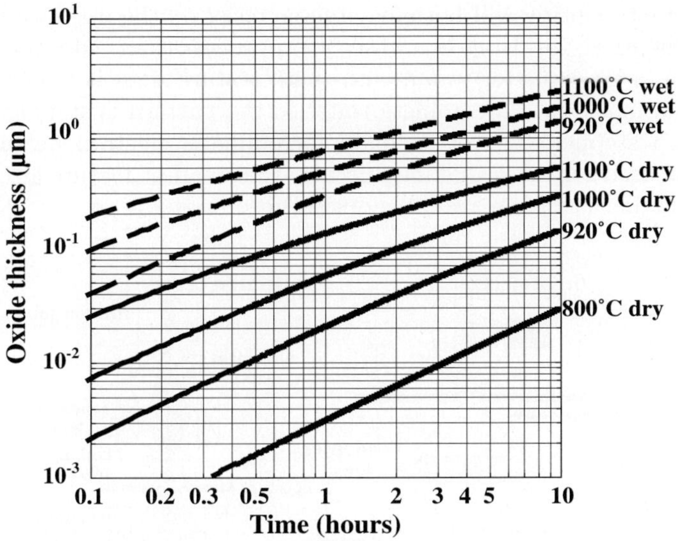

Fig. 10.9. The thickness of oxides as a function of growth time for a number of different temperatures. Both wet and dry oxidation is shown

figure clear demonstrates the faster oxide growth in wet oxidation compared to dry.

10.2.2 Lithography

Lithography is the process of producing a predefined pattern onto a substrate or layer on a substrate. The major manufacturing technology is optical lithography with electron beam lithography used for small feature sizes and also for writing optical lithography masks. A general lithography process involves the spinning of a resist onto the substrate. This resist is selectively exposed to the radiation, typically either photons or electrons. Then the resist is developed in a chemical solvent with the areas which have been irradiated being more soluble than those without radiation. This positive process leaves behind the resist where it has not been exposed and then an etch process is used to transfer that pattern into the substrate or the top layer or layers on the substrate. Negative processes can also be used where the exposed resist becomes less soluble (for example by cross linking long chain polymers) and the exposed sections remain after chemical development.

Optical lithography is the dominant technology as it allows fast parallel processing. It comes in two main forms, shadow printing in contact or proximity mode (Fig. 10.10(a)) where the mask is in contact or very close to the resist covered substrate and projection lithography (Fig. 10.10(b)) where a lens is used to focus the radiation down to a fraction of the feature size on the mask. The second technique is the dominant for integrated circuit manufacture as contact printing will both leave some resist on the mask and over time damage the mask resulting in a lower yield process than reduction lithography. For the large diameter wafers and small feature sizes it is now impossible to focus a mask onto a complete wafer so the pattern is stepped using a tool called a stepper. The other main technique is electron beam lithography which is used for mask making and also the smallest feature size devices which are smaller than the available optical resolution.

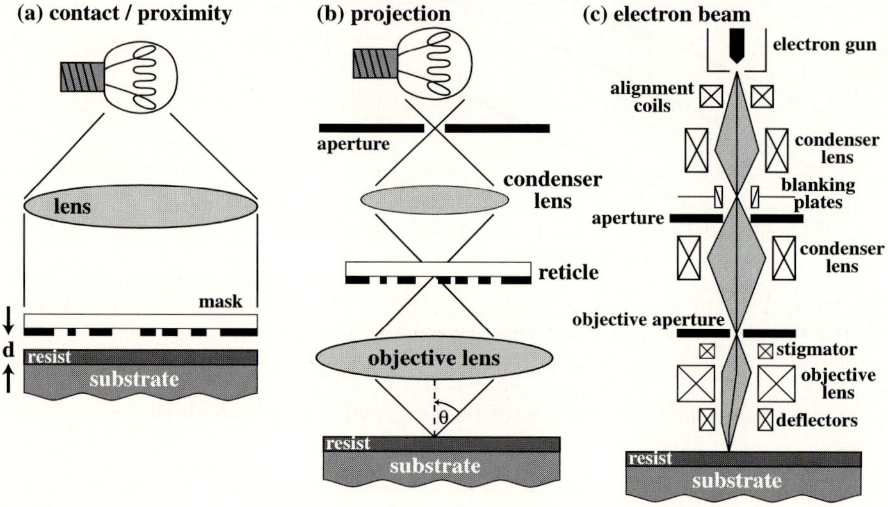

Fig. 10.10. Schematic diagrams of (a) shadow printing using contact (d=0) or proximity (d= finite number) mode (b) projection lithography and (c) an electron beam lithography system

There are a number of important parameters in optical lithography which determine what devices can be fabricated. The resolution is the minimum feature size or critical dimension which can be accurately transferred from a mask to the resist. If multiple layers are being patterned then each mask must be aligned to the previous layer. Registration is the measure of the accuracy of alignment between different lithographically defined layers. For manufacture, the number of wafers which can be processed every hour or throughput is another important parameter.

Shadow printing (Fig. 10.10(a)) is still used for many small wafer processes such as power electronics or in universities. It is a relatively straightforward process with the mask being a direct replica at the same scale as the pattern to be transferred to the resist on the substrate. Typically the whole wafer is exposed at once to provide a large throughput of wafers. The highest resolution may be obtained by using the technique in contact mode ($d = 0$ in Fig. 10.10(a)) but any dust or particles captured on the resist surface may damage the mask. Also over time some resist will build up on the mask and reduce the accuracy of the transferred pattern. Therefore frequent mask cleaning is required to maintain accurate transfer of patterns. With all these problems, proximity printing is frequently used where the mask is kept at some finite distance, d away from the wafer. This is not ideal and reduces the resolution since the critical dimension, CD is approximately

$$CD \simeq \sqrt{\lambda d} \tag{10.19}$$

for a wavelength, λ of light. Higher resolution can be achieved by reducing the wavelength with the ultimate resolution being limited by the diffraction from the edge of the mask pattern. The more significant issue as the feature size is reduced is that it becomes more and more difficult and expensive to produce accurate masks which are 1:1 replicas of the pattern required. The solution to this problem comes from projection lithography (Fig. 10.10(b)).

Projection lithography uses two lenses to focus the light from a source through a mask or reticle and then reduce the image onto the resist covered substrate (Fig. 10.10(b)). To increase resolution, only a small part of the wafer is exposed, typically one chip or die at a time. This also has the advantage of allowing larger wafer diameters to be used in processing which are useful for increasing yield. Reduction ratios of 10:1 or 5:1 are common which makes mask making substantially easier as larger features are required. The resolution of a projection system with an objective lens of numerical aperture, NA is

$$CD \propto \frac{\lambda}{NA} \tag{10.20}$$

where the numerical aperture is given by

$$NA = \hat{n} \sin\theta \tag{10.21}$$

θ is the half angle of the cone of light converging to an image point on the wafer (see Figure 10.10(b)) and $\hat{n}$ is the refractive index of the air or vacuum above the resist which is approximately 1. Wide lenses are required to increase θ and hence increase the NA but costs also increase as the NAs grow larger. For a projection system, the wafer must be at the focal point for the image to appear sharp and at the correct reduction factor. There is a depth of focus, DOF inside which the required critical dimension can be met which is defined as

$$DOF \propto \frac{\lambda}{(NA)^2} \tag{10.22}$$

The problem as the wavelength of the lithography is reduced is that the depth of focus is also reduced. Also it should be clear from the above equations that to reduce the CD, one needs to reduce the NA of the objective lens but this also decreases the DOF. The ideal situation would be a small CD but a large depth of focus but this unfortunately cannot be achieved simultaneously.

As will be shown later in the chapter when complete fabrication processes are discussed, one of the major problems with aggressively scaled devices is that the surface may be quite rough with many different heights relating to devices, contacts, gates and insulating regions. Planarisation techniques frequently have to be used on sub-micron processes to allow sufficient depth of focus for the lithography.

A number of different optical sources of light are used in optical lithography. Many shadow and projection systems use mercury-arc lamps which has a spectrum with emission at a number of lines. These are named G-line for 436 nm, H-line for 405 nm and I-line for 365 nm. I-line lithography projection systems typically have minimum resolutions of around 0.3 μm. For smaller features new sources are required. KrF excimer lasers emit at 248 nm and are now one of the main stepper projection sources. There are also systems with ArF excimer lasers operating at 193 nm and fluorine excimer lasers operating at 157 nm. When these systems are combined with phase shifting technology which uses the interference of waves to produce smaller feature sizes, smaller CDs than those from (10.20) can be achieved and the KrF, ArF and F_2 systems are now used for 180 nm, 100 nm and 70 nm lithography processes respectively.

Electron beam lithography is predominantly used for the direct write of photomasks. Figure10.10(c) shows the schematic diagram of an electron beam system. Electromagnetic lens are used to focus the electrons into a beam. The electrons require to be in vacuum so the whole system requires either a high vacuum for a LaB_6 electron gun or ultra high vacuum if a field emission source is used for electrons. For research systems, electron beam lithography can pattern using a single beam spot which is rastered to produce features below 10 nm, substantially below that achievable using optical techniques. The problem with such systems is that the throughput is significantly lower than optical lithography as this is a serial rather than parallel patterning approach. For a number of mask making systems where the minimum feature size is above 100 nm, extra electromagnetic lens are used to produced a patterned beam with either a square or rectangular aperture to speed up the mask writing process.

10.2.3 Etching

For the high yields required in manufacture, the standard method of fabrication is the deposition of a two dimensional film onto the substrate followed by the patterning of this film using lithography and etching. The other main process is the direct etching of the substrate using lithography and a mask. One of the main reasons silicon is the dominant semiconductor technology is that selective etches exist with respect to the insulators SiO_2 and Si_3Ni_4 along with the major metals used for contacts and interconnects. This allows layers to be patterned and the etch to stop on oxides or silicon as appropriate. There is also the possibility of epitaxially depositing or growing an etch stop layer into a structure which allows flexibility in manufacturing processing. Such layers are commonly used in the fabrication of SOI wafers with a thin top silicon layer.

There are two main etching techniques, the first chemical etching and the second is dry or plasma etching which for ULSI is typically reactive ion etching (RIE). RIE involves the electrical excitation of a low pressure gas which then dissociates into a plasma of electrons and ions. The advantage of RIE over chemical etching is that anisotropic or directional etching can be easily achieved along with the selectively between layers. Chemical etching is typically isotropic, that is the same rate in all directions. As feature sizes get smaller, it is more difficult to fabricate small devices using isotropic etching although it can be used to over etch a feature to produce smaller features than the lithography that has been used. Therefore there is a strong move in ULSI processing towards RIE.

The field of etching for semiconductor fabrication is a substantial branch of chemistry and not appropriate for this text. In Table 10.2 a number of etch chemistries are listed for many of the standard substrates, metals and silicides in silicon processing. The list is not exhaustive but will give the reader an idea of groups of chemicals required to etch different materials. Some etches are used for polishing while others are selective to other materials allowing etch stops to be used in a process.

Table 10.2. A number of the standard chemical etches for some of the materials used in silicon process technology

Material	Etchant	Temp.	Etch Rate
Si(100)	126 HNO_3: 60 H_2O : 5 NH_4F	25 °C	150 nm/min
Si(100)	10 NHO_3 (65%): 1 HF (50%)	25 °C	12 μm/min
Si(100)	15g KOH : 50 ml H_2O : 15 ml C_3H_7OH	60 °C	1.1 μm/min
Ge(100)	50% wt. HF: 50% wt. H_2O_2	25 °C	21.2 μm/min
SiO_2	15 H_2O : 1 HF	25 °C	16 nm/min
SiO_2	BHF (58.6% wt. H_2O : 6.8% HF : 34.6% NH_4F)	25 °C	16 nm/min
Si_3N_4	H_3PO_4 (85%)	180 °C	10 nm/min
Al	H_3PO_4	25 °C	10 nm/min
Al	4 ml HNO_3 : 3.5 ml CH_3COOH : 73 ml H_3PO_4 : 19 H_2O	25 °C	30 nm/min
Cu	5 HNO_3 : 1 H_2O		
Ni	3 HNO_3 : 1 H_2SO_4 : 1 H_3PO_4 (98%) : 5 CH_3COOH	90 °C	1 μm/min
Ti	9 H_2O : 1 HF	32 °C	12 μm/min

Most of the etch processes now used in ULSI processing are RIE processes. This is predominantly because there are very few anisotropic chemical etches unlike RIE where numerous anisotropic etches have been developed. Figure

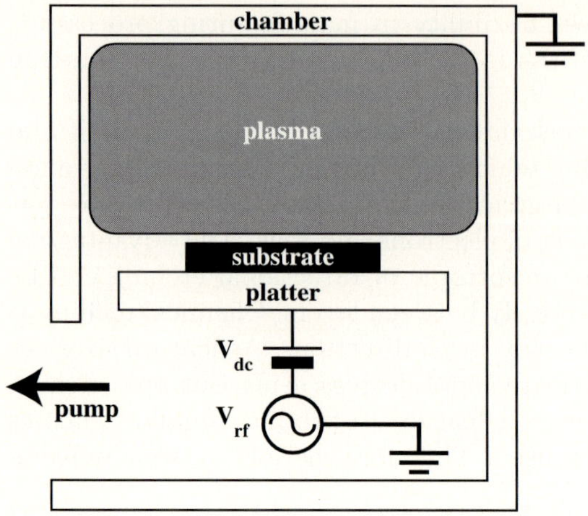

Fig. 10.11. A schematic diagram of a reactive ion etch (RIE) chamber

10.11 shows a schematic diagram of a typical RIE system. The chamber is pumped out to a low pressure before the etch gas is allowed into the chamber using mass flow controllers and kept at a low pressure using a butterfly or similar valve. The outer chamber is grounded and both a dc and rf voltage signal are applied to the electrode on which the substrate to be etched sits. The rf signal dissociates electrons from the gas molecules producing reactive ions for etching while the dc bias attracts these ions to the substrate which is to be etched. By controlling the dc bias, the etch rate and anisotropy can be controlled in many systems. In addition to the rf generator, a matching network is also required to impedance match the 13.56 MHz signal to the electrode and chamber.

There are many different etch chemistries. Most RIE systems consist of both chemical and physical (sputtering) components. The chemical component is normally used to define selectivity while the physical part is used to produce anisotropic etching. By varying the gas flow rate, the chamber pressure and the dc bias, the ratio between the chemical and physical etching processes can be varied. Table 10.3 list some of the standard plasma chemistries used in the semiconductor industry.

A number of new developments have been used on modern RIE systems. Electron cyclotron resonance (ECR) systems were developed to produce a plasma outside the main process chamber. This allowed the dc bias to be used to filter only low energy ions for the etching process, thereby reducing the damage to substrate by removing the bombardment of high energy ions. Such ECR sources also allow much higher density plasmas which can etch at high rates but also make etching small features especially in deep trenches significantly easier. More recently inductively coupled plasma (ICP) sources have replaced ECR as they only require a rf coil rather than the complicated

Table 10.3. A number of the standard plasma etch chemistries for some of the materials used in silicon process technology

Material being etched	RIE chemistry
Deep Si trench	$HBr/NF_3/O_2/SF_6$
Shallow Si trench	$HBR/Cl_2/O_2$
Poly-Si	SF_6, BCl_3/Cl_2, HBr/O_2
SiO_2	CH_2F_2, C_2F_6, C_3F_8, $CF_4/CHF_3/Ar$
Si_3N_4	CHF_3/O_2, CH_2F_2
Al	$SiCl_4/Cl_2$, BCL_3/Cl_2, HBr/Cl_2
$TiSi_2$	CF_4Cl_2

magnet systems and are therefore much cheaper to buy, operate and maintain while providing similar etch conditions.

10.3 CMOS

CMOS is still the most predominant of all the microelectronic technologies and commands the most research resources. Present technology uses 0.13 μm lithography and through diffusion of the Ohmic implants achieves 70 nm gate-lengths. The technology is being aggressively scaled to lower dimensions and 90 nm technology will be available by the time this book is published. The typical finished cross section of a n- and p-MOS transistor are shown in Fig. 10.12. Each transistor is isolated by shallow trench isolation in the lateral dimensions and by doping with a punch stop (heavily doped layer) in the vertical direction. The advanced designs use a retrograde doping profile which has reduced doping density towards the Si/SiO_2 interface to reduce impurity scattering in the channel. These profiles therefore provide good source-drain isolation while reducing the Coulombic scattering in the inversion layer to increase the mobility. The Ohmic contacts use a high doped (HDD) region with a self-aligned silicide (this is termed salicide) and a second section with a lightly doped drain (LDD). This is achieved by the formation of a spacer layer at the side of the gate to allow a low-doped region to be implanted. This low doped region is to reduce the diffusion, damage and more importantly the electric field across the device as high energy electrons can easily be excited into the oxide to form defects.

A typical CMOS fabrication process is shown in Figs. 10.13 to 10.15. The typical silicides which are used depends on the technology node being fabricated but as device sizes are reduced, there is demand for reduction in the resistivity of the silicide and the temperature of formation. For 0.25 μm processes, $TiSi_2$ was predominantly used while many 0.13 μm processes have moved to $CoSi_2$. Research is presently concentrating on NiSi for 90 nm and below.

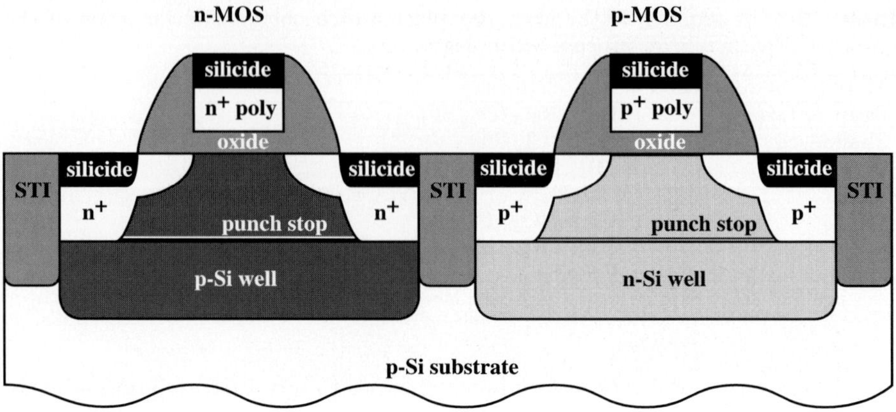

Fig. 10.12. A schematic diagram of a standard n- and p-MOS transistor system used in a typical modern CMOS fabrication process

The interconnect metal has also changed as the dimensions have been reduced. For decades all interconnects were made of Al with a small quantity of Cu to reduce electron migration. This is the transportation of atoms of the interconnect by electrons used for conduction which eventually results in voids in the interconnect and failure of the line. Most present processes now use Cu interconnects which have a higher conductivity and therefore reduce the interconnect delay on chips. In addition the insulator between the metal interconnects has also changed in an attempt to reduce the capacitance between metal layers. Fluorine has been added to SiO_2 to reduce the dielectric constant and hence the capacitance.

The processing for all devices manufactured on CMOS production lines follows a fairly standard form of two dimensional blanket deposition, lithography and then pattern transfer through etching. While photo resists can be used to pattern many layers, it is frequently advantageous to deposit a SiO_2 or Si_3N_4 layer and pattern that first. There are specific chemical processes with high selectivity between Si and the insulators thereby allowing selective etching and the ability to stop on selected layers.

The starting wafer for most CMOS processes is p-type (100) doped around 10 to 20 Ω-cm. Most manufacturers then grown an epitaxial Si layer on top of the wafer before starting the CMOS processing. This provides higher quality material than a polished starting surface by burying any defects created by polishing the surface. The first stage to electrically isolate each transistor involves the growth of a thermal oxide as a protection layer and then the deposition of Si_3N_4. Lithography is then used to pattern the Si_3N_4 before the pattern is transfered down into the Si substrate to form an isolation trench (Fig. 10.13 (b)). The whole wafer is coated in a CVD deposited SiO_2 layer (Fig. 10.13 (c)) before chemical mechanical polishing (CMP) is used to remove the patterned nitride and protective oxide layers. This leaves behind

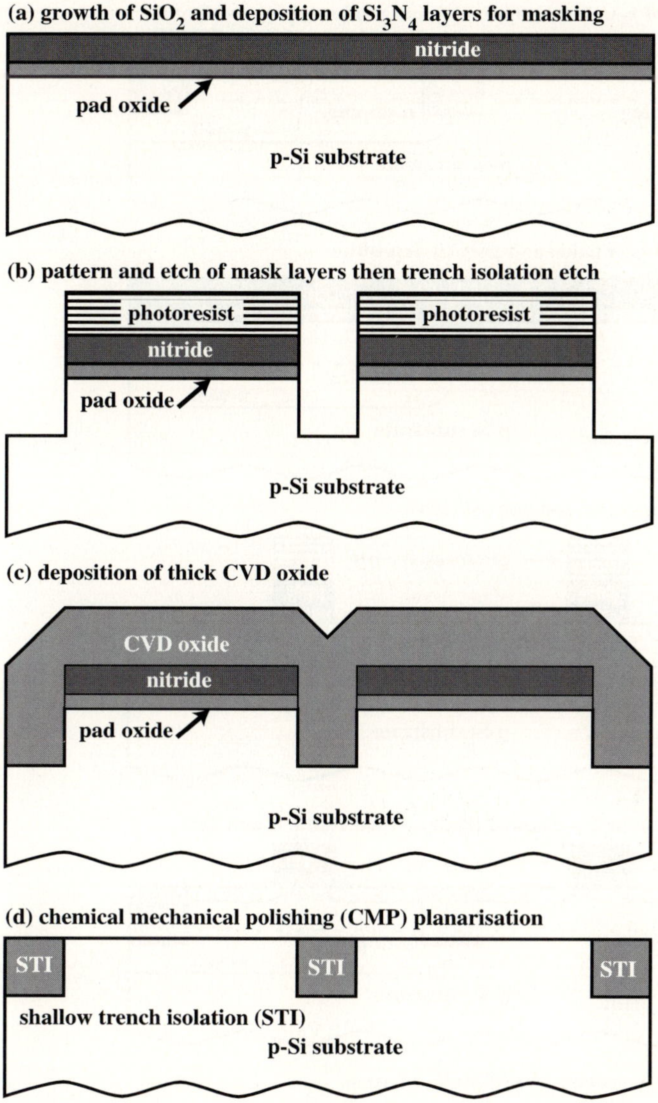

Fig. 10.13. The first four stages in a CMOS process

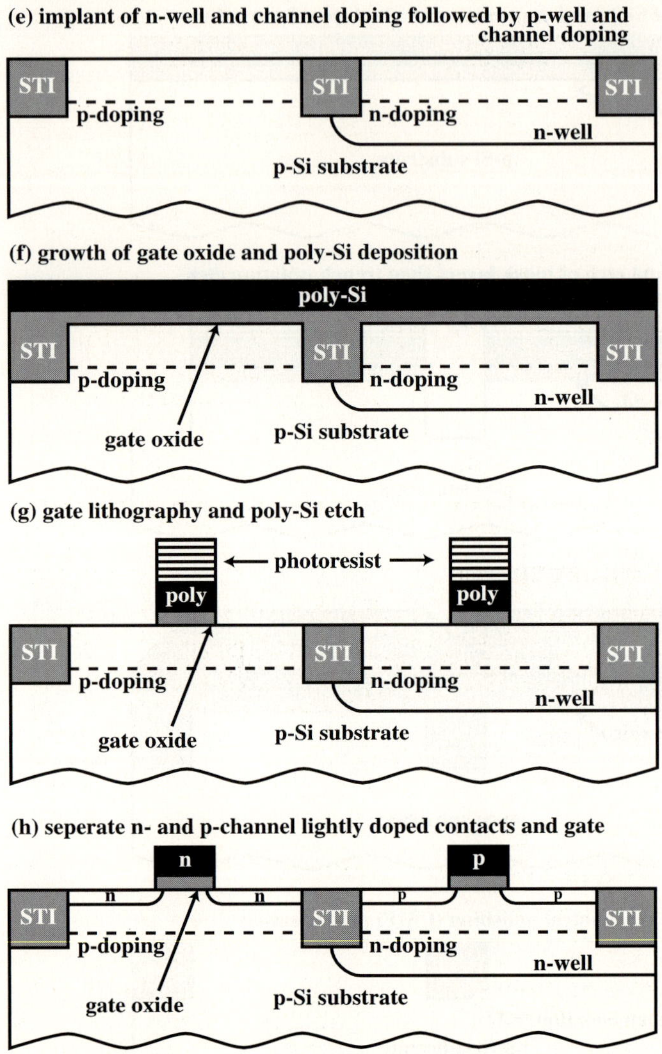

Fig. 10.14. The middle stages of a CMOS process

trenches filled with SiO_2 which is termed shallow trench isolation (Fig. 10.13 (d)).

The next stage is the implantation of the wells for the n- and p-MOS transistors. Resist is used to mask all the n-MOS regions and a n-type dopant such as P or As is implanted for the p-MOS wells. Implantation is normally through a thin oxide to prevent sputtering of the Si surface. Frequently a high doped δ-layer is also designed to act as a punch stop layer - this is a well defined potential plane to isolate the transistor vertically. Then the resist is

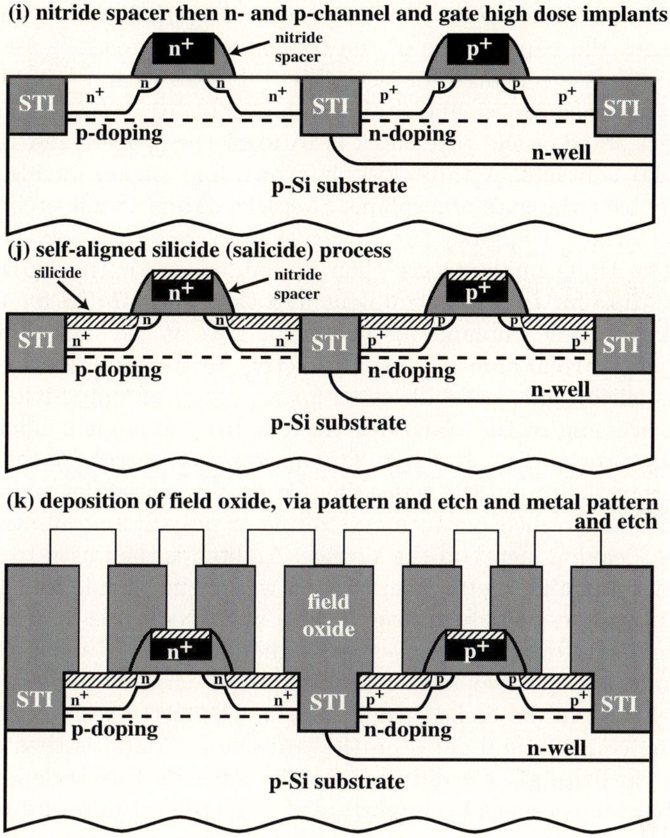

Fig. 10.15. The final stages in a CMOS process

stripped, the p-MOS regions are protected by resist and the correct p-doping for the n-MOS is implanted. The well implants are used to set the threshold voltages of the n- and p-MOS transistors (Fig. 10.13 (e)).

The next stage is the deposition and patterning of the gate stack. The thermal gate oxide is first grown in a furnace tube before amorphous Si or amorphous SiGe is deposited on top. SiGe is now used by some manufacturers as it allows better matching of the threshold voltages by the potential alignments of the gates from the vacuum level. Lithography is then used for the gate pattern and reactive ion etching will selectively etch the poly gate, stopping on the gate oxide (Fig. 10.13 (g)). For some CMOS processes the gate oxide would be stripped at this stage but with many modern processes the oxide is left to prevent sputter of the Si surface during the subsequent implantation stage. The resist is then stripped and lithography is used to open first a window around the whole of the n-MOS transistors. A n-type implant is then used to implant both the gate and the low doped drain region of the implanted Ohmic contact. This produces a self-aligned Ohmic contact which reduces the capacitance of the transistor. The resist is stripped and the same process is repeated for the p-MOS devices with a p-implant of the gate and low doped source and drain regions (Fig. 10.13 (h)).

The next stage of the gate stack is to form the spacers and high doped drain (HDD) contacts. A thin oxide is grown and silicon nitride deposited conformally over the gate amorphous Si stacks. An isotropic etch and CMP are used to form a triangular like spacer at the side of the poly-Si as in Fig. 10.13(i). The HDD implants are then implanted as for the LDD, masking first the n-MOS for the boron implant and then the p-MOS for the n-type implant. Again these implants also dope the gate at the same time. At this point a rapid thermal anneal is used to active all the implantations and the amorphous silicon gate crystallises to form poly-Si or poly-SiGe. The final stage of processing of the transistor itself is the self-aligned silicide known as a salicide process Fig. 10.13(j). The appropriate metal for the silicide is sputtered onto the top of the wafer (Ti, Co or Ni). The wafer is annealled so that silicides are formed between crystalline or poly-Si and all other regions do not react leaving metal on the surface. An etch is then used to selectively remove any remaining metal from regions where the silicide has not formed (Figure 10.13(j)), especially on top of SiO_2 or Si_3N_4 layers.

The final stage of CMOS processing is the formation of the interconnects. A field oxide is deposited before lithography is used to pattern the areas where via holes will be etched using fluorine chemistry reactive ion etching down to the contacts and gates on the transistors. Metal is then sputtering into these via holes after a diffusion barrier such as TaN is deposited. The diffusion barrier is especially important if a deep level impurity metal such as copper is used as the interconnect metal. This is to prevent the metal damaging the transistors. The metal will cover the top of the field oxide and will then be patterned by lithography and etching to form interconnects (Fig. 10.13(k)). This stage of interconnect formation can be repeated many times to form different layers of interconnects. 0.13 μm technology node CMOS has up to seven levels of metal. The insulator between the metal layers above the first metal layer is sometimes spin-on-glass. This is basically similar in chemistry to silicon dioxide but is significantly easier to planarise the surface of the wafer when spun onto the surface.

The process described above is a generic description of a typical CMOS process. There are, however, many differences that different companies may use in CMOS processing. The above description should therefore be considered as a guide rather than the exact process used in every CMOS fab..

10.4 Heterolayer Integration Issues

Before moving on to discuss fabrication processes for SiGe HBTs or other SiGe strained layers into processes, there are a few important issues which need to be discussed. The major issue is diffusion of Ge. From the previous section on CMOS processing, it should be clear that there are a significant number of high temperature steps. These include the activation of the Ohmic implants, the growth of the gate oxide and the formation of the silicides for the contacts. There are many more stages with thermal budgets but all provide no significant diffusion compared to these processes.

Diffusion of Ge or impurity atoms is very similar to the diffusion of carriers discussed in Chap. 3. If the concentration of Ge is given by C and the flux of Ge atoms passing through unit area in unit time is F then

$$F = -D\frac{\partial C}{\partial x} \tag{10.23}$$

where D is the diffusion constant or diffusivity. This equation demonstrates that it is the concentration gradient $\frac{\partial C}{\partial x}$ which drives the diffusion process. Remembering the relationship between the diffusion current and the diffusion constant for electrons from (5.1) we obtain

$$\frac{\partial C}{\partial t} = -\frac{\partial F}{\partial x} = \frac{\partial}{\partial x}\left(D\frac{\partial C}{\partial x}\right) = D\frac{\partial^2 C}{\partial x^2} \tag{10.24}$$

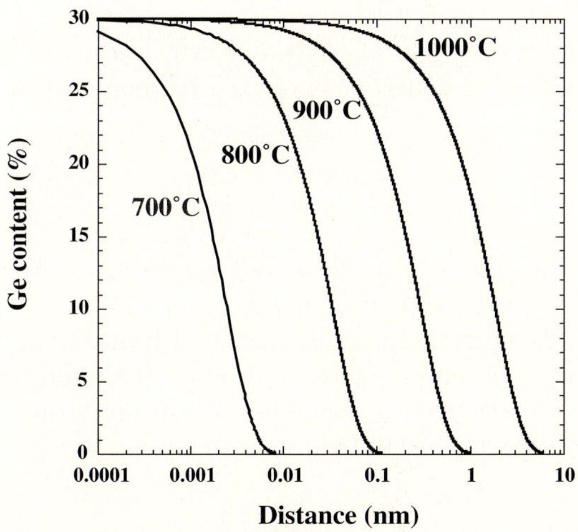

Fig. 10.16. The diffusion of Ge into strained-Si from a $Si_{0.7}Ge_{0.3}$ layer at different temperatures. All anneals were for 180 s

When written in the form

$$\frac{\partial C}{\partial t} = D\frac{\partial^2 C}{\partial x^2} \tag{10.25}$$

this is called Fick's diffusion equation. The solution to this equation is given by

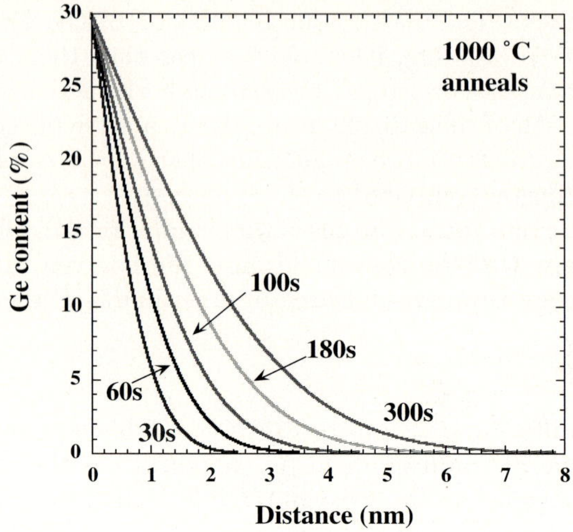

Fig. 10.17. The diffusion of Ge into strained-Si from a $Si_{0.7}Ge_{0.3}$ layer at 1000 °C for different times

$$C(x,t) = C_s \mathrm{erfc}\left[\frac{x}{2\sqrt{Dt}}\right] \tag{10.26}$$

where C_s is the surface concentration of Ge at x=0 and erfc is the complementary error function. The diffusion constant is related to temperature variations through

$$D = D_0 \exp\left(\frac{-E_A}{k_B T}\right) \tag{10.27}$$

where D_0 is the diffusion coefficient and E_A is the activation energy of the process. For diffusion of Ge into Si D_0 is $0.04\,\mathrm{m^2/s}$ and $E_A = 4.7\,\mathrm{eV}$.

Ge diffusion from a $Si_{0.7}Ge_{0.3}$ layer into silicon is simulated from (10.26) in Figs. 10.16 and 10.17 for a number of temperatures and times. It should be clear that for many of the temperatures used in standard silicon processing, Ge diffusion is a serious issue if SiGe layers are to be integrated into a process.

10.5 Bipolar and HBT Fabrication Processes

The bipolar transistor is still used for many applications particularly in analogue applications where low noise is important. It is also used in some areas of high speed logic where speed is more important than power dissipation.

The standard bipolar fabrication process used today is basically similar in many ways to the original fabrication of bipolar transistors at least with the fabrication techniques which are used. The reduction of the device dimensions, however, has substantially changed both the sizes, the transistor design

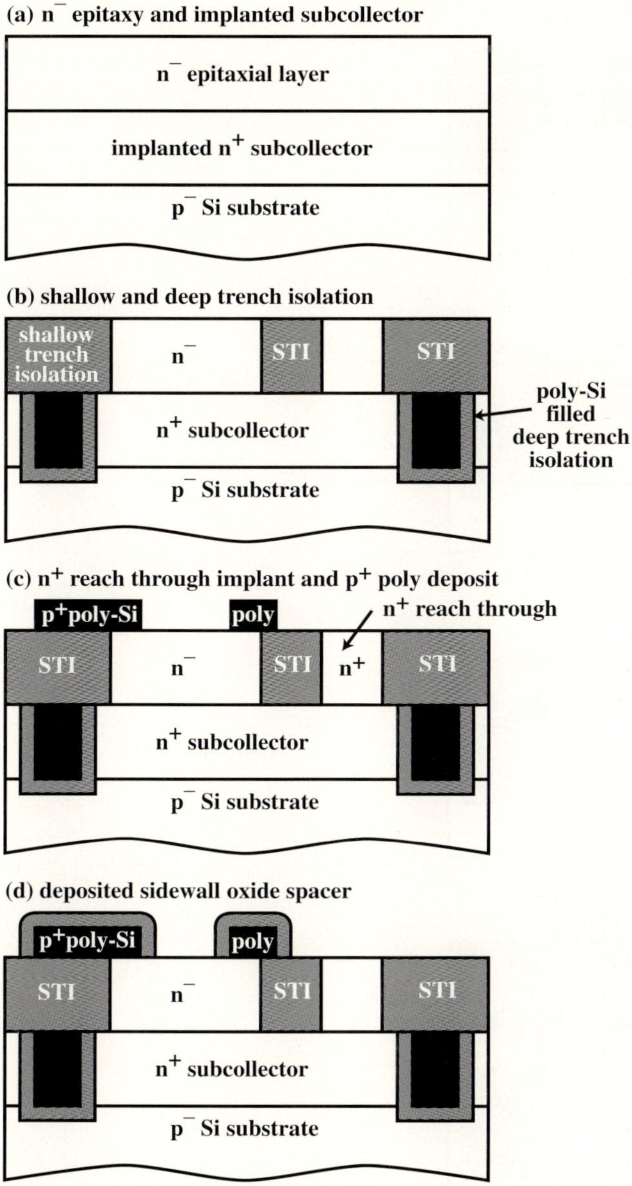

Fig. 10.18. The fabrication processes involves in silicon bipolar technology

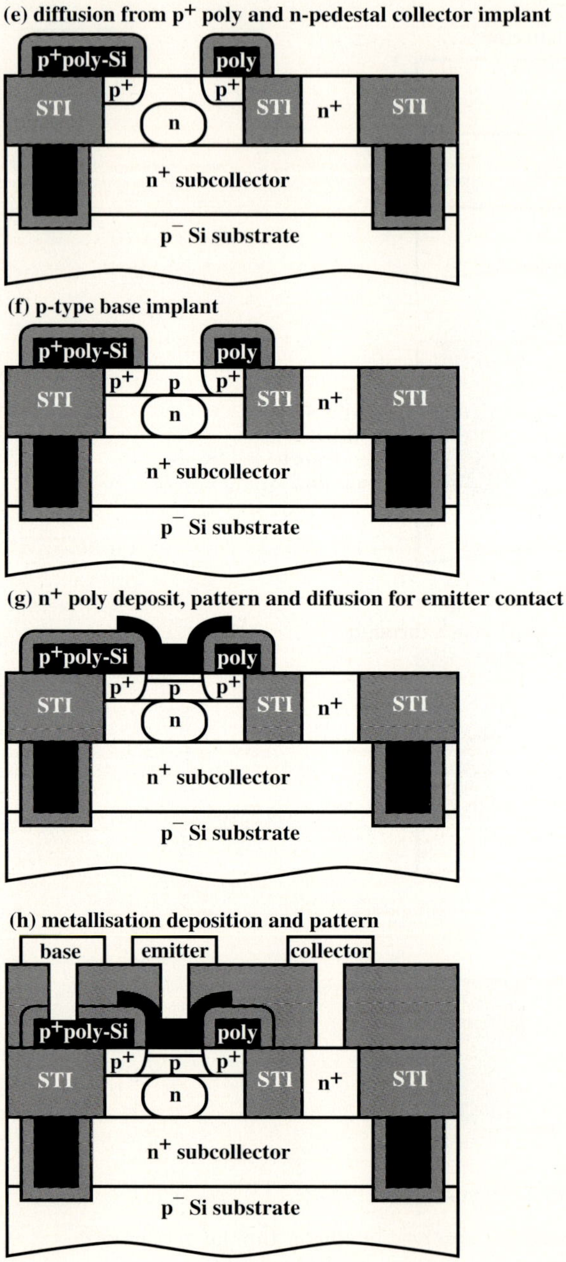

Fig. 10.19. The fabrication processes involves in silicon bipolar technology

and the specific fabrication processes to achieve ever smaller dimensions. Figures 10.18 and 10.19 show the stages involved with a modern implanted base bipolar transistor. Modern bipolars use a poly-Si emitter which reduces the emitter -base width to increase performance. For the following sections a n-p-n bipolar will be discussed. For p-n-p the polarities need to be reversed.

The starting wafers for bipolar transistors are typically p-type Si(100) wafers doped around 10 to 20 Ω-cm. On top of the substrate are grown two silicon epitaxial layers by CVD with different doping densities. Many manufacturers implant the doping at the two different levels shown in Figure 10.18(a) as this is presently a more accurate process than trying to achieve the doping levels through epitaxy. The lower heavily doped layer forms the subcollector for the bipolar or HBT transistor after a RTA to active the implants. At this stage lithography and etching is used to create trenches to electrical isolate the transistors laterally on the wafer. Both shallow and deep trench isolation is used with the deep trenches also filled with poly-Si.

The next stage involves the deposition of p^+ poly-Si which will be used to form contacts to the base of the transistor. After being patterned by lithography and etched (Fig. 10.18(e)) the poly is oxidised to form electrical isolation to subsequent layers before a n-type pedestal is implanted to reduce the base-collector resistance before diffusing the p-doping from the poly-Si base contacts. For an implanted base silicon bipolar the base is implanted at this stage. For HBTs, numerous situations exists at this point in the process. For an implanted base, both Ge and B can be implanted to form a p-SiGe base. Options exist to epitaxially grow a SiGe base with boron doping or to grown an undoped SiGe layer and then implant the p-type doping.

The final stage to produce the transistor before fabricating interconnects involves the deposition and patterning of a poly-Si emitter. After implantation and diffusion of the n-type dopant to create the emitter-base interface, a field oxide is deposited before the interconnect metal is deposited and patterned. As in the CMOS case, advanced bipolars use copper interconnects with TaN as a diffusion barrier.

10.6 BiCMOS

BiCMOS is a mixture of both bipolar and CMOS on the same chip. It is used in a number of different applications, either for mixed signals or where high speed is required for parts of the circuit. Good examples include some digital to analogue converters, analogue to digital converters along with the microelectronic parts of optical or ethernet switching circuits which require high speed. While normal Si bipolar is the standard form of the bipolar device with an implanted base, the IBM SiGe BiCMOS process flow is shown as it demonstrates a clever method of reducing the thermal budget after the selective deposition of the SiGe base of the bipolar transistor. This is important to prevent Ge and B outdiffusion from the base which is the main limitation in the ultimate speed of the technology.

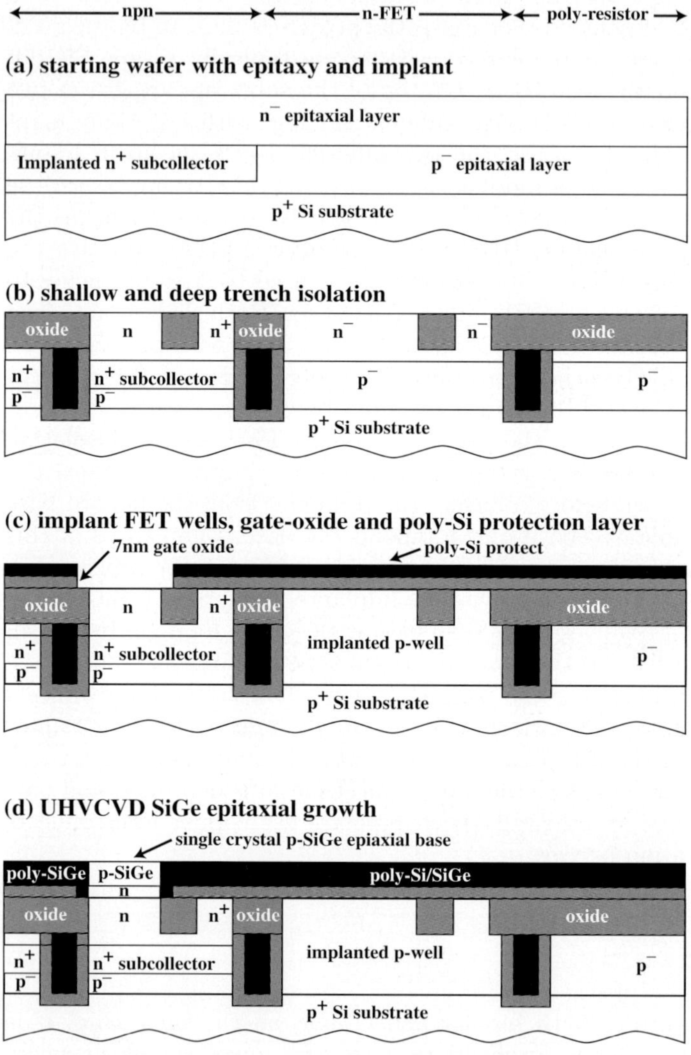

Fig. 10.20. The first stages of the IBM BiCMOS process

The process starts in a similar manner to a standard bipolar process of Sect. 6.4 with a subcollector implanting into epitaxial starting material (Fig. 10.20(a)). After lithographic patterning, trench isolation is etched into the wafer to electrically isolate each device in both lateral and also vertical directions (Fig. 10.20(b)). The deep trench isolation is filled with oxide and

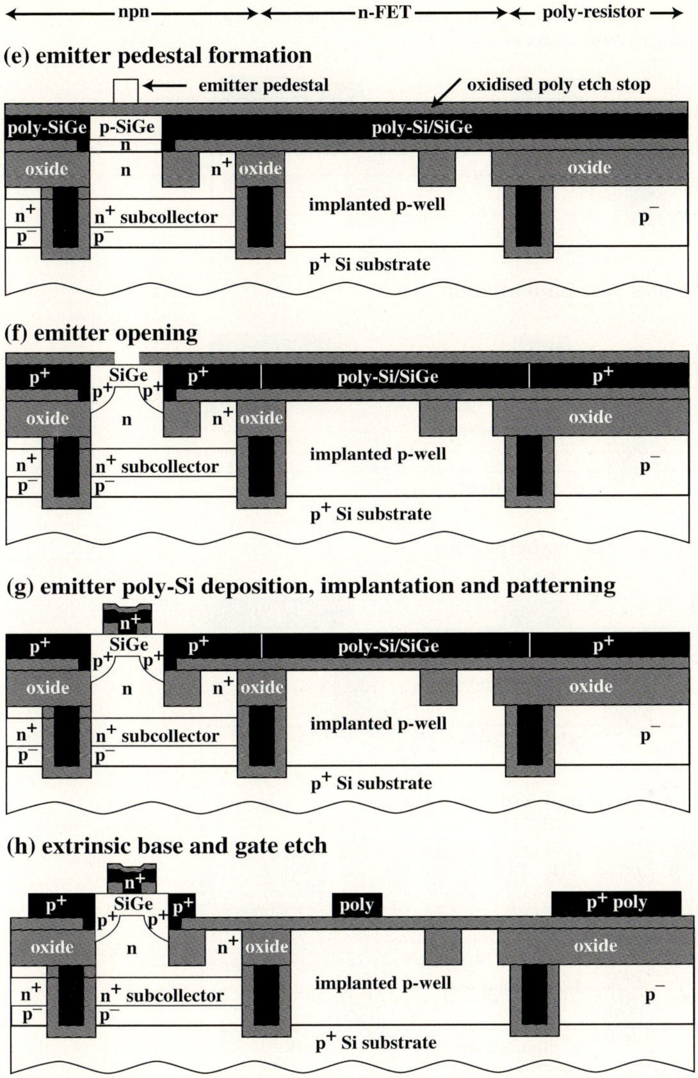

Fig. 10.21. The middle stages of the IBM BiCMOS process

poly-Si while the shallow trench isolation (STI) is filled with just silicon dioxide. At this stage the gate oxide is grown which is the main thermal budget in the process.

The IBM process demonstrates how the highest thermal budget parts of a process are achieved before any SiGe strained layers are deposited. On top of the oxide a poly-Si protection layer is grown. Lithography and selective etching are then used to open a window in the poly-Si and oxide layers leaving behind single crystal silicon above the bipolar collector (Fig. 10.20(c)). The

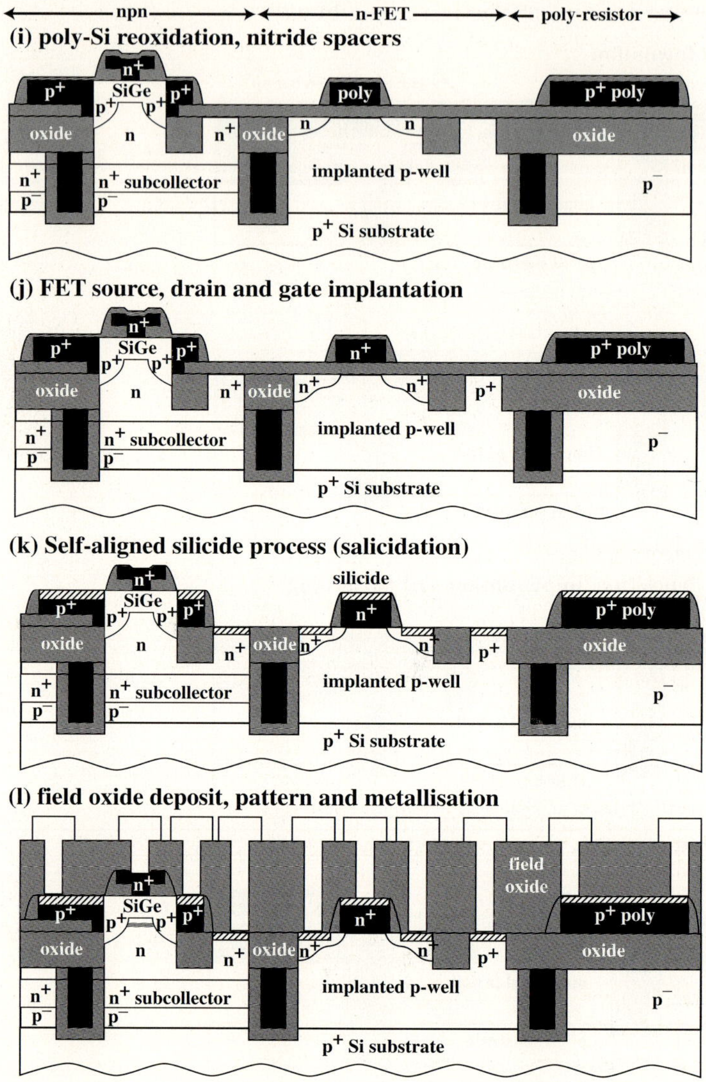

Fig. 10.22. The final stages of the IBM BiCMOS process

wafer is now cleaned with a hydrofluoric acid dip as the final stage to remove any oxide and hydrogen passivate the surface before placing the wafer in a CVD tool for the growth of the SiGe base of the bipolar transistor. Single crystal SiGe only grows on top of the single crystal silicon and poly-SiGe is deposited elsewhere as a crystal is required to seed the single crystal growth (Fig. 10.20(d)).

The next stages are related to the formation of the emitter pedestal of the HBT. First the poly and single crystal SiGe is oxidised before lithography and selective etching are used to open a hole where the emitter will contact the base. Next the poly-Si emitter is deposited and implanted with an n-type dopant. This is then oxidised before lithography and etching are used to pattern the emitter pedestal (Fig. 10.21(g)). Next the MOSFET gates, the poly-SiGe base contacts to the HBT and poly-SiGe resistors are patterned and etched, stopping on the gate oxide (Fig. 10.21(h)).

The self-aligned contacts to the MOSFETs are now formed, first implanting the low doped drain (LDD) before reoxidising the poly and then depositing silicon nitride. The n- and p-type implants to the n-MOS and p-MOS transistors respectively have to be done independently while masking the other parts of the wafer with resist. This implant is also used to dope the poly-SiGe gate of the transistors. The nitride is chemical mechnically polished (CMP) down to the top of the poly-Si before an isotropic etch is used to form spacers on the sides of the MOSFET transistor gates. Now the high doped drain (HDD) implants are formed with a higher doping to allow low contact resistivity Ohmic contacts to be formed (Fig. 10.22(j)). Again the n- and p-type dopants are implanted separately while the remaining parts of the wafer are masked with resist. The implants are activated with a two stage thermal process. First a short rapid thermal anneal (RTA) at high temperature ($\sim$ 950 °C) of a few seconds followed by a longer furnace anneal ($\sim$ 700 °C) for 30 minutes are used to activate the implants and diffuse the n-doping from the poly-Si emitter pedestal to form the base-emitter junction in the HBT.

The gate oxide is now etched away from the Ohmic contact regions of the FETs and HBTs where it was used to prevent the sputtering of the surface of the wafer during the previous implantation stages. Metal is then sputtered and annealled onto the surface. Where the metal is on top of single crystal silicon or poly-Si, a silicide will form. In the IBM process, cobalt is sputtered and annealled to form $CoSi_2$ but titanium or nickel are alternatives. Where the metal was above oxide, no reaction occurs and a selective etch is used to remove the unwanted metal. This leaves silicided contacts and gates to reduce the contact resistivities (Fig. 10.22(k)). The final stages involve the formation of interconnects. First a field oxide is deposited before lithography and etching are used to form via holes. These are then filled with metal, typically copper after a TaN diffusion barrier is deposited to prevent copper reacting with any silicon or silicide. The metal is finally patterned and the process repeated a number of times to build up multiple layers of interconnects on the chip (Fig. 10.22(l)).

There are many other examples of BiCMOS processes in the literature with SiGe HBT transistors in addition to the one described above. This was the first production BiCMOS process which used a SiGe base and still provides a good insight into techniques for reducing the thermal budget for strained-SiGe layers.

10.7 Strained-Si CMOS

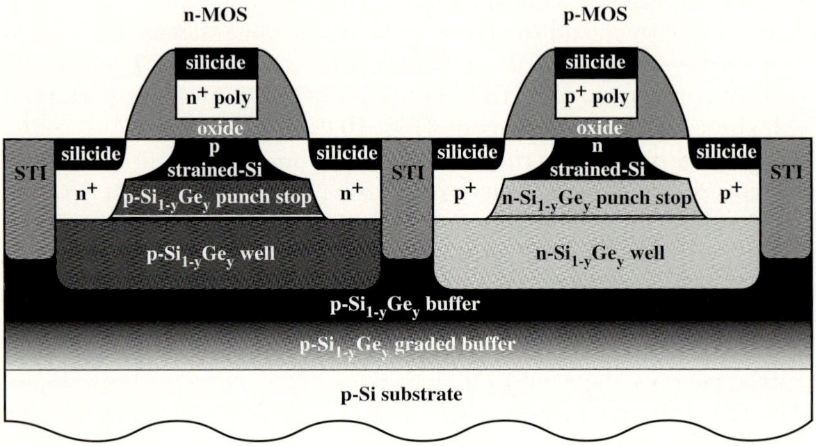

Fig. 10.23. A schematic diagram of strained-Si n- and p-MOS transistors. The Ge content in the substrate (y) can be chosen to optimise the performance of either the n- or p-MOS transistor or both

Strained-Si transistors where discussed in detail in Sect. 7.2. The major advantage of such a technique is that MOSFETs with the same design as that which has been discussed above in Sect. 10.3 may be fabricated using the same CMOS processing tools on fabrication plants which already exist. Slight modifications must be made to the fabrication process as implantation into a strained layer will result in an amorphous layer which when annealled may not return to its originally grown level of strain. In addition the problems of Ge diffusion discussed in Sect. 10.4 must also be addressed as no strained-Si layer will remain if too high a thermal budget is used. Therefore clever designs must be implemented to circumvent such problems.

Figure 10.24 shows one possible way of circumventing many of the fabrication issues of strained-Si CMOS. First the virtual substrate is grown (Fig. 10.24(a)) and the n- and p-wells for the CMOS are implanted (Fig. 10.24(b)). Shallow trench isolation (STI) can also be implemented before the well implants are implanted. The wafer can now be annealled at high temperature to active the well implants. The wafer is cleaned, placed in a CVD system and the strained-Si layer is grown on top of the wafer (Fig. 10.24(c)). This can now be processed almost as a normal CMOS process (Sect. 10.3).

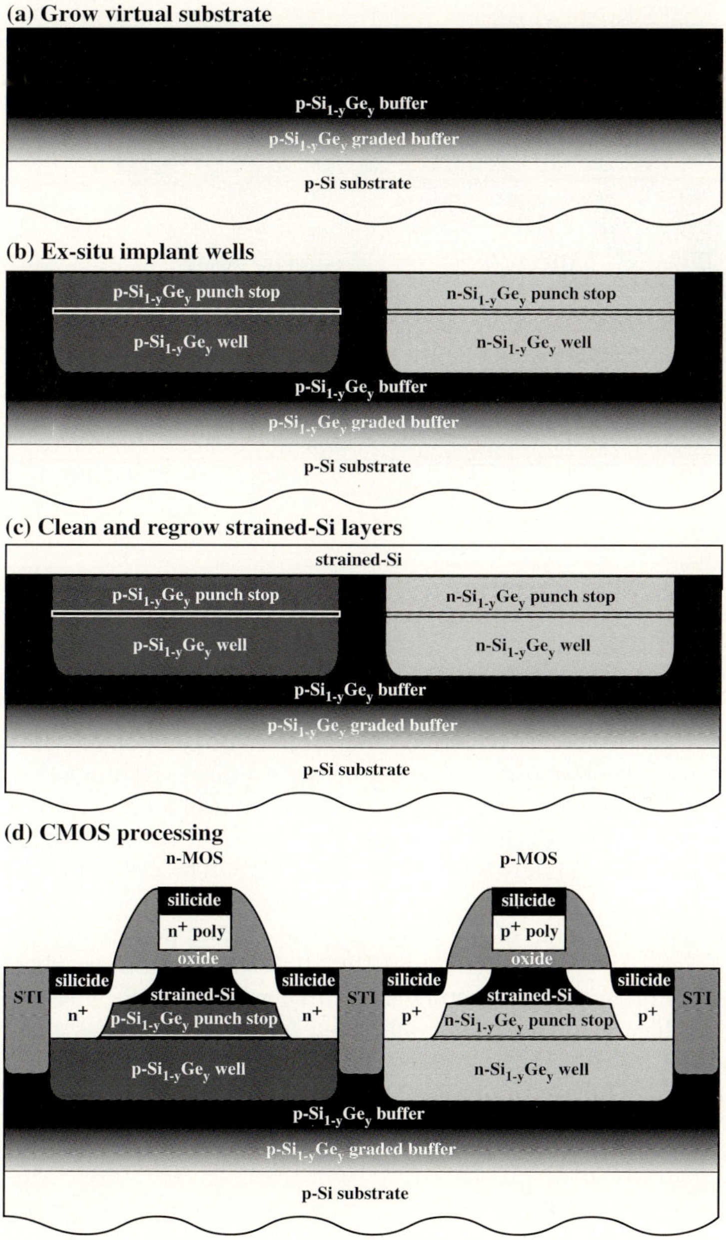

Fig. 10.24. A schematic diagram of the different fabrication stages of producing a strained-Si CMOS circuit using regrowth

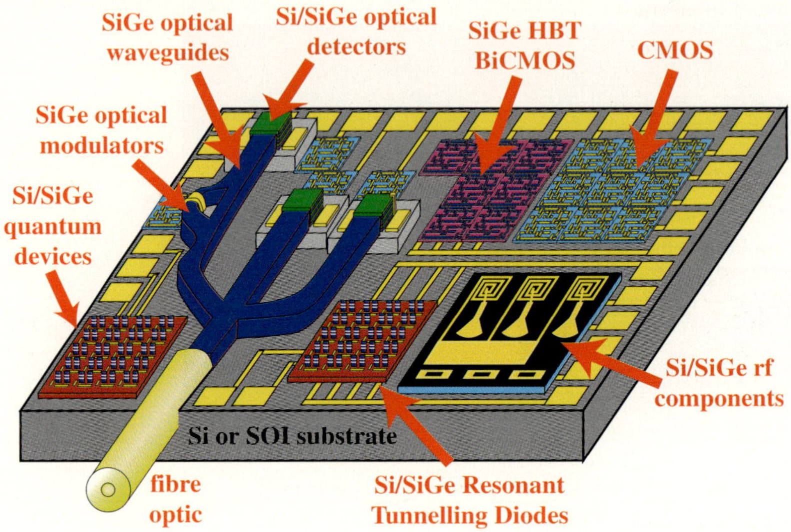

Fig. 10.25. A schematic of a system-on-a-chip using many different SiGe devices

10.8 The System on a Chip

The drive in microelectronics is to reduce the number of components in a system because costs increase for fabrication of the complete system as the number of components increase. Therefore there is substantial economic drive for a system-on-a-chip. SiGe devices have great potential in this area of research as ways of integrating many SiGe devices onto CMOS or silicon bipolar wafers can be envisaged. Figure 10.25 shows a schematic diagram of a very idealised view of the silicon chip of the future. Already rf and passive components are integrated with CMOS, HBT and BiCMOS chips. In optics, there are many examples of Si photodetectors integrated with CMOS. The real challenge is to produce a Si-based laser and then full optoelectronics could be integrated, either for interchip communications using optical fibres or for optical interconnects. A full set of optoelectronic components including modulators, waveguides and efficient detectors would be required for such chips and many researchers are already working in such areas.

10.9 Fault Tolerant Architectures

The main reason for the high cost of CMOS fabrication plants is that the top-down architecture required from lithographic fabrication techniques is not defect tolerant. Therefore any transistor or interconnect which fails on the circuit potentially destroys the whole chip. Some redundancy can be built in but this is expensive and not ideal. Almost all nanoelectronic devices still rely on architectures which have no fault tolerance. This problem is substantially worse for applications specific integrated circuits (ASICS) where each application has a different circuit design unlike the mass repetition of memory or processor manufacture. Testing has become a substantial cost (>60% of the total chip cost) as the new circuit design must be tested to check that no mistakes in circuit design or mask fabrication has taken place. With up to 20 mask levels in some chips and with up to 300 million transistors in designs, the potential for errors at some point in the design and fabrication process is substantial.

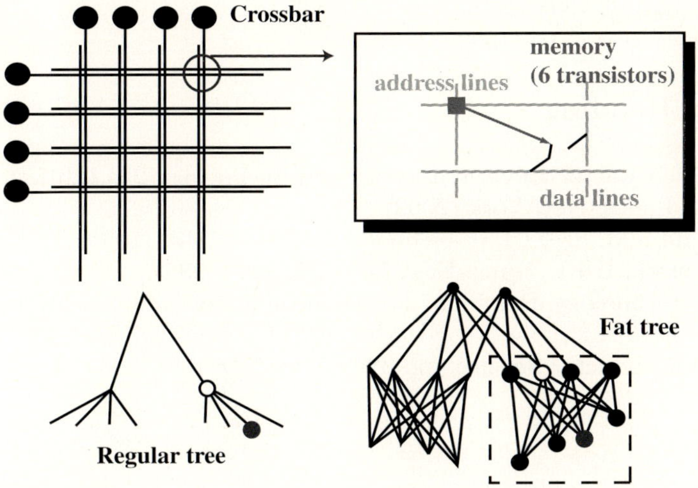

Fig. 10.26. A schematic diagram of the Teramac architecture

If an architecture could be found for which not all the transistors or interconnects are required to correctly function for the operation of a complete chip then this would substantially reduce costs. Most molecular ideas may require a fault tolerant architecture if they are ever going to be successful. One of the few defect tolerant architectures which has appeared is the Teramac architecture. This concept is shown in Fig. 10.26 and depends on the ability to have a large number of interconnects in a system so that some path may always be found around defects and non-functional parts of a circuit. Before the system is run, a map of all the defects is found and the system is configured so that the defects can be circumvented. Therefore the Teramac paradigm is to build a cheap and defective computer, find the defects, configure the resources with software, compile the programme and then run the computer. The major disadvantage with the Teramac concept is that the biggest problem with both CMOS and molecules is the interconnects. With CMOS as interconnects are reduced in size, the conductivity is constant with the resistance increasing faster than the decrease in capacitance from reduced size. Therefore the RC time constants increase for reduced interconnect size and the speed of the circuit correspondingly is reduced. High quality interconnects are a serious problem in many of the molecular schemes. Therefore using larger interconnect bandwidth to provide defect tolerance is not an option with most of the known nanoelectronic schemes.

Architecture is one area that could have a major impact on the design and fabrication of silicon chips if new more efficient architectures could be found. To date little progress has been made in finding efficient fault-tolerant architectures but if Moore's law runs out, there would be a substantial economic driver to find ways of improving the performance of silicon chips without scaling the gate length of MOSFETs.

10.10 Further Reading

1. S.M. Sze, Semiconductor Devices: Physics and Technology 2nd Edition, John Wiley and Sons, New York (2002)
2. C.Y. Chang and S.M. Sze, ULSI Technology, McGraw Hill (1996)
3. D.L. Harame et al., IEEE Trans. Elec. Dev. **24**, 324 (1995)
4. International Technology Roadmap for Semiconductors 2003 Edition, (http://public.itrs.org)
5. D.J. Paul, Semiconductor Science and Technology **19**, R59 (2004)

11. Outlook

This book on silicon based quantum electronics concentrated in its device section strongly on developments which in a short or medium time scale either rival dramatically existing silicon microelectronics or deliver added value to them at such a substantial scale that the high barriers for the introduction of new structures in production can be overcome. Microelectronics production now and probably for the next 10 to 15 years will be dominated by CMOS technology and hence we have to consider the market significance of quantum electronics against the background of CMOS products. The technical details of future CMOS generations are laid down in International Technology Roadmap for Semiconductors (ITRS) which are revised annually and therefore, give a good picture of the near future developments, the obvious roadblocks and the already known solutions. The reader will find roadmap information in the introduction and the conventional device chapters. To understand what technologies have the potential to gain market share requires analysis of the basic microelectronic circuit application assumptions and the inherently connected economic conditions of the dominant CMOS technology. The main basic assumptions in a 10 year forecast are described by

- Electronic devices must operate at room temperature. In this context room temperature is more broadly defined to include the extremes of climate in which any technology must be able to operate, namely from -20 °C to 60 °C, or -40 °C to 80 °C for the automotive applications. The temperatures in devices can be further increased by electron heating, self-heating and power dissipation of the devices up to 150 °C.
- Monolithic integration of the devices into circuits will continue with further expansion from the giga-scale to the tera-scale. This includes not only more devices for the standard functions (logic, processor, memory, amplifer, phase back loop, converter) but also integration of new and different functions onto the chip. This approach is termed system on a chip (SOC) and new popular examples include embedded memories, mixed analogue/digital circuits, rf front-end/intermediate signal processing, single chip radio, smart power electronics and optoelectronic switching. The whole system is complicated and time consuming to design, charaterise and test. Hence the re-use of building blocks and easy integration of these building blocks is an essential demand of the SOC approach. As the testing

of such systems is now the major cost, reuse of proven building blocks or cores is essential to keep costs to a minimum.
- Compatibility with CMOS technology. The acceptance of new technologies increases steeply by the proof or demonstration of at least a sound technical assessment of its compatibility with CMOS technology. By that assessment a further scaling of the lateral dimensions, different gate materials, silicide contacts and a significant reduction of processing temperatures may be anticipated.

There are a number of economic conditions which must also be considered. A general rule for any semiconductor technology is that provided the required performance for an application can be met (including speed, power dissipation, functionality, lifetime of product, etc.) the cheapest technology will dominate. There are also strong economic drivers which includies specific market volumes, price per chip, price per transistor and the cost of the factory. Figure 11.1 summarises some of the economic data for the past and forecast them up to 2010.

On a long term consideration the basic microelectronic circuit assumptions and some of the economic conditions could change and then potentially other quantum effects which are now not in the main stream silicon based devices may be of stategic importance. There are too many possible scenarios that we only consider the following personal thoughts to have some chance of fruition:

Scenario A: Novel cooling concepts allow device operation at cryogenic temperatures. Nowadays, small and compact cryogenic coolers have been developed driven by thermal imaging demands to reduce the size and power consumption along with improvement of reliability. This is only one step toward a breakthrough. The most important step could come from artificial semiconductor materials composed from superlattices with nanometer or sub-nanometer periodicity (Fig.11.2).
A superlattice (SL) with monolayer peridicity (five monolayers (ML) would have a period of 0.7 nm) can be considered as an artificial semiconductor with its own electronic and thermal properties. The thermal properties of semiconductors depends on the phonon spectrum which is changed in superlattices by the so called Brillouin zone folding effect. In metals the thermal conductivity is coupled to the electronic conductivity through the Wiedemann-Franz law. In uniform semiconductors the thermal conductivity is decoupled from the electronic conductivity resulting in high dissipation (fortunately for normal heat) in hard semiconductors like Si. In the superlattice direction, the heat conduction is partly decoupled from the lattice properties by the additional superlattice periodicity. Micro-Peltier elements can shift the heat from the device layer to the underlying substrate when the thermal conductivity of the SL layer is low. Within this concept only the thin device layer is cooled by Peltier cooling. It is worthwhile mentioning that by cooling the device layer, leakage

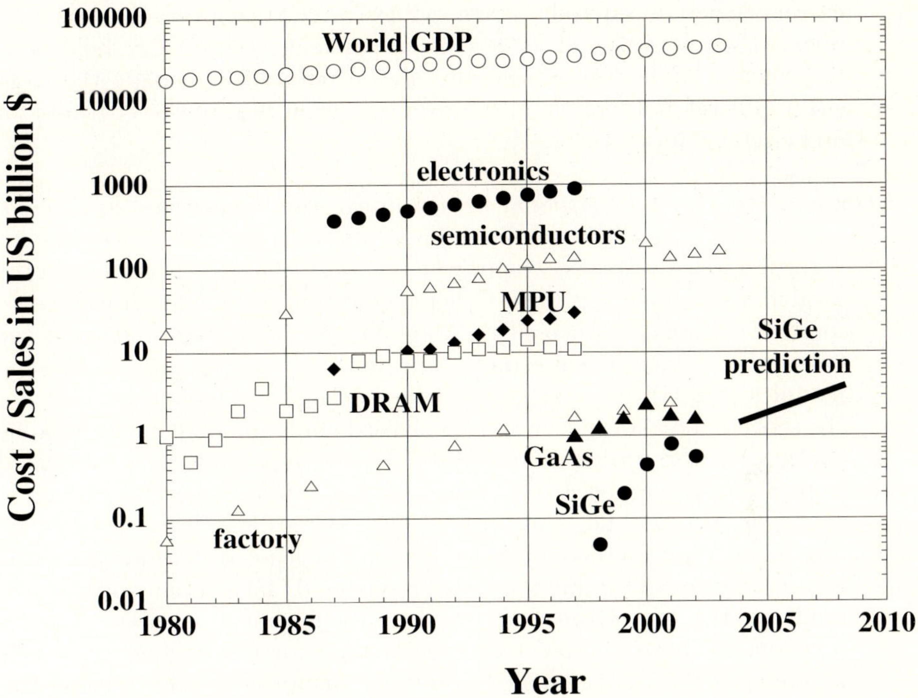

Fig. 11.1. The economics of the semiconductor market as a function of year. GDP is gross domestic product of the world economy, MPU is sales of microprocessor units, DRAM is sales of dynamic random access memory, factory is the cost of building a semiconductor foundry, GaAs is sales of GaAs devices in the rf market and SiGe is the sales of SiGe HBT and BiCMOS products (predominantly in the rf market). The SiGe growth prediction is for 30% per annum growth up to 2008

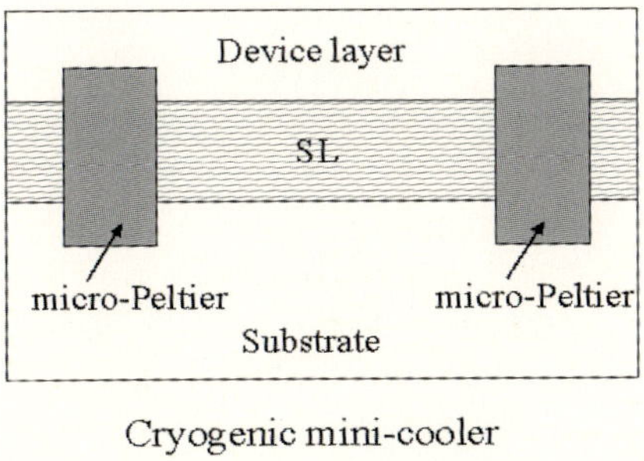

Fig. 11.2. A superlattice substrate structure for cryogenic chip cooling. The thin device layer is separated from the silicon substrate by a superlattice (SL) region. The thickness of the SL region is a few microns, the periodicity of the monolayer-superlattice is nanometer or sub-nanometer. When built up from strained hetero-layer materials (strained layer superlattice SLS) the superlattice can additionally have the function of a virtual substrate with strain adjustment in the device layer

currents are reduced and voltage swings can also be reduced thereby reducing the selfheating of the circuit. In particular the subthreshold slope of present CMOS transistors can only be reduced by cooling the transistor. Therefore lower circuit power dissipation requires lower transistor and circuit temperatures.

Scenario B: Three dimensional hybrid integration techniques could compete successfully with monolithic integration. In the past monolithic integration was always the winner when it was technically feasable and the production volume was high. Other integration techniques proceeded to improve automatic processing, reliability and performance. Two of the most intersting technical routes are wafer level packaging and multichip modules (MCM), the first lowering the package and mounting costs and the second one allowing several specialised chips to be mounted into one package. In the present state of MCMs the chips preferably fabricated from Si are placed on a base which for reduction of thermal expansion problems can also be made from silicon. Chips which are stacked on top of the lower ones could deliver three dimensional (3D) arrays in the future. Figure 11.3 shows the concept of the 3D integration with the example of the top of the wafers being thinned but a standard wafer is used on the bottom. Key technologies used for this technique include handling of the thinned wafers using, for instance, a wafer bonded stabiliser and interchip-via connections between the metallisation systems. For complex and large systems the whole variety of improved packaging, multichip modules and on board mounting techniques will compete with monolithic system -on chip solutions.

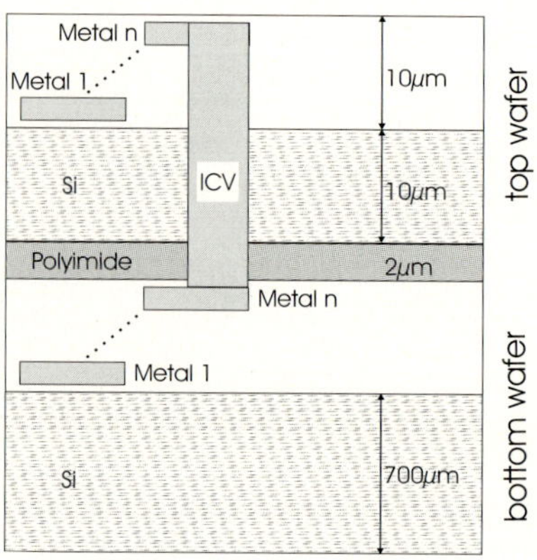

Fig. 11.3. 3D integration with interchip-via (ICV) connects. The top wafer is thinned to about 10 μm to improve ICV footprint at a given aspect ratio. After P. Ramm (U.S. patent application 950 123)

Scenario C: Self-organised quantum dot device circuits with tera-scale complexity are more economically to fabricate compared to deterministically positioned devices in conventional circuits of the same complexity. The cost of factories in microelectronics industry is steadily increasing and they will hit the ten billion dollar mark within ten years. The costs are necessary because the numerous devices of small dimension are placed on predetermined positions using processes with incredibly small alignment tolerances. In comparison, it is amazing to see how fast nature produces $10^9 - 10^{11}$ islands per square centimeter by a self-organised process like Stranski-Krastanov growth. In a typical epitaxy experiment with strained Ge/Si this huge number of islands is created during ten seconds assuming a growth rate of around one monolayer per second. Opponents of such a technology who argues that neither the position nor the size of the islands is determined, that the quantum dot island is not a functional device and that the interconnection strategy is not viable are correct. Within the last five years, however, considerable progress has been made in demonstrating uniform size, positioning and functionality of the islands. Figure 11.4 shows a scheme of a quantum dot tunnel device which has been positioned along an oxide window.

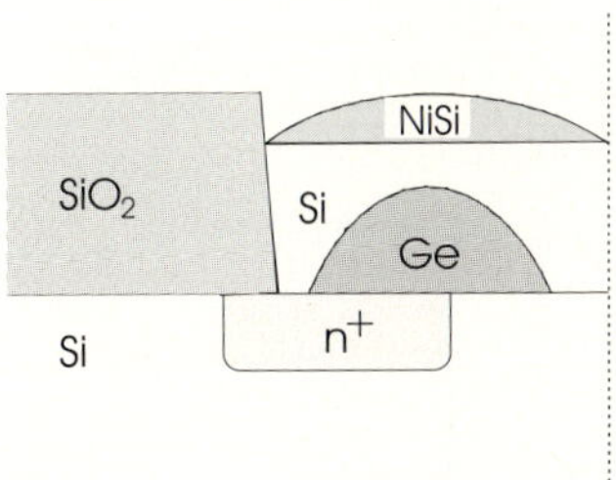

Fig. 11.4. Cross section through a quantum dot row of Ge islands postioned along the edge of a SiO_2 window. The islands are overgrown with silicon. The top contact is grown by self-aligned silicide nucleation, the botton contact is from a buried n^+-channel

On a long term time scale these cost and thermal load driven technical developments could offer new chances to quantum effects now not considered as serious contenders in silicon electronics because of the need of cryogenic operation or a problematic compatibility with CMOS. The effect which might then come to larger importance in silicon based quantum devices include

- electron wave interference devices / Aharonov-Bohm effect
- quantum computing
- superconducting single flux quantum logic
- DNA and other molecular conduction mechanisms
- silicon-based interband light emission or lasing

A. List of variables

a_0 = lattice constant (nm)
$\mathbf{a}$ = lattice basis vector
a_B = Bohr radius = $\frac{4\pi\varepsilon_0\hbar^2}{e^2 m_0}$ (m)
a_B^* = effective Bohr radius = $\frac{4\pi\varepsilon_0\varepsilon_r\hbar^2}{e^2 m^*}$ (m)
a_c = conduction band hydrostatic deformation potential (eV)
a_v = valence band hydrostatic deformation potential (eV)
A = area (m^2)
$\mathrm{Ai}[x]$ = Airy's function
$\mathbf{b}$ = Burger's vector
b_c = conduction band uniaxial deformation potential (eV)
b_v = valence band uniaxial deformation potential (eV)
c = speed of light = 3.00×10^8 (ms^{-1})
C = capacitance (F)
C_{BC} = base-collector capacitance (F)
C_{BE} = base-emitter capacitance (F)
C_{ox} = oxide capacitance (F)
C_s = capacitance of depletion layer (F)
D = diffusivity of atoms (m^2s^{-1})
D^* = detectivity of a photodetector ($m(Hz)^{1/2}/W$)
D_h = diffusion coefficient of holes (m^2s^{-1})
D_n = diffusion coefficient of electrons (m^2s^{-1})
D_S = diffusion coefficient for adatoms on a surface (m^2s^{-1})
D_0 = diffusion coefficient of atoms (m^2s^{-1})
E = energy (J or eV)
E_{ad} = potential energy for the surface absorption of an adatom (J or eV)
E_F = Fermi energy (J or eV)
E_{Fi} = intrinsic Fermi energy (J or eV)
E_g = band gap energy (J or eV)
E_S = the energy gain from the incorporation of an adatom onto a surface (J or eV)
f = lattice mismatch
F = electric field
F = atomic or molecular flux for epitaxial growth
F_{des} = the desorbing flux of adatoms on a surface

$\mathbf{F}$ = flux of atoms (in growth)
$\mathbf{F_{des}}$ = desorbing flux
$f(E)$ = Fermi-Dirac function
f_{max} = maximum oscillatory frequency = frequency at which the unilateral power gain of a transistor becomes unitary (Hz)
f_T = cutoff frequency = frequency at which the gain is unitary (Hz)
g_m = transconductance (S)
G_n = electron generation rate (s)
G_S = surface generation rate for adatoms (s)
h = Planck's constant = 6.63 x 10^{-34} (Js)
h = thickness of a monolayer of atoms (m)
h_c = the critical thickness for a stained heterolayer (m)
$\hbar$ = Planck's constant / 2π = 1.05 x 10^{-34} (Js)
H_n = Hermite polynomials
I = dc current (A)
i = ac current (A)
$\mathrm{i} = \sqrt{-1}$
I_B = dc base current (A)
i_B = ac base current (A)
I_C = dc collector current (A)
i_C = ac collector current (A)
I_{ds} = drain-source current (A)
$I_{d,sat}$ = saturation drain-source current (A)
I_E = dc emitter current (A)
i_E = ac emitter current (A)
I_{th} = threshold current in a laser (A)
J = current density (Am^{-2})
J_p = current density of holes (Am^{-2})
J_P = peak current density in tunnel diode (Am^{-2})
$\mathbf{k}$ = wave vector of an electron (m^{-1})
k_B = Boltzmann's constant = 1.38×10^{-23} (JK^{-1})
L = length of device (m)
L_D = Debye length (m)
L_g = gate length (m)
L_n = diffusion length of electrons (m)
M = molecular weight (kg)
m^* = effective mass of electrons (kg)
m_0 = electron rest mass = 9.11 $\times 10^{-31}$ (kg)
m_l^* = effective mass of longitudinal electrons (kg)
m_{lh}^* = effective mass of light holes (kg)
m_{hh}^* = effective mass of heavy holes (kg)
m_{so}^* = effective mass of split-off holes (kg)
m_t^* = effective mass of transverse electrons (kg)
N_A = density of acceptors (m^{-3})

N_{Av} = Avogadro's number = 6.02×10^{23} (mole^{-1})
N_D = density of donors (m^{-3})
n_e = density of electrons (m^{-3})
n_i = intrinsic carrier density (m^{-3})
n_{ie} = effective intrinsic carrier density (m^{-3})
n_h = density of holes (m^{-3})
n_s = sheet carrier density (m^{-2})
N_S = density of atoms on a surface (m^{-2})
n_0 = electron concentration at thermal equilibrium (m^{-3})
p = vapour pressure (Pa)
P_{opt} = the optical power at wavelength, λ incident on a photodetector
$\mathbf{p}$ = momentum (kg.m)
q = electronic charge = 1.6 x 10^{-19} (C)
Q = macroscopic charge density (m^{-3})
$\tilde{r}$ = reflected amplitude
$\tilde{R} = |\tilde{r}|^2$ = reflection coefficient
r_B = base resistance (Ω)
R = resistance (Ω)
R = epitaxial growth rate (ms^{-1})
R^* = reciprocal lattice vector (m^{-1})
R_C = contact resistance (Ω)
r_d = channel resistance in a FET (Ω)
R_H = Hall coefficient (Ω)
R_n = electron recombination rate (s)
R_S = surface recombination rate for adatoms (s)
S = subthreshold slope (mV dec^{-1})
$\boldsymbol{S}$ = surface flux vector
t = time (s)
$\tilde{t}$ = transmission amplitude
$\tilde{T} = |\tilde{t}|^2$ = transmission coefficient
T = temperature (K)
t_{ox} = oxide thickness (m)
U_S = activation barrier to surface diffusion (J or eV)
$u(\nu; T)$ = the spectral energy density as a function of frequency, ν and temperature, T
v = electron velocity (ms^{-1})
V = applied voltage (V)
v_d = drift velocity (ms^{-1})
V_{ds} = source-drain voltage (V)
V_g = gate voltage (V)
v_E = exit velocity for electrons exiting a base of a bipolar transistor
V_P = voltage of peak in tunnel diode (V)
v_{sat} = saturation velocity of electrons (ms^{-1})
V_t = thermal voltage (V)

V_T = threshold voltage (V)
W = width of device (m)
W_B = base width (m)
W_D = depletion width (m)
W_m = maximum depletion width (m)
x = length in x-direction (m)
x_n = edge of depletion region in n-doped material (m)
x_p = edge of depletion region in p-doped material (m)
y = length in y-direction (m)
z = length in z-direction (m)
α = optical absorption coefficient
β = gain or amplification
ΔE_c = conduction band discontinuity (eV)
ΔE_g = effective band gap narrowing (eV)
ΔE_v = valence band discontinuity (eV)
Δ_S = segregation length (m)
ϵ = strain tensor = $(\epsilon_x, \epsilon_y, \epsilon_x)$
ε_0 = permittivity in a vacuum = 8.85 x 10^{-12} (Fm^{-1})
ε_r = relative dielectric constant
ε_s = dielectric constant of semiconductor
η = quantum efficiency of a photodetector
λ = wavelength of photon (m)
λ_F = Fermi wavelength (m)
λ_S = surface diffusion length (m)
μ = shear modulus
μ_H = Hall mobility ($\mathrm{m^2V^{-1}s^{-1}}$)
μ_n = drift mobility of electrons ($\mathrm{m^2V^{-1}s^{-1}}$)
μ_p = drift mobility of holes ($\mathrm{m^2V^{-1}s^{-1}}$)
ν = frequency of photon (Hz)
ν = Poisson's ratio
ω_q = phonon wave vector (m^{-1})
Φ_B = Schottky barrier height (eV)
Φ_0 = flux of incident photons per unit area
ψ = wavefunction
ψ = potential (V)
ψ_B = potential in the bulk (V)
ψ_s = surface potential (V)
$\Re$ = responsivity of a photodetector (A/W)
ρ = resistivity (Ω m)
σ = stress tensor = $(\sigma_x, \sigma_y, \sigma_z)$
σ_h = hole conductivity (S)
σ_n = electron conductivity (S)
σ_S = the surface supersaturation
Σ = surface tension (N)

Σ_i = the interface tension (N)
τ_B = base transit time (s)
τ_C = collector charging time of the base-collector junction capacitance (s)
τ_{Des} = desorption lifetime (s)
τ_{inc} = incorporation time (s)
τ_n = lifetime of an electron (s)
χ = electron affinity (eV)

B. Physical Properties of Important Materials at 300 K

Table B.1. Physical Properties of Si, Ge, SiO_2 and Si_3N_4 at room temperature

Property	Si	Ge	SiO_2	Si_3N_4
Atomic / molecular weight	28.09	72.60	60.08	140.28
Atoms, molecules (m^{-3})	5.0×10^{28}	4.42×10^{28}	2.3×10^{28}	1.23×10^{28}
Breakdown field ($V\,m^{-1}$)	3×10^{7}	10^{7}	$>10^{9}$	10^{9}
Debye temperature (K)	645	360	-	-
Deformation potential, a_v (eV)	2.46	1.24	-	-
Deformation potential, a_c (eV)	4.18	2.55	-	-
Deformation potential, b_v (eV)	-2.35	-2.55	-	-
Deformation potential, b_c (eV)	9.16	9.42	-	-
Density ($kg\,m^{-3}$)	2370	5326	2533	2887
Dielectric constant	11.9	16.0	3.9	7.9
Effective longitudinal electron mass, m_l^* (m_0)	0.1905	0.082	-	-
Effective transverse electron mass, m_t^* (m_0)	0.9163	1.58	-	-
Effective heavy hole mass, m_{HH}^* (m_0)	0.537	0.284	-	-
Effective light hole mass, m_{LH}^* (m_0)	0.153	0.044	-	-
Effective split-off hole mass, m_{SO}^* (m_0)	0.234	0.095	-	-
Elastic moduli, C_{11} (GPa)	165.8	128.5	-	-
Elastic moduli, C_{12} (GPa)	63.9	48.3	-	-
Elastic moduli, C_{44} (GPa)	79.6	66.8	-	-
Electron affinity, χ (eV)	4.05	4	0.9	
Electron mobility ($cm^2\,V^{-1}\,s^{-1}$)	1430	3900	-	-
Energy gap, E_g (eV)	1.12	0.66	9	5
Hole mobility ($cm^2\,V^{-1}\,s^{-1}$)	470	1900	-	-
Intrinsic carrier density (cm^{-3})	1.45×10^{10}	2.4×10^{13}	-	-
Intrinsic Debye length (μm)	24	0.66	-	-
Intrinisic resistivity (Ω-m)	2.3×10^{3}	0.47	$\sim10^{14}$	$\sim10^{12}$
Lattice constant (nm)	0.543095	0.564613	amorphous	amorphous
Melting point (°C)	1415	935	>1713	1900
Minority carrier lifetime (s)	2.5×10^{-3}	10^{-3}	-	-
Optical phonon energy (meV)	63	37	-	-
Poisson's ratio, ν	0.280	0.273	0.167	0.270
Refractive index (@ 633 nm)	3.94	5.62	1.46	2.05

Property	Si	Ge	SiO_2	Si_3N_4
Specific heat ($J\,g^{-1}\,K^{-1}$)	0.7	0.31	1.0	-
Thermal conductivity ($W\,m^{-1}K^{-1}$)	140	60	1.1	30
Thermal diffusivity (cm^2s^{-1})	0.9	0.36	0.006	0.07
Thermal expansion coeff. (K^{-1})	2.56×10^{-6}	5.9×10^{-6}	0.5×10^{-6}	3.3×10^{-6}
Velocity of sound ($m\,s^{-1}$)	2329	5323	5759	-
Youngs Modulus (GPa)	47	30	75	85

C. Fundamental Physical Constants

Table C.1. Physical Constants

Property	Symbol	Value	Units
Avogadro constant	N_{Av}	6.02214×10^{23}	mol^{-1}
Bohr radius	a_B	0.052917	nm
Boltzmann constant	k_B	1.38066×10^{-23}	$\mathrm{J\,K}^{-1}$
Elementary charge	q	1.602177×10^{-19}	C
Electron rest mass	m_e	9.109389×10^{-31}	kg
Electron volt	eV	1.602177×10^{-19}	J
Intrinsic impedance of free space		376.7	Ω
Molar gas constant	$N_{Av}k_B$	8.314510	$\mathrm{J\,mol}^{-1}\mathrm{K}^{-1}$
Permeability in a vacuum	μ_0	12.566370×10^{-7}	$\mathrm{H\,m}^{-1}$
Permittivity in a vacuum	ε_0	8.854187×10^{-12}	$\mathrm{F\,m}^{-1}$
Planck constant	h	6.626075×10^{-34}	J s
Planck constant/2π	$\hbar$	1.05457266×10^{-34}	J s
Proton rest mass	M_P	1.672623×10^{-27}	kg
Speed of light in a vacuum	c	2.99792458×10^{8}	$\mathrm{m\,s}^{-1}$
Standard atmosphere		1.01325×10^{5}	Pa
Thermal energy @ 4.2 K	$4.2k_B$	0.362	meV
Thermal energy @ 300 K	$300k_B$	25.85	meV
Thermal voltage at 300 K	$\frac{k_BT}{q}$	0.025852	V

Printing: Krips bv, Meppel
Binding: Litges & Dopf, Heppenheim